Computer Algebra
Symbolic and
Algebraic Computation

Second Edition

Edited by
B. Buchberger, G. E. Collins, and R. Loos
in cooperation with R. Albrecht

Springer-Verlag
Wien New York

Prof. Dr. Bruno Buchberger
Institut für Mathematik
Johannes Kepler Universität Linz, Austria

Prof. Dr. George Edwin Collins
Computer Science Department
University of Wisconsin-Madison
Madison, Wis., U.S.A.

Prof. Dr. Rüdiger Loos
Institut für Informatik I
Universität Karlsruhe
Federal Republic of Germany

Prof. Dr. Rudolf Albrecht
Institut für Informatik und Numerische Mathematik
Universität Innsbruck, Austria

With 5 Figures

ISBN 3-211-81776-X Springer-Verlag Wien-New York
ISBN 0-387-81776-X Springer-Verlag New York-Wien
ISBN 3-211-81684-4 1. Aufl. (= Computing, Supplementum 4) Springer-Verlag Wien-New York
ISBN 0-387-81684-4 1st ed. (= Computing, Supplementum 4) Springer-Verlag New York-Wien

Foreword to the Second Edition

We are pleased to observe that the first edition of this volume has already obtained wide distribution, which might be an indication that a book like this fills a gap. We are therefore grateful to Springer-Verlag for the decision to make the volume available in a second, and less expensive, edition. The text of the first edition has not been changed except for the correction of errors which have been brought to our attention.

The Editors

Preface

The journal Computing has established a series of supplement volumes the fourth of which appears this year. Its purpose is to provide a coherent presentation of a new topic in a single volume. The previous subjects were Computer Arithmetic 1977, Fundamentals of Numerical Computation 1980, and Parallel Processes and Related Automata 1981; the topic of this 1982 Supplementum to Computing is Computer Algebra. This subject, which emerged in the early nineteen sixties, has also been referred to as "symbolic and algebraic computation" or "formula manipulation".

Algebraic algorithms have been receiving increasing interest as a result of the recognition of the central role of algorithms in computer science. They can be easily specified in a formal and rigorous way and provide solutions to problems known and studied for a long time. Whereas traditional algebra is concerned with constructive methods, computer algebra is furthermore interested in efficiency, in implementation, and in hardware and software aspects of the algorithms. It develops that in deciding effectiveness and determining efficiency of algebraic methods many other tools – recursion theory, logic, analysis and combinatorics, for example – are necessary. In the beginning of the use of computers for symbolic algebra it soon became apparent that the straightforward textbook methods were often very inefficient. Instead of turning to numerical approximation methods, computer algebra studies systematically the sources of the inefficiency and searches for alternative algebraic methods to improve or even replace the algorithms.

The remarkable results of computer algebra so far are scattered over conference proceedings, journal papers and technical reports. Until now no comprehensive treatment has been undertaken. The present volume is a first attempt at removing

this gap. In sixteen survey articles the most important theoretical results, algorithms and software methods of computer algebra are covered, together with systematic references to literature. In addition, some new results are presented. Thus the volume should be a valuable source for obtaining a first impression of computer algebra, as well as for preparing a computer algebra course or for complementary reading.

The preparation of some papers contained in this volume has been supported by grants from the Austrian "Fonds zur Förderung der wissenschaftlichen Forschung" (Project No. 3877), the Austrian Ministry of Science and Research (Department 12, Dr. S. Höllinger), the United States National Science Foundation (Grant MCS-8009357) and the Deutsche Forschungsgemeinschaft (Lo-231-2). The work on the volume was greatly facilitated by the opportunity for the editors to stay as visitors at the Department of Computer and Information Sciences, University of Delaware, at the General Electric Company Research and Development Center, Schenectady, N.Y., and at the Mathematical Sciences Department, Rensselaer Polytechnic Institute, Troy, N.Y., respectively. Our thanks go to all these institutions. The patient and experienced guidance and collaboration of the Springer-Verlag Wien during all the stages of production are warmly appreciated.

The editors of the   Cooperative editor of
Supplementum   Computing

B. Buchberger   **R. Albrecht**
G. Collins
R. Loos

Contents

Introduction

R. Loos, Karlsruhe

Abstract

In this introduction we first give a working definition of computer algebra. We then describe the organization of research activities in this field. Finally the overall structure and the intention of the present volume on computer algebra is explained. Some technical information (basic references, notation etc.) about the volume is given.

1. What Is Computer Algebra?

1.1 An Attempt at a Definition

Computer algebra is that part of computer science which designs, analyzes, implements and applies algebraic algorithms. We may consider this as a working definition for the scope of the following chapters but we do not claim that this is the only possible one. The definition is based on the view that the study of algorithms is at the heart of computer science [1, 6, 13]. Compared with other algorithms, algebraic algorithms have simple formal specifications, have proofs of correctness and asymptotic time bounds which can be established on the base of a well developed mathematical theory. Furthermore, algebraic objects can be represented exactly in the memory of a computer, so that algebraic computations can be performed without loss of precision and significance. Usually, algebraic algorithms are implemented in software systems allowing input and output in the symbolic notation of algebra. For all these reasons, computer algebra has attracted an increasing interest in computer science, mathematics, and application areas.

Negatively, one could say that computer algebra treats those subjects which are too computational to appear in an algebra text and which are too algebraic to be presented in a usual computer science text book. Another negative description says that computer algebra deals with non-numeric computations, a description which had only some content in the very early days of computing. Somewhat in between is Knuth's definition of seminumerical algorithms [12]. The chapter of this name in his series of books is the best introduction to parts of our subject.

1.2 The Relation to Algebra

It is not easy to separate computer algebra from mathematical disciplines like algebra, analysis (calculus) and numerical analysis. Algebra has a long tradition of constructive tools, going back to Euclid's algorithm (at least), to Kronecker's [14] concept of constructions "in endlich vielen Schritten", to H. Weyl's [17] intuitionistic insistence on constructions, to Grete Hermann's [11] actual bounds for the number of steps in ideal theory algorithms. Even the influential book of van der Waerden on Modern Algebra [16] contains a bulk of constructive algebraic methods – at least in its early editions. Birkhoff [2, 3] emphasizes the role of "modern (i.e. symbolic) algebra" in computing. Many of the concepts and results of his universal algebra [4] belong to the basis of a rigorous theory of simplification (see this volume). Many-sorted term algebras correspond most precisely to the name symbolic algebra which is sometimes used instead of computer algebra. Already the Arabian word algebra means to shuffle a term from one side to the other side of an equation and Hasse [10] defines algebra as the theory of solving equations built up from the four arithmetical operations. Surprisingly, he points out already in 1957 that "the process (the general formal rules, formalisms, algorithms) of solving the equations, but not the objects" are at focus in algebra. Therefore, from this point of view there is very little difference between algebra and computer algebra as we described it. The same point of view is elaborated by Buchberger and Lichtenberger [7]. However, we would like to maintain formally a difference even if the same subject and the same algebraic algorithm is considered: in algebra the general rules underlying an algorithm are the primary object of study whereas in computer science the tool itself, the algorithm in all its aspects, including its efficiency and implementation, is the central theme. A little synopsis of constructive proofs presented in a sample of algebra books reveals quickly that very different views are possible and that, for example, the books of Lang represent the other end of the spectrum of views on algebra. We refer the reader to [3] and the following chapters to find his own point of view.

1.3 The Relation to Analysis

The reader may be surprised to see in a text on computer algebra algorithms for integration, for computing with elementary transcendental functions or even for solving differential equations. There are several reasons for this inclusion. First: the algorithms just mentioned operate on terms and formulas and they produce symbolic output. Second: a closer look at these algorithms shows that all of them achieve their solution by some kind of algebraization of the problem. We mention only the family of Liouville type theorems. Third: there are algebraic counterparts of analytical concepts. The derivative of a polynomial can be defined in a pure combinatorial manner. The Newton iterations for computing zeros can be replaced by p-adic expansions. Both cases do not involve limit considerations. Finally, there are methods to represent exactly entities defined as limits and having infinite numerical representations, which leads us to the next section.

1.4 The Relation to Numerical Analysis

Many computer algebra computations produce formula output which, in turn, is used as input to numerical procedures, e.g. for integration of rational functions in several dimensions, the first and perhaps the second integration is done symbolically and then the remaining ones numerically. Numerical procedures use finite precision arithmetic and are based on the theory of approximation. A numerical root finder working with finite precision, for example, cannot always isolate the roots, but only isolate clusters of roots the diameter of which depends on the precision and many other parameters. In principle, it is possible and desirable to specify numerical algorithms with the same formal rigor as algebraic algorithms; however, the amount of detail required is much higher and resembles less transparently the specification of the mathematical problem. On the other side, an algebraic algorithm pays for the exactness with — in general much — higher execution times and storage requirements compared to its numerical counterpart. However, there are many instances of problems where an approximation does not make much sense, for example for decision procedures for elementary algebra, for counting real roots of polynomials or even for the degree of a polynomial in one indeterminate, defined as the exponent of the highest non-vanishing term. Since 0 numerically is defined only within the given precision, consequently the degree is also only defined within this precision — a questionable concept of degree. Having seen that Sturm sequences with floating point coefficients count real zeros, giving sometimes more zeros in the subintervals than in the entire interval, the price for a reliable answer seems acceptable. To overstate the point, an algebraic algorithm will never trade the mathematical specification of the output for efficiency to get it approximately.

A consequence of this consideration is that the analytic and empirical analysis of the efficieny of algebraic algorithms is necessary for a responsible use of time and memory. This includes the analysis of the growth of the outputs; they may be so large that any content of information is lost.

Let us consider the relation between numerical and algebraic algorithm more closely in the case of real algebraic numbers. In the numerical case the irrational number α may be represented by two floating point numbers α_m and $\varepsilon > 0$ such that $\alpha_m - \varepsilon < \alpha < \alpha_m + \varepsilon$. In the algebraic case d is the zero of a positive, primitive squarefree integral polynomial $A(x)$ of positive degree. In order to indicate which zero of $A(x)$ is α, one may use an interval I_α with rational endpoints r and s such that $r < \alpha < s$. In addition, one requires that α is the only zero of A with this property. Conceptually, α is represented by A, I_α and a bisection algorithm which for any inputs A, $I_\alpha = (r, s)$ computes $m = (r + s)/2$ and selects (r, m) if $A(r)A(m) < 0$ or (m, s) if $A(m)A(s) < 0$ or (m, m) if $A(m) = 0$. In this way $|r - s|$ can be made arbitrarily small at any desired point in an algebraic computation with α, whereas in the finite precision representation the interval length will grow very quickly.

An idea reconciling the efficiency of numerical arithmetic with the reliability of algebraic computations is k-precision arithmetic where k can dynamically be adjusted such that the algebraic specifications are still met. This has to be carefully checked for every algorithm implemented.

To use integers of arbitrary large size "is natural since computers are now fast enough that any restrictions imposed by the word length of the computer are artificial and unnecessary" (Collins [8]).

2. The Organization of Research Activities in Computer Algebra

From the early beginnings of computer algebra, research activities in this field have been coordinated by SIGSAM (the *Special Interest Group on Symbolic and*

*A*lgebraic *M*anipulation) of ACM (*A*ssociation for *C*omputing *M*achinery). SIGSAM publishes a Bulletin that reports regularly on all events and developments in computer algebra. Information about membership in SIGSAM and subscription to the Bulletin may be obtained from

> Association for Computing Machinery, 1133 Avenue of the Americas, New York, NY 10036, U.S.A.

Furthermore, many other organizations in mathematics and computer science have helped to develop computer algebra by including special sessions on this subject into their meetings and conferences.

The European research activities in computer algebra are not centrally organized so far. There are several independent research groups at various universities and research institutes. The following professional organizations loosely coordinate the activities:

NIGSAM (Nordic Interest Group for Symbolic and Algebraic Manipulation):
> Address: Y. Sundblad, Department of Numerical Analysis and Computer Science, KTH, S-100 44 Stockholm, Sweden.

The group publishes the "NIGSAM-News".

SAM-AFCET (Le Group "Calcul Formel" d'Association Française pour la Cybernétique Economique et Technique):
> Address: M. Bergman, IREM, Faculté de Science de Luminy, Case 901, F-13288 Marseille Cédex 9, France; J. Calmet, IMAG, Laboratoire d'Informatique et de Mathématiques Appliquées de Grenoble, BP 53, F-38041 Grenoble Cédex, France.

The group publishes the CALSYF-Bulletin.

SEAS-SMS (SHARE European Association, Symbolic Mathematical Computation Group):
> Address: J. H. Davenport, Cambridge University, Computer Laboratory, Corn Exchange Street, Cambridge CB2 3QG, England.

Recently, the European groups have established a steering committee. All persons and groups interested in computer algebra are invited to contact this committee. The coordinator is

> H. van Hulzen, Twente University of Technology, Department of Applied Mathematics, P.O. Box 217, 7500 AE Enschede, Netherlands.

SIGSAM, partly in collaboration with the European professional organizations, has organized a series of symposia whose proceedings form a significant portion of what constitutes computer algebra:

SYMSAM 1966:
> ACM Symposium on Symbolic and Algebraic Manipulation, Washington, D.C., March 29 – 31, 1966. Proceedings (Floyd, R. W., ed.): Commun. ACM **9**, 547 – 643 (1966).

SYMSAM 1971:
  ACM Symposium on Symbolic and Algebraic Manipulation, Los Angeles, March 23 – 25, 1971. Proceedings (Petrick, S. R., ed.): published by ACM. [Some of the papers published also in Commun. ACM **14**, 509 – 560 (1971) or in J. ACM **18**, 477 – 565 (1971).]

EUROSAM 1974:
  European Symposium on Symbolic and Algebraic Manipulation, Stockholm, August 1 – 2, 1974. Proceedings (Jenks, R. D., ed.): SIGSAM Bull. **8**/3 (1974).

SYMSAC 1976:
  ACM Symposium on Symbolic and Algebraic Computation, Yorktown Heights (New York), August 10 – 12, 1976. Proceedings (Jenks, R. D., ed.): published by ACM.

EUROSAM 1979:
  European Symposium on Symbolic and Algebraic Manipulation, Marseille, June 26 – 28, 1979. Proceedings (Ng, E. W., ed.): Lecture Notes in Computer Science, Vol. 72 (1979).

SYMSAC 1981:
  ACM Symposium on Symbolic and Algebraic Computation, Snowbird (Utah), August 5 – 7, 1981. Proceedings (Wang, P. S., ed.): published by ACM.

In addition, various conferences on special topics within computer algebra, symposia and summer schools organized by regional organizations, and special sessions on computer algebra as a part of more general conferences have contributed to consolidate the field. Among them are:

IFIP 1966:
  IFIP Working Conference on Symbol Manipulation Languages and Techniques, Pisa, September 5 – 9, 1966. Proceedings (Bobrow, D. G., ed.): Amsterdam: North-Holland 1968.

OXFORD 1967:
  Conference on Computational Problems in Abstract Algebra, Oxford, August 29 – September 2, 1967. Proceedings (Leech, J., ed.): Oxford: Pergamon Press 1970.

MAXIMIN 1977:
  Conference on Symbolic Computational Methods and Applications, St. Maximin (France), March 21 – 23, 1977. Proceedings (Visconti, A., ed.): available from A. Litman, Luminy, 70 Route Léon-Lachamp, F-13288 Marseille Cédex 2, France.

MACSYMA 1977:
  MACSYMA Users' Conference, Berkeley, July 27 – 29, 1977. Proceedings (Fateman, R. J., ed.): published by MIT.

MACSYMA 1979:
  MACSYMA Users' Conference, Washington, D.C., June 20 – 22, 1979. Proceedings (Lewis, V. E., ed.): published by MIT.

ANN ARBOR 1980:
    Short Course: Computer Algebra – Symbolic Mathematical Computation,
    Ann Arbor (Michigan), August 16 – 17, 1980. Synopses of the lectures (Yun,
    D. Y. Y., ed.): AMS Notices, June 1980.

EUROCAM 1982:
    European Computer Algebra Conference, Marseille, April 5 – 7, 1982.
    Proceedings: available from J. Calmet, Grenoble.

The above conference abbreviations will be used in the reference parts of this
volume. For information (including abstracts of papers) about other computer
algebra events (for instance, Boston 1968, Charkov 1972, Paris 1973, Rimini 1973,
Leuven 1973, Purdue 1974, San Diego 1974, Jülich 1975, Amsterdam 1978, Rome
1978, Riken Wakoshi 1978, Kent 1979, Brighton 1979, Antwerp 1981, Stara Lesna
1981) see the respective issues of the SIGSAM Bulletin. References to the special
conferences covering computational group theory are given in the respective
chapter of this volume.

We present a list of bibliographies on computer algebra that have been compiled in
the past (including two surveys by Sammet 1971 and Yun/Stoutemyer 1980, which
contain extensive bibliographies).

Sammet, J. E.: Revised Annotated Descriptor Based Bibliography on the Use of
    Computers for Non-Numerical Mathematics. IFIP **1966**, 358 – 484.
    [Updated in SIGSAM Bull. **10** (1968), **12** (1969), **15** (1970).]
Sammet, J. E.: Software for Nonnumerical Mathematics. In: Mathematical
    Software (Rice, J., ed.), pp. 295 – 330. Academic Press 1971.
Loos, R.: SIGSAM KWIC Index. SIGSAM Bull. **8**/1, 17 – 44 (1974).
Loos, R.: KWIC Index of SIGSAM Bulletin Contributions and Abstracts
    1972 – 1977. SIGSAM Bull. **11**/3, 21 – 61 (1977).
Felsch, V.: A Bibliography on the Use of Computers in Group Theory and Related
    Topics: Algorithms, Implementations, and Applications. Kept current and
    obtainable from Lehrstuhl D für Mathematik, Rheinisch-Westfälische
    Technische Hochschule Aachen, D-5100 Aachen, Federal Republic of
    Germany [see also SIGSAM Bull. **12**/1, 23 – 86 (1978)].
Yun, D. Y. Y., Stoutemyer, D. R.: Symbolic Mathematical Computation. In:
    Encyclopedia of Computer Science and Technology (Belzer, J., Holzman, A.
    G., Kent, A., eds.), Vol. 15, 235 – 310 (1980).
Miola, A. M., Pottosin, I. V.: A Bibliography of Soviet Works in Algebraic
    Manipulations. SIGSAM Bull. **15**/1, 5 – 7 (1981).

(Surveys on computational group theory containing bibliographies are cited in the
respective chapter of this volume. For the whole field of computational number
theory, which is not covered by the present volume, the following book may serve as
a first reference:

Zimmer, H. G.: Computational Problems, Methods, and Results in Algebraic
    Number Theory. Lect. Notes Math., Vol. 268. Berlin-Heidelberg-New York:
    Springer 1972.)

For those interested in installing a computer algebra software system at their computing center, addresses of persons who may be contacted are given below. (This list is drawn from the above survey by Yun/Stoutemyer 1980. It constitutes only a small portion of the whole scale of approximately 60 computer algebra systems.) Most of the systems are available to universities and research organizations for modest handling charges. A description of the systems is given in the chapter entitled "Computer Algebra Systems".

ALTRAN (based on ANSI FORTRAN):
> W. S. Brown, Computing Information Library, Bell Telephone Laboratory, 600 Mountain Ave., Murray Hill, NJ 07974, U.S.A.

CAMAL (based on a portable system implementation language):
> J. P. Fitch, School of Mathematics, University of Bath, Claverton Down, Bath BA2 7AY, England.

FORMAC (host languages FORTRAN or PL/I):
> K. A. Bahr, Gesellschaft für Mathematik und Datenverarbeitung G/IFV, Rheinstrasse 75, D-6100 Darmstadt, Federal Republic of Germany.

MACSYMA (available on the VAXIMA operating system):
> J. Moses, Department of Electrical Engineering and Computer Science, M.I.T., 545 Technology Square, Cambridge, MA 02139, U.S.A.

muMATH (for Intel 8080 and Zilog Z-80 based microcomputers):
> The Soft Warehouse, P.O. Box 11174, Honolulu, HI 96828, U.S.A. (D. R. Stoutemyer).

REDUCE (bootstraped from a LISP kernel):
> A. C. Hearn, The Rand Corporation, 1700 Main Street, Santa Monica, CA 90406, U.S.A.

SAC-2 (based on ANSI FORTRAN):
> G. E. Collins, Computer Science Department, University of Wisconsin, 1210 W. Dayton Street, Madison, WI 53706, U.S.A. (or: R. Loos, Institut für Informatik I, Universität Karlsruhe, Zirkel 2, D-7500 Karlsruhe, Federal Republic of Germany.)

SCRATCHPAD (LISP-based):
> D. Jenks, D. Y. Y. Yun, IBM T. J. Watson Research Center, P.O. Box 218, Yorktown Heights, NY 10598, U.S.A.

SYMBAL (based on an implementation language):
> M. Engeli, FIDES Trust Company, P.O. Box 656, CH-8027 Zürich, Switzerland.

3. About This Volume

3.1 Prerequisites

First we list some prerequisites from computer science. We expect the reader to be familiar with the design and analysis of algorithms. The algorithms in this volume will be presented in informal mathematical notation. The reader should have some programming experience and a basic insight in list processing and storage management in order to see in advance which algebraic programs he can expect to run on his computer. This implies that he should have full control of the algorithms actually used in his computer algebra system, assuming that full documentation and analysis of the algorithms are at hand.

The analysis of algorithms forms an essential part of computer algebra. Thus, for many algorithms computing time analyses are included in this volume. However, for some very important algorithms no analysis is available at present. Also, because of the limited space the question of optimal algorithms (lower bounds, algebraic complexity theory [5]) is not considered in this volume. The basic concepts of worst-case, average and minimal time and space complexity (in dependence on various parameters characterizing classes of inputs) as contained, for instance, in [1, 12, 8] are presupposed. The calculus of O-notation or dominance, respectively, is used throughout the volume, see [1, 12, 8]. (We say that f is *dominated* by g, and write $f \leq g$, iff $f(n) = O(g(n))$. If $f \leq g$ and $g \leq f$ we say that f and g are *codominant* and write $f \sim g$. If $f \leq g$ but not $g \leq f$ we say that f is *strictly dominated* by g and write $f < g$).

The material contained in the books of Knuth [12], in particular chapter 4, is assumed and only shortly repeated when appropriate. Thus, [12], chapter 4, is considered as a "default reference" for algebraic algorithms. The default reference for algebra is van der Waerden's book [16].

For the correctness proofs and computing time analysis of computer algebra algorithms results from various branches of mathematics other than algebra are required as, for instance, analysis, probability theory and statistics, topology and number theory.

3.2 The Intention of the Volume

The exposition in this volume is not intended as a textbook for classroom use, but rather as a condensed skeleton for giving a course in computer algebra. By providing extensive references, it should help to find an access to the literature, which so far is scattered in conference proceedings, technical reports and journals. Hence, this volume may be considered as a step to a systematic treatment of the field. However, it cannot be a substitute for a comprehensive presentation of the subject in a book, for which the need has been felt for a long time [9]. Actually, Lipson [15] is the only algebra text book that covers at least part of computer algebra. In some parts of the volume new material is presented in order to stimulate research in our area.

The material in this volume should be useful for supplementing a traditional course on algebra and as a basic reference for a special computer algebra course as well.

3.3 The Arrangement of the Volume

There are several possible methods for structuring a computer algebra course. In algebra, the material usually is arranged according to structures: semigroups, groups, lattices, rings, fields etc. In computer science, algorithms are often grouped together according to their underlying strategy: divide and conquer methods, greedy algorithms, dynamic programming, transformation and mapping strategies etc. Both approaches would be possible and, indeed, attractive also for computer algebra. In this volume we have used a third alternative: we followed a problem oriented top-down approach, different from the data type approach of algebra and the strategy approach of computer science. This alternative is based on the view that there is a fairly small number of problem types, as for instance the problem type of "solving equations", according to which computer algebra can be partitioned in a natural way. Roughly, the structure of the volume is such that one chapter is devoted to each of the basic problem types and that the sequence of chapters starts with high-level problem types whose motivation is immediately apparent and proceeds to more elementary problems whose solution is needed for the problems at higher levels.

Thus, the volume starts with a chapter on the problem of (canonical) *simplification* of expressions, a subject that pervades nearly all subfields of computer algebra. The next chapter is on the *analysis of* (finite) *algebraic structures*, in particular groups. In fact, this is one of the most successful areas of computer algebra. Closed form solutions for *sums* and *integrals*, a topic that relies on results of nearly all the other areas of computer algebra, is the subject of the next two chapters. Similarly, *decision problems* in algebra, in particular the method of *quantifier elimination*, is another important problem type that stimulated many of the research activities in computer algebra. The *determination of zeros of polynomials*, *factorization* (of polynomials), and the computation of *greatest common divisors* and *generalized remainder sequences* are important subproblems for the solution of the above problems. However, they are interesting and demanding problems also in their own right. A chapter is devoted to each of them. The basic layer of this top-down problem hierarchy is formed by the technique of *computing by homomorphic images*, *computing in transcendental and algebraic extensions* and the *arithmetical operations in various basic algebraic domains*. Correspondingly, these topics are treated in the final four problem chapters. Two chapters on *applications* of computer algebra and *computer algebra systems* form an independent part of the volume. A short chapter on various *bounds* at the end of the volume provides some useful background for the analysis of algorithms presented in the other chapter.

References

[1] Aho, A. U., Hopcroft, J. E., Ullman, J. D.: The Design and Analysis of Computer Algorithms. Reading, Mass.: Addison-Wesley 1974.
[2] Birkhoff, G.: The Role of Modern Algebra in Computing. SIAM Proc. **1971**, 1 – 47.
[3] Birkhoff, G.: Current Trends in Algebra. Am. Math. Mon. **80**/7, 760 – 762 (1973).
[4] Birkhoff, G., Lipson, J.: Heterogeneous Algebra. J. Comb. Theory **8**, 115 – 133 (1970).
[5] Borodin, A. B., Munro, I.: The Computational Complexity of Algebraic and Numeric Problems. New York: American Elsevier 1975.
[6] Brainerd, W. S., Landweber, L. H.: Theory of Computation. New York: J. Wiley 1974.
[7] Buchberger, B., Lichtenberger, F.: Mathematik für Informatiker I (Die Methoden der Mathematik), 2nd ed. Berlin-Heidelberg-New York: Springer 1981.
[8] Collins, G. E.: Computer Algebra of Polynomials and Rational Functions. Am. Math. Mon. **80**/7, 725 – 755 (1973).
[9] Griesmer, J. H.: The State of Symbolic Computation. SIGSAM Bull. **13**/3, 25 – 28 (1979).

[10] Hasse, H.: Höhere Algebra I. Berlin: De Gruyter 1957.
[11] Hermann, G.: Die Frage der endlich vielen Schritte in der Theorie der Polynomideale. Math. Ann. **95**, 736 – 788 (1926).
[12] Knuth, D. E.: The Art of Computer Programming, Vol. I – III. Reading, Mass.: 1968 – 1981.
[13] Knuth, D. E.: Algorithms. Scientific American **236**/4, 63 – 80 (1977).
[14] Kronecker, L.: Werke (Hensel, K., ed.). Leipzig: 1845.
[15] Lipson, J. D.: Algebra and Algebraic Computing. London: Addison-Wesley 1981.
[16] van der Waerden, B. L.: Modern Algebra, Vol. I and II. New York: Frederick Ungar 1953.
[17] Weyl, H.: Randbemerkungen zu Hauptproblemen der Mathematik II: Fundamentalsatz der Algebra und Grundlagen der Mathematik. Math. Z. **20**, 131 – 150 (1924).

Prof. Dr. R. Loos
Institut für Informatik I
Universität Karlsruhe
Zirkel 2
D-7500 Karlsruhe
Federal Republic of Germany

Algebraic Simplification

B. Buchberger, Linz, and **R. Loos,** Karlsruhe

Abstract

Some basic techniques for the simplification of terms are surveyed. In two introductory sections the problem of canonical algebraic simplification is formally stated and some elementary facts are derived that explain the fundamental role of simplification in computer algebra. In the subsequent sections two major groups of simplification techniques are presented: special techniques for simplifying terms over numerical domains and completion algorithms for simplification with respect to sets of equations. Within the first group canonical simplification algorithms for polynomials, rational expressions, radical expressions and transcendental expressions are treated (Sections 3–7). As examples for completion algorithms the Knuth-Bendix algorithm for rewrite rules and an algorithm for completing bases of polynomial ideals are described (Sections 8–11).

1. The Problem of Algebraic Simplification

The problem of simplification has two aspects: obtaining *equivalent but simpler* objects and computing *unique representations for equivalent objects.*

More precisely, let T be a class of (linguistic) objects ("*expressions*") as, for instance, (first-order) terms, (restricted classes of) logical formulae, or (restricted classes of) programs and let $\sim$ be an *equivalence relation* on T, for instance functional equivalence, equality derivable from axioms, equality valid in certain models of axioms, congruence modulo an ideal etc.

The *problem of obtaining equivalent but simpler objects* consists in finding an effective procedure S that maps T in T and meets the following specification: For all objects t in T

(SE) $\qquad\qquad\qquad\qquad S(t) \sim t$ and

(SS) $\qquad\qquad\qquad\qquad S(t) \leqslant t.$

Here, $\leqslant$ is the concept of *simplicity* one is interested in (for instance, "$s \leqslant t$" might be "s is shorter than t", "s needs less memory for representation in computer storage than t", "the evaluation of s is less complex than that of t", "numerically, s is more stable than t", or "the structure of s is more intelligible than that of t" etc.). Of course, $S(t) < t$ will be required for "most" t.

The *problem of computing unique representations for equivalent objects* (= the problem of *canonical simplification*) consists in finding an effective procedure S (a "*canonical simplifier*" or "*ample function*" for $\sim$ (on T)) that maps T in T and meets the following specifications: For all objects s, t in T

(SE) $S(t) \sim t$ and

(SC) $s \sim t \Rightarrow S(s) = S(t)$,

(i.e. S singles out a unique representative in each equivalence class. $S(t)$ is called a *canonical form* of t).

A well known example of a canonical simplifier is the transformation of elementary arithmetical expressions into "polynomial" form: $x^2 - 1$ is the polynomial form of $(1/2)(2x + 2)(x - 1)$, for instance.

The two types of problems are not totally independent: a consequent iterative reduction of the size of expressions with respect to some measure of "simplicity" may, sometimes, well establish a canonical simplifier or at least give an idea how to construct a canonical simplifier. Conversely, a canonical simplifier trivially defines a corresponding notion of simplicity: the canonical form of an object may be called "simpler" than the object itself. On the other hand, practical procedures that intuitively "simplify" expressions may lead to distinct simplified forms for two equivalent expressions and, conversely, canonical forms of simple expressions may be quite "complex" (for instance, the canonical form of $(x + 2)^5$ is $x^5 + 10x^4 + 40x^3 + 80x^2 + 80x + 32$).

In this paper we shall exclusively be concerned with *canonical simplification* of first-order *terms* (= *algebraic* simplification) and only briefly survey the literature on simplification of other types of expressions and on non-canonical simplification at the end of this section.

For characterizing a canonical simplification problem a careful analysis of the notion of *equivalence* involved is vital. The phrases set in italics in the following paragraphs represent a few examples of frequent equivalence concepts.

As linguistic objects, the elementary arithmetical expressions $(x + 2)^5$ and $x^5 + 10x^4 + 40x^3 + 80x^2 + 80x + 32$ are not *identical*. They are equivalent, however, in the sense that they *represent the same polynomial* in $\mathbb{R}[x]$. $x^6 - 1$ and $x - 1$ do not represent the same polynomial in $\mathbb{R}[x]$ nor in $GF(2)[x]$. They are, however, *functionally equivalent* (i.e. describe the same function) over $GF(2)$. The rational expression $(x^6 - 1)/((x - 3)(x^2 - 1))$ is not functionally equivalent to $(x^4 + x^2 + 1)/(x - 3)$ over $\mathbb{Q}$ (consider the values of the two expressions at $x := \pm 1$). These expressions *denote the same element* in $\mathbb{Q}(x)$, however, and are functionally *equivalent as meromorphic functions*. Similarly, the radical expressions $(x - y)/(\sqrt{x} + \sqrt{y})$ and $\sqrt{x} - \sqrt{y}$ are not functionally equivalent over $\mathbb{Q}$ in a strong sense, but are equivalent as meromorphic functions and may be conceived as denoting the same element in the extension field $\mathbb{Q}(x, y)[\sqrt{x}, \sqrt{y}]$, which is isomorphic to $\mathbb{Q}(x, y)[z_1, z_2]$ modulo the polynomial ideal generated by $z_1^2 - x$ and $z_2^2 - y$. Here *congruence* modulo an ideal appears as yet another notion of equivalence. The *equality* $\text{Append}(\text{Cons}(A, \text{Cons}(B, \text{Null})), \text{Cons}(C, \text{Null})) = \text{Cons}(A, \text{Cons}(B, \text{Cons}(C, \text{Null})))$ *may be derived* from the two axioms $\text{Append}(\text{Null}, x) = x$, $\text{Append}(\text{Cons}(x, y), z) = \text{Cons}(x, \text{Append}(y, z))$ by mere "equational reasoning" and is valid in all models of these axioms. In this sense the expressions on the two sides of the equality might be called equivalent. On the other hand, the equality $\text{Append}(x, \text{Append}(y, z)) = \text{Append}(\text{Append}(x, y), z)$ is not derivable from the above axioms by equational calculus alone (some additional induction principle is needed), though the *equality holds* in all computationally "interesting" models (the "initial" models of the axioms). Thus, the two terms $\text{Append}(x, \text{Append}(y, z))$ and $\text{Append}(\text{Append}(x, y), z)$ are equivalent in a sense that is distinct from the preceding one.

For computer algebra, the problem of constructing *canonical simplifiers* is basic, because it is intimately connected with

> the problem of effectively carrying out operations ("*computing*") in algebraic (factor) domains and

> the problem of effectively *deciding equivalence*.

In fact, an effective canonical simplifier immediately yields an algorithm for deciding equivalence and for computing in the factor structure defined by the equivalence. Conversely, a decision procedure for equivalence guarantees the existence of an effective (though, in general, not an efficient) canonical simplifier:

Theorem (Canonical simplification and decidability of equivalence). *Let T be a decidable set of objects and $\sim$ an equivalence relation on T. Then: $\sim$ is decidable iff there exists a canonical simplifier S for $\sim$.* ∎

Example. In the commutative semigroup defined by the generators a, b, c, f, s and the defining relations $as = c^2 s$, $bs = cs$, $s = f$ the rewrite rules $s \to f$, $cf \to bf$, $b^2 f \to af$ constitute a canonical simplifier (this can be established by the methods given in Section 11). $a^5 bc^3 f^2 s^3$ and $a^5 b^2 c^2 s^5$ represent the same element of the semigroup because their canonical forms are identical, namely $a^7 f^5$, whereas cs^2 and $c^2 s$ represent different elements because their canonical elements bf^2 and af are distinct.

Proof of the Theorem. "$\Leftarrow$": By (SE) and (SC): $s \sim t \Leftrightarrow S(s) = S(t)$.

"$\Rightarrow$": Let g be a computable function that maps $\mathbb{N}$ onto T. Define: $S(s) := g(n)$, where n is the least natural number such that $g(n) \sim s$. S is computable, because g is. It is clear that (SE) and (SC) are satisfied. ∎

Theorem (Canonical simplification and computation). *Let T be a decidable set of objects, R a computable binary operation on T, and $\sim$ an equivalence relation on T, which is a congruence relation with respect to R. Assume we have a canonical simplifier S for $\sim$. Define:*

$$\text{Rep}(T) := \{t \in T \mid S(t) = t\} \quad (\text{set of "canonical representatives", ample set}),$$

$$R'(s, t) := (S(R(s, t)) \quad (\text{for all } s, t \in \text{Rep}(T)).$$

Then, $(\text{Rep}(T), R')$ is isomorphic to $(T/\sim, R/\sim)$, $\text{Rep}(T)$ is decidable, and R' is computable. (Here, $R/\sim (C_s, C_t) := C_{R(s,t)}$, where C_t is the congruence class of t with respect to $\sim$). ∎

This theorem shows that, having a canonical simplifier for an equivalence relation that is a congruence with respect to a computable operation, one can algorithmically master the factor structure. Of course, for concrete examples computationally more efficient ways of realizing R' will be possible. The theorem is proven by realizing that $i(t) := C_t$ $(t \in \text{Rep}(T))$ defines an isomorphism between the two structures.

Example. The rewrite rules $x^3 \to x^2$ and $x^2y \to x^2$ constitute a canonical simplifier S on $T := \mathbb{Q}[x, y]$ for the congruence relation modulo the ideal I generated by the two polynomials $x^3y - x^3$ and $x^3y - x^2$. (Again, this can be established by the methods developed in Section 11.) $\text{Rep}(T) =$ set of all polynomials f in $\mathbb{Q}[x, y]$, such that no monomial in f is a multiple of x^3 or x^2y. Multiplication of residue classes modulo I, for instance, may be isomorphically modeled in $\text{Rep}(T)$ by $R'(s, t) := S(s \cdot t)$. Thus, $R'(xy + 1, xy - 1) = S((xy + 1) \cdot (xy - 1)) = S(x^2y^2 - 1) = x^2 - 1$. ∎

It is not always possible to find a canonical simplifier for an equivalence relation on a given set of expressions. In addition to the practical desire of system users to simplify with respect to various intuitive and "local" simplification criteria, this is a theoretical motivation for dealing with non-canonical simplification, see bibliographic remarks below. Actually, it has been proven that some rather simple classes of expressions have an undecidable (functional) equivalence problem and, hence, no canonical simplifier can be found for them (see Section 5). Accordingly, various notions weaker than, but approximating canonical simplification have been introduced in the literature. The two most common ones are *zero-equivalence* simplification and *regular* simplification.

Zero-equivalence simplification may be defined in a context where the given set of expressions contains an element that plays the role of a zero-element:

A computable mapping from T to T is called *zero-equivalence* (or *normal*) *simplifier* for the equivalence relation $\sim$ on T iff for all t in T:

(SE) $S(t) \sim t$ and

(SZ) $t \sim 0 \Rightarrow S(t) = S(0)$.

Of course, canonical simplifiers are normal ones. On the other hand, if there exists a computable operation M on T such that for all s, t in T

(ME) $s \sim t \Leftrightarrow M(s, t) \sim 0$,

then the existence of a normal simplifier implies the existence of a canonical simplifier for $\sim$: Assume M satisfies (ME) (for instance, M might be an operation that represents subtraction) and S is a normal simplifier for $\sim$. Then $\sim$ is decidable, because $s \sim t \Leftrightarrow M(s, t) \sim 0 \Leftrightarrow S(M(s, t)) = S(0)$ (use (ME), (SE) and (SZ)). Now, the existence of a canonical simplifier is guaranteed by the above theorem on canonical simplification and decidability of equivalence.

The notion of *regular simplification* is used in the context of expressions involving transcendental functions (for instance exp, log etc.). Roughly, a regular simplifier guarantees that all non-rational terms in the simplified expression are algebraically independent. (For details see the chapter on computing in transcendental extensions in this volume).

The conceptual framework concerning *canonical simplification*, decidability of equivalence and effective computation in algebraic structures, outlined above is based on various analyses given in Caviness [17], Moses [94], Caviness, Pollack, Rubald [19], Musser [95], Lausch, Nöbauer [77], Loos [79], and Lauer [76].

The simplification of linguistic objects embraces the simplification of *logical formulae* and the "optimization" of *programs*. Both problems are extremely relevant for the computer-aided design and verification of software. Following common usage, we do not treat these topics under the heading of "computer algebra" in this paper, although we believe that a unified treatment of these subjects will be inevitable and promising in the future. King, Floyd [63], Polak [110], Luckham et al. [83], Gerhart et al. [42], Boyer, Moore [8], Suzuki, Jefferson [133], Reynolds [112], Loveland, Shostak [82] are some sources that give an impression of the wide range of simplification techniques for logical formulae in the field of computer-aided program verification and synthesis.

Simplification procedures for algebraic terms that do not explicitly aim at canonicality, but emphasize actual "simplification" of expressions with respect to various, intuitively appealing criteria of simplicity (*non-canonical simplification*) are of great practical importance in computer algebra systems. Since the user of a computer algebra system will be interested in simplification with respect to quite distinct simplification objectives in different contexts, most computer algebra systems provide various simplification procedures for one and the same class of expressions with a possibility for the user to *choose* the most adequate type of simplification *interactively*. Some of the systems include the possibility for the user to *teach the system* new simplification rules.

An overview on existing (non-canonical) simplification facilities, design objectives and simplification criteria applied in computer algebra systems may be obtained by consulting (the introductory sections of) the following papers: Moses [94], Fateman [35], Fitch [40], Brown [11], Hearn [48], Fateman [37] and Pavelle et al. [104] and the manuals of the computer algebra systems, see also the chapter on computer algebra systems in this volume. Fateman [36] and Hearn [47] are illustrative summaries of two of the most advanced simplifiers in existing systems. Yun, Stoutemyer [140] is the most thorough treatment of (non-canonical) simplification.

Substitution (Moses [94]) and *pattern matching* (Fateman [35]) are two basic general purpose techniques for (non-canonical) simplification of expressions. Substitution is used in both directions: substituting expressions for variables in other expressions and recognizing identical subexpressions in a complex expression by pattern matching. An example of applying these techniques is Stoutemyer's simplifier (Stoutemyer [132]) for expressions involving the absolute value function and related functions. Mathematical and software engineering details of general and special purpose (non-canonical) simplifiers are treated in a big number of articles, some typical of them being Fateman [34], Griss [45], Jenks [60], Lafferty [67]. Still, basic techniques are used from the pioneer works of Fenichel [38], Korsvold [66], Tobey [135] and others.

Some special and challenging types of non-canonical "simplifications" are standardization of expressions for *intellegible output* (see, for instance, Martin [87], Foderaro [41]), and simplification in the sense of "*extracting interesting informations*" from expressions (for instance, information about boundedness, continuity, monotonicity, convexity etc.), see Stoutemyer [131].

2. Terms and Algebras: Basic Definitions

A *signature* (or *arity-function*) is a family of non-negative integers (i.e. a function a: $\text{Def}(a) \to \mathbb{N}_0$). The index set of a (i.e. $\text{Def}(a)$) may be called the set of *function symbols* of a, in short $F(a)$. Furthermore, $F(a, n) := \{f \in F(a)/a(f) = n\}$ (set of n-ary function symbols). (We often shall omit the arguments of operators like F if they are clear from the context, compare C, Term etc. in the sequel).

An *a-algebra* is a family A (with index set $F(a)$) of functions ("*operations*") on a set M, such that for all $f \in F(a)$ the function $A(f)$ "denoted by the function symbol f" is a mapping from $M^{a(f)}$ into M. ($A(f)$ is an element in M if $a(f) = 0$). M is called the *carrier* $C(A)$ of A. By abuse of language, sometimes A will be written in contexts where $C(A)$ should be used.

We assume that the reader is familiar with the following notions: *generating set* for an *a*-algebra, *homomorphic* (*isomorphic*) *mapping* between two *a*-algebras and *congruence* relation on an *a*-algebra.

In the class of all *a*-algebras, for every set X there exists a "*free a-algebra A_0 with generating set X*" with the following property: $X \subseteq C(A_0)$, A_0 is generated by X, and for all *a*-algebras A and all mappings $m: X \to C(A)$ there exists a unique homomorphism h from A_0 to A which extends m (*universal property* of free algebras). Up to isomorphisms, A_0 is uniquely determined.

Let X be a (denumerable) set disjoint from $F(a)$ (X is called a set of *variables*). The following *term algebra* $\text{Term}(X, a)$ is an example of a free *a*-algebra with generating set X: The carrier of $\text{Term}(X, a)$ is the minimal set T' such that

$$x \in X \Rightarrow x \in T',$$

$$t_1, \ldots, t_n \in T', \qquad f \in F(a, n) \Rightarrow ft_1 \cdots t_n \in T'$$

($x, t_1, \ldots, t_n, ft_1 \cdots t_n$ denote strings of symbols here!). Furthermore, for every function symbol f in $F(a, n)$ the function $\text{Term}(X, a)(f)$, i.e. the function denoted by f in the term algebra, is defined by

$$\text{Term}(X, a)(f)(t_1, \ldots, t_n) = ft_1 \cdots t_n \qquad (t_1, \ldots, t_n \in C(\text{Term}(X, a))).$$

(For example, $t := \cdot + xy7$ is a *term* from $\text{Term}(X, a)$, where $X := \{x, y\}$, $F(a) = \{\cdot, +, 7\}$, $a(\cdot) = a(+) = 2$, $a(7) = 0$. More readable notations for terms (infix, parentheses etc.) will be used in the subsequent examples). Concrete representations of terms in computers (for instance as threaded lists) are other examples of free *a*-algebras. Terms in $\text{Term}(X, a)$ that do not contain variables are called *ground terms* over a.

A homomorphic mapping σ from Term into Term with the property that $\sigma(x) = x$ for almost all $x \in X$ is called *substitution*. If A is an *a*-algebra, an *assignment* in A is a homomorphic mapping v from Term into A. For $t \in$ Term, $v(t)$ is called the *value* of t at v. Because of the universal property of Term, a substitution (an assignment) is totally determined by fixing its values for all $x \in X$.

Suppose that X is denumerable and is bijectively enumerated by $x : \mathbb{N} \to X$. Let A be an a-algebra, $t \in$ Term, and n be such that all variables occurring in t appear in $\{x_1, \ldots, x_n\}$. The n-ary *function* $\text{Fun}(A, n, t)$ *described by* t on A is defined by $\text{Fun}(A, n, t)(a_1, \ldots, a_n) = v(t)(a_1, \ldots, a_n \in A)$, where v is an assignment that satisfies $v(x_1) = a_1, \ldots, v(x_n) = a_n$. t_1 and t_2 are *functionally equivalent* on A iff $\text{Fun}(A, n, t_1) = \text{Fun}(A, n, t_2)$ (where n is such that the variables occurring in t_1 or t_2 appear in $\{x_1, \ldots, x_n\}$).

Let $E \subseteq$ Term $\times$ Term. (A pair (a, b) in E will be conceived as the left-hand and right-hand side of an "*equation*" in the subsequent context). Two important equivalence relations on Term are associated with E, which are basic for all algebraic theories: $E \models s = t$ (s and t are "*semantically equal* in the theory E") and $E \vdash s = t$ (s and t are "*provably equal* in the theory E").

Definition of $E \models s = t$. If A is an a-algebra, we say that the equation $s = t$ is *valid* in A (or that A is a *model* for $s = t$) iff for all assignments v in A $v(s) = v(t)$. The set of all a-algebras A, in which all equations $a = b$ of E are valid is called the *variety* $\text{Var}(a, E)$ of E. Now,

$$E \models s = t \text{ iff } s = t \text{ is valid in all } a\text{-algebras of Var}(a, E). \quad \blacksquare$$

Definition of $E \vdash s = t$. $E \vdash s = t$ iff the formula $s = t$ can be derived in finitely many steps in the following calculus ("*equational calculus*"; a, b, c etc. $\in$ Term):

$$\frac{\quad}{a = b} \qquad \text{for all } (a, b) \in E \text{ ("elements of } E \text{ are } axioms\text{"),}$$

$$\frac{\quad}{a = a}, \quad \frac{a = b}{b = a}, \quad \frac{a = b, b = c}{a = c} \quad (reflexivity, symmetry, transitivity \text{ of } =),$$

$$\frac{a = b}{\sigma(a) = \sigma(b)} \qquad (\text{for all substitutions } \sigma; \; substitution \text{ rule}),$$

$$\frac{a_1 = b_1, \ldots, a_n = b_n}{fa_1 \cdots a_n = fb_1 \cdots b_n} \qquad (\text{for all } f \in F(a, n); \; replacement \text{ rule}). \quad \blacksquare$$

The equational calculus is correct and complete for the above defined notion of semantical equality by the following theorem.

Theorem (Birkhoff 1935).

$$E \models s = t \qquad iff \qquad E \vdash s = t. \quad \blacksquare$$

By this theorem, semantical equality can be semi-decided if E is at least a recursively enumerable set. For more details about the above concepts see the monographies on universal algebra (for instance Cohn [21], Lausch, Nöbauer [77]) and the survey on equational theory (Huet, Oppen [58]) which covers also the extensions of the terminology to many-sorted algebras, introduced in Birkhoff, Lipson [7], and the literature on abstract data types, see for instance Musser [96]. The proof of Birkhoff's theorem is given in Birkhoff [6], compare also O'Donnell [101].

In the sequel, for fixed E, $=_E$ denotes the equivalence relation defined by: $s =_E t$ iff $E \vdash s = t$.

3. Canonical Simplification of Polynomials and Rational Expressions

The General Concept of Polynomial

Let E be a set of equations and A an a-algebra in $\mathrm{Var}(E)$. All terms with "constants denoting elements in A", which can be proven equal by E and the definition of the operations in A, are said to "represent the same E-polynomial over A". More precisely, the algebra of E-*polynomials* over A (with variable set X) $\mathrm{Pol}(X, a, A, E)$ is the a-algebra that has as its carrier the factor set $C(\mathrm{Term}(X, a'))/\sim$, where a' and (the congruence) $\sim$ is defined as follows: $F(a') := F(a) \cup \mathrm{Name}(A)$, where $\mathrm{Name}(A)$ (a set of *names* for elements in $C(A)$) is a set that is disjoint from $F(a)$ and X and has the same cardinality as $C(A)$. $a'(f) := a(f)$ for $f \in F(a)$ and $a'(f) := 0$ for $f \in \mathrm{Name}(A)$. Furthermore, for $s, t \in \mathrm{Term}(X, a')$

$$s \sim t \qquad \text{iff} \qquad (E \cup \mathrm{Op}(A)) \vdash s = t,$$

where $\mathrm{Op}(A)$ is the set of axioms that completely describe the operations of A, i.e.,

$$\mathrm{Op}(A) := \{(f i_1 \cdots i_n, i)/f \in F(a, n), A(f)(c(i_1), \ldots, c(i_n)) = c(i),$$

$$i_1, \ldots, i_n, i \in \mathrm{Name}(A), n \in \mathbb{N}\}$$

(c is some bijection from $\mathrm{Name}(A)$ onto $C(A)$).

Example. $E :=$ set of axioms for commutative rings, $A :=$ ring of integers.

$$\mathrm{Op}(A) = \{1 \cdot 1 = 1, \ 1 \cdot 2 = 2, \ldots, 5 \cdot 7 = 35, \ldots, 1 + 1 = 2, \ldots,$$

$$(-5) + 7 = 2, \ldots\}.$$

$(3x + 1) \cdot (2 - 3x \cdot x) = -9 \cdot x \cdot x \cdot x - 3 \cdot x \cdot x + 6 \cdot x + 2$ is derivable from E and $\mathrm{Op}(A)$ in the equational calculus, hence, the two sides of the equation denote the same E-polynomial over A. If $A := GF(2)$, then $\mathrm{Op}(A) = \{0 \cdot 0 = 0, 0 \cdot 1 = 0, \ldots, 1 + 1 = 0\}$. The equation $x \cdot x + 1 = x + 1$ is not derivable from E and $\mathrm{Op}(A)$ (proof!). $x \cdot x + 1$ and $x + 1$ do not represent the same E-polynomial over A (although $\mathrm{Fun}(A, 1, x \cdot x + 1) = \mathrm{Fun}(A, 1, x + 1)$). For E as above, $\mathrm{Pol}(\{x_1, \ldots, x_n\}, a, R, E)$ is denoted by $R[x_1, \ldots, x_n]$, usually. ∎

Canonical Simplifiers for Polynomials

Let $E :=$ set of axioms for *commutative rings* with 1 and R be an arbitrary commutative ring with 1. It is well known (see, for instance, Lausch, Nöbauer [77], p. 24), that the set of all terms of the form $a_n x^n + \cdots + a_0$ ($a_i \cdots$ (names for) elements in R, $a_n \neq 0$) and the term 0 form a system of canonical forms for $R[x]$ (x^n abbreviates $x \cdot x \cdot \cdots \cdot x$, n times). The set of terms of the form $a_{i_1} \cdot x^{i_1} + \cdots + a_{i_m} x^{i_m}$ ($a_{i_j} \neq 0 \cdots$ (names) of elements from R, $i_j \cdots$ "multi-indices" $\in \mathbb{N}_0^n$, $i_1 > \cdots > i_m$ in the lexicographic ordering, for instance) and the term 0 form a system of canonical forms for $R[x_1, \ldots, x_n]$ (x^i is an abbreviation for the term $x_1^{i_1} \cdot \cdots \cdot x_n^{i_n}$, if i is a multi-index). It is clear, how a canonical simplifier $S : \mathrm{Term}(X, a') \to R[x]$, respectively $\to R[x_1, \ldots, x_n]$, may be constructed by iteratively "applying" the axioms of E and the "operation table" $\mathrm{Op}(A)$ as "rewrite rules" (supposing that the operations in R are computable). An alternative system of canonical forms for $R[x_1, \ldots, x_n]$, often used in computer algebra systems, is the "recursive" representation: $(3x^2 + 5)y^2 + (x - 3)y + (x^3 - 2x + 1)$, for instance, is a "*recursive*" *canonical form* of a polynomial (see Collins [22]).

In Lausch, Nöbauer [77], pp. 27 canonical forms of polynomials over *groups*, *distributive lattices* with 0 and 1 and *boolean algebras* for $X := \{x_1, \ldots, x_n\}$ are given. Again, the respective canonical simplifiers essentially consist in iteratively "applying" the axioms defining the respective variety.

Canonical Simplifiers for Rational Expressions

Informally, rational expressions over an integral domain R are those formed from the constants and variables by the symbols for the ring operations and a symbol for "division". The equivalence of such expressions cannot be defined along the above pattern, because the special role of the zero element in division cannot be formulated by an equational axiom. Therefore, rational expressions are considered as representing elements in the quotient field formed from $R[x_1, \ldots, x_n]$. Thus, the simplification problem consists in finding a canonical simplifier S for the equivalence relation $\sim$ defined on $R[x_1, \ldots, x_n] \times (R[x_1, \ldots, x_n] - \{0\})$ by:

$$(f_1, g_1) \sim (f_2, g_2) \qquad \text{iff} \qquad f_1 \cdot g_2 = f_2 \cdot g_1.$$

In the case when R is a field, the quotient field defined above is isomorphic to the field obtained by transcendentally extending R by $x_1, \ldots, x_n$. Note that $r \sim r'$ does not imply that r and r' are functionally equivalent, compare Brown [11], p. 28. For example $(x^2 - 1)/(x - 1) \sim x + 1$, but $(x^2 - 1)/(x - 1)$ is undefined for $x = 1$ whereas $x + 1$ yields 2. However, one can define division of functions in the way it is done in the theory of meromorphic functions (where, in the above example, the value of $(x^2 - 1)/(x - 1) = (x - 1)(x + 1)/(x - 1)$ for $x = 1$ is *defined* to be $x + 1 = 2$). We, then, have $(f_1, g_1) \sim (f_2, g_2)$ iff (the function denoted by f_1/g_1) = (the function denoted by f_2/g_2). Rational expressions t that are not quotients of two polynomials, for instance $(x - x/x^2)/(x + 1)$, are simplified by using rational arithmetic (see the chapter on arithmetic in algebraic domains) on the simplified quotients that appear as subterms in t:

$$(x - x/x^2)/(x + 1) \to (x/1 - 1/x)/((x + 1)/1) \to ((x^2 - 1)/x)/((x + 1)/1)$$
$$\to (x - 1)/x.$$

Let us suppose that $R[x_1, \ldots, x_n]$ is a unique factorization domain with 1, for which there exist three computable functions $G, /, l$ with the following properties:

$G(f, g) =$ a greatest common divisor (GCD) of f and g
  (i.e. G satisfies: $G(f, g)|f$, $G(f, g)|g$ and
  for all h: if $h|f$ and $h|g$, then $h|G(f, g)$),

$f|g \Rightarrow f \cdot (g/f) = g$ (division),

$l(f)$ is a unit and $f \equiv g \Rightarrow l(f) \cdot f = l(g) \cdot g$

(i.e. the function $s(f) := l(f) \cdot f$ is a canonical simplifier for the relation $\equiv$ defined by: $f \equiv g$ iff f and g are associated elements, i.e. $f = u \cdot g$ for some unit u.

A unit is an element for which there exists an inverse).

In this case the above simplification problem can be solved by the following *canonical simplifier*:

$$S((f,g)) := (c \cdot (f/G(f,g)), \; c \cdot (g/G(f,g))), \quad \text{where} \quad c = l(g/G(f,g)).$$

This solution of the simplification problem for rational expressions is essentially based on procedures for a GCD (see the chapter on polynomial remainder sequences). Since these procedures are relatively time consuming it is important to carefully avoid superfluous calls of GCD in the arithmetic based on the above canonical forms (see the chapter on arithmetic in algebraic domains, algorithms of Henrici [49]). Also, it may be recommendable to store the numerator and denominator of rational expressions in factored or partially factored form, if such factors are known in special contexts (see Brown [11], Hall [46]).

4. Canonical Simplification of Radical Expressions

Roughly, *radical expressions* are terms built from variables $x_1, \ldots, x_n$, constants (for elements in $\mathbb{Q}$), the arithmetical function symbols $+$, $-$, $\cdot$, $/$ and the radical sign $\sqrt[q]{}$ or, equivalently, rational powers ("*radicals*") s^r ($r \in \mathbb{Q}$) (s in s^r is called a *radicand*). A natural (modified) notion of functional equivalence for these expressions is as follows:

$$s \sim t \; (s \text{ is } meromorphically \text{ } equivalent \text{ to } t) \quad \text{iff} \quad \mathrm{Fun}'(s) = \mathrm{Fun}'(t).$$

Here, Fun' essentially is defined as Fun, but $\mathrm{Fun}'(s/t) = \mathrm{Fun}'(s)/'\mathrm{Fun}'(t)$ where $/'$ denotes division of meromorphic functions (see the preceding section) and, furthermore, $\mathrm{Fun}'(s^r)$ (where r denotes a positive rational) is a meromorphic function that satisfies

(R) $\mathrm{Fun}'(s^r)^v - \mathrm{Fun}'(s)^u = 0$

(where u, v are relatively prime integers with $r = u/v$). Let T be a set of radical terms built from the radicals $s_1^{r_1}, \ldots, s_k^{r_k}$. Suppose that for all $1 \leq i \leq k$ the equation (R) for $s_i^{r_i}$ is irreducible over $\mathbb{Q}(\mathrm{Fun}'(x_1), \ldots, \mathrm{Fun}'(x_n), \mathrm{Fun}'(s_1^{r_1}), \ldots, \mathrm{Fun}'(s_{i-1}^{r_{i-1}}))$. Then all possible *interpretations* Fun' of the expressions in T satisfying the above conditions are "differentially isomorphic", i.e. the structure of the field of functions described by expressions in T does not depend on the chosen "branches" of the meromorphic functions defined by (R) ("*regular algebraic field description*", see Risch [116], Caviness, Fateman [18], Davenport [25]).

Thus, $\sim$ on T may be decided by the following procedure:

1. Construct an algebraic extension field $K := \mathbb{Q}(x_1, \ldots, x_n)[\alpha_1, \ldots, \alpha_m]$ such that all terms in T may be conceived as pure arithmetical terms in K, i.e. terms that are composed from constants for elements in K and the arithmetical function symbols $+$, $-$, $\cdot$, $/$.

2. For deciding $s \sim t$ evaluate s and t by applying rational simplification and an effective arithmetic in K. $s \sim t$ iff both evaluations yield the identical result. The construction of minimal irreducible polynomials for $\alpha_1, \ldots, \alpha_m$ is a means for having an effective arithmetic available in K.

An effective evaluator for ground terms over K, hence, is a *canonical simplifier* for $\sim$ on T. (Note, however, that this simplifier depends on the field K constructed in step 1 of the procedure).

For unambiguously specifying the element of K denoted by a given radical term t it is still necessary to state the *conventions* one wants to use for relating subterms of the form $(s \cdot t)^r$ with s^r and t^r. Note that the rule $(s \cdot t)^r = s^r \cdot t^r$ is a convention that attributes a special interpretation to $(s \cdot t)^r$. Other conventions are possible that imply other interpretations. For instance, $\sqrt{(x \cdot x)}$ may equally well be meant to denote $\sqrt{x} \cdot \sqrt{x} = x$ or $\sqrt{(-x)} \cdot \sqrt{(-x)} = -x$ (this corresponds to computing in $\mathbb{Q}(x)[y]$ modulo the two irreducible factors $y - x$ and $y + x$ of $y^2 - x^2$ respectively). Conventions of the kind $\sqrt{(x \cdot x)} = \mathrm{abs}(x)$, though desirable in certain contexts, would lead outside the range of interpretations, for which an isomorphism theorem holds.

A simplification algorithm of the above type has been developed in Caviness [17], Fateman [35], Caviness, Fateman [18] for the special case of unnested radical expressions, i.e. radical expressions whose radicals do not contain other radicals as subterms:

Algorithm (Canonical Simplification of Radical Expressions, Caviness, Fateman 1976).

Input:

$t_1, \ldots, t_l$ (unnested radical expressions in the variables $x_1, \ldots, x_n$).

Output:

1. $M_1, \ldots, M_m$ (minimal irreducible polynomials for algebraic entities $\alpha_1, \ldots, \alpha_m$, such that $t_1, \ldots, t_l$ may be conceived as pure arithmetical ground terms with constants from $K := \mathbb{Q}(x_1, \ldots, x_n)[\alpha_1, \ldots, \alpha_n])$.

2. $s_1, \ldots, s_l$ (s_i is the result of evaluating t_i in K. s_i is taken as the *canonically simplified* form of t_i).

1. Construction of K:

1.1.   All $s_i := t_i$.

1.2.   Transform all s_i
        by rationally *simplifying all radicands* and applying the rule
        $(p/q)^{-r} \twoheadrightarrow (q/p)^r$.

1.3. $R :=$ set of all "*radicals*"
        in the s_i, i.e. of subterms of the form $(p/q)^r$, where p and q are relatively prime elements in $\mathbb{Z}[x_1, \ldots, x_n]$, q is positive (i.e. the leading coefficient of q is positive), r is a positive rational number $\neq 1$.

1.4. $P :=$ set of all "*polynomial radicands*"
        in the s_i, i.e. polynomials p, q for which $(p/q)^r$ is in R.

1.5. $B := $ a "*basis*" for P,

        i.e. a set of irreducible positive polynomials in $\mathbb{Z}[x_1, \ldots, x_n]$ of degree
$\geqslant 0$ such that

        if b is in B then $b|p$ for a p in P and

        if p is in P then $p = \pm \prod b_i^{e_i}$ for certain b_i in B and e_i in $\mathbb{N}$.

        (A basis B for P can be obtained by factorizing all p in P and collecting all irreducible factors).

1.6.   If P contains a non-positive element: Add the element -1 to B.

        $m := $ *number of elements* in B.

1.7.   Transform all s_i by applying the following *conventions*:

        $(p/q)^r \rightarrow p^r/q^r$,

        $(\prod b_i^{e_i})^r \rightarrow \prod (b_i^{e_i \cdot r})$, $(- \prod b_i^{e_i})^r \rightarrow (-1)^r \cdot \prod (b_i^{e_i \cdot r})$,

        $b^r \rightarrow b^w \cdot b^{u/v}$, if $r = w + u/v$ ($w \in \mathbb{N}_0$, $0 \leqslant u/v < 1$, u, v relatively prime).

1.8.   For all b in B determine the *radical degree* $d(b)$:

        $d(b) := \text{lcm}(v_1, \ldots, v_k)$, where $u_1/v_1, \ldots, u_k/v_k$ are the reduced rational powers to which b appears in the s_i.

1.9.   Transform all s_i by

        expressing all $b_j^{u/v}$ ($b_j \in B$) as *powers of* $b_j^{1/d(b_j)}$ and

        replacing all $b_j^{1/d(b_j)}$ by new variables y_j.

1.10. Determine *minimal irreducible polynomials*:

        $M_j := $ an irreducible factor of $y_j^{d(b_j)} - b_j$ over

        $\mathbb{Q}(x_1, \ldots, x_n)[\alpha_1, \ldots, \alpha_{j-1}]$, where $\alpha_j := $ "$b_j^{1/d(b_j)}$".

        (If -1 is in B, say $b_1 = -1$, then $\alpha_1 := \omega_{2d(b_1)}$).

2. Determine the canonical forms s_i:

2.1. Iterate:

        Rationally simplify all s_i and

        execute the arithmetical operations in K, i.e. in

        $\mathbb{Q}(x_1, \ldots, x_n)[y_1, \ldots, y_m]$ modulo the polynomials M_j. ∎

Example. $t = (((2x - 2)/(x^3 - 1))^{-7/3} + (2/(x + 1))^{1/2})/(24x + 24)^{1/4}$.

1.2.: $s = (((x^2 + x + 1)/2)^{7/3} + (2/(x + 1))^{1/2})/(24x + 24)^{1/4}$.

1.3., 1.4.: $R = \cdots$, $P = \{x^2 + x + 1, 2, x + 1, 24x + 24\}$.

1.5.: $B = \{2, 3, x + 1, x^2 + x + 1\}$.

1.7.: $s = ((((x^2 + x + 1)^2(x^2 + x + 1)^{1/3})/(2^2 2^{1/3})$

        $+ (2^{1/2}/(x + 1)^{1/2}))/(2^{3/4} 3^{1/4}(x + 1)^{1/4})$.

1.8.: $d(2) = 12$, $d(3) = 4$, $d(x + 1) = 4$, $d(x^2 + x + 1) = 3$.

1.9.: $s = (((x^2 + x + 1)^2 \cdot y_4)/(4 \cdot y_1^4) + y_1^6/y_3^2)/(y_1^9 \cdot y_2 \cdot y_3))$.

1.10.: $M_1 = y_1^{12} - 2$, $M_2 = y_2^4 - 3$, $M_3 = y_3^4 - (x + 1)$, $M_4 = y_4^3 - (x^2 + x + 1)$.

2.1.: $s = ((x^2 + x + 1)^2 \cdot y_3^2 \cdot y_4 + 4 \cdot y_1^{10})/(4 \cdot y_1^{13} \cdot y_2 \cdot y_3^3)$

        $= ((x^4 + 2x^3 + 3x^2 + 2x + 1)/(48x + 48)) \cdot 2^{11/12} \cdot 3^{3/4}(x + 1)^{3/4}$

        $\cdot (x^2 + x + 1)^{1/3} + (1/(6x + 6)) \cdot 2^{9/12} \cdot 3^{3/4} \cdot (x + 1)^{1/4}$. ∎

Note that the above construction, in general, does not lead to the minimal extension field K of $\mathbb{Q}(x_1, \ldots, x_n)$ in which the above simplification procedure may be carried out.

Several computational improvements of the algorithm and a complexity analysis based on work of Epstein [32] can be found in Caviness, Fateman [18].

(Theoretically, the algorithm is exponential in the number of variables. However, it is feasible for expressions of moderate size.)

An algorithm similar to the Caviness-Fateman algorithm has also been developed in Fitch [39, 40]. The simplification of radicals is also treated in Zippel [142]. In Shtokhamer [127] ideas for "local" simplification of unnested radical expressions are given (i.e. simplification in the style of step 2 prior to constructing the field K for the whole expression). For treating nested radicals effective computation in residue class rings modulo arbitrary polynomials in $\mathbb{Q}[x_1, \ldots, x_n]$ is an adequate framework. Proposals in this direction have been made in Kleiman [64], Buchberger [12], Shtokhamer [125, 126], see also Section 11.

5. Canonical Simplification of Transcendental Expressions: Unsolvability Results

Further extension of the expressive power of the term classes considered may lead to term classes whose simplification problem is algorithmically unsolvable. An example is the class $R2$ of terms generated from one variable x, constants for the rationals, π, and the function symbols $+$, $\cdot$, sin, abs, whose simplification problem with respect to functional equivalence $\sim_{\mathbb{R}}$ on $\mathbb{R}$ is algorithmically unsolvable. (It should be clear how the definition of $R2$ and $\sim_{\mathbb{R}}$ can be made precise following the pattern given in Section 2. Expressions involving function symbols for transcendental functions like exp, log, sin, erf are called *transcendental expressions*. The class of rational expressions over $\mathbb{C}$ extended by exp and log is called the class of *elementary transcendental* expressions).

Theorem (Caviness, Richardson, Matijasevic 1968, 1970). *The predicate "$t \sim_{\mathbb{R}} 0$" is undecidable (for $t \in R2$). (Hence, $\sim_{\mathbb{R}}$ is undecidable and, by the theorem on canonical simplification and equivalence, there cannot exist a zero-equivalence or canonical simplifier for $\sim_{\mathbb{R}}$ on $R2$).* ∎

Proof (Sketch). The proof proceeds by a number of problem reductions via recursive translators ("m-reductions" in the sense of algorithm theory, Rogers [119]. The following variant of the famous result of Matiyasevic [89] concerning the undecidability of Hilbert's tenth problem serves as a basis of the proof. For a certain n the predicate

(1) "There exist $z_1, \ldots, z_n \in \mathbb{Z}$ such that $\mathrm{Fun}_{\mathbb{Z}}(t)(z_1, \ldots, z_n) = 0$"

is undecidable for $t \in P := \mathbb{Z}[x_1, \ldots, x_n]$. (Actually $n = 13$ is the minimal such n known so far, see Manin [84].)

First problem reduction. There exists a recursive "translator" $T1 : P \twoheadrightarrow R3$ such that for all $t \in P$:

(2) there exist $z_1, \ldots, z_n \in \mathbb{Z}$ such that $\mathrm{Fun}_{\mathbb{Z}}(t)(z_1, \ldots, z_n) = 0$ $\Leftrightarrow$
    there exist $b_1, \ldots, b_n \in \mathbb{R}$ such that $\mathrm{Fun}_{\mathbb{R}}(T1(t))(b_1, \ldots, b_n) < 0$.

Here, $R3$ essentially is defined as $R2$ with the difference that we allow n variables $x_1, \ldots, x_n$ and that abs does not appear in the terms of $R3$. Using sin and π, a suitable $T1$ may be defined as follows ($t \in P$):

(3) $T1(t) := (n+1)^2 \cdot [t^2 + \sin^2(\pi x_1) \cdot D(1, t)^2 + \cdots + \sin^2(\pi x_n) \cdot D(n, t)^2] - 1$.

Here $D(i, t)$ is a "dominating" function for $\partial/\partial x_i (t^2)$. It is clear that, given $t \in P$, a term $t' \in P$ may be found effectively, such that t' describes $\partial/\partial x_i(t^2)$. Thus, for constructing $D(i, t)$ it suffices to have a recursive function $T: P \to P$ such that, for all $t \in P$, $T(t)$ describes a "dominating" function for t, i.e. a function that satisfies for all $b_1, \ldots, b_n \in \mathbb{R}$, and for all $\varDelta_1, \ldots, \varDelta_n \in \mathbb{R}$ with $|\varDelta_i| < 1$:

(4) $\text{Fun}_{\mathbb{R}}(T(t))(b_1, \ldots, b_n) > 1$ and

$\quad \text{Fun}_{\mathbb{R}}(T(t))(b_1, \ldots, b_n) > |\text{Fun}_{\mathbb{R}}(t)(b_1 + \varDelta_1, \ldots, b_n + \varDelta_n)|.$

A suitable T may be defined by induction on the structure of the terms in P. Knowing (4), the proof of (2) is easy.

Second problem reduction. There exists a recursive "translator" $T_2: R3 \to R3$ such that for all $t \in R3$:

(5) there exist $b_1, \ldots, b_n \in \mathbb{R}$ such that $\text{Fun}_{\mathbb{R}}(t)(b_1, \ldots, b_n) < 0 \quad \Leftrightarrow$

$\quad$ there exists $b \in \mathbb{R} \qquad$ such that $\text{Fun}_{\mathbb{R}}(T2(t))(b) \qquad < 0.$

The construction of $T2$ is tricky: it simulates the application of "pairing functions", which normally are used in algorithm theory for the reduction of n-dimensional problems to one-dimensional ones. Since pairing functions are not available in $R3$, the sinus-function is used to partially approximate the effect of pairing functions.

Third problem reduction. There exists a recursive "translator" $T3: R3 \to R2$ such that for all $t \in R3$:

(6) there exists $b \in R$ such that $\text{Fun}_{\mathbb{R}}(t)(b) < 0 \quad \Leftrightarrow$

$\quad\quad\quad\quad \Leftrightarrow \quad \text{not } (T3(t) \sim_{\mathbb{R}} 0).$

A suitable $T3$ is:

(7) $T3(t) := \text{abs}(t) - t.$

Given the unsolvability of the problem stated in (1) and the problem reductions in (2), (5), and (6), it is thus shown that $\sim_{\mathbb{R}}$ is undecidable. ∎

The proof method for the above unsolvability result has been developed in Richardson [113]. It has been applied to Matiyasevic's result in Caviness [17]. Matiyasevic's theorem is the culmination of earlier work by M. Davis, J. Robinson and H. Putnam, for instance Davis, Putnam, Robinson [27]. A clear and detailed presentation of Matiyasevic's result is given in Davis [26]. Other unsolvability results concerning transcendental terms in computer algebra may be found in Fenichel [38], Risch [116], Moses [93].

6. (Canonical) Simplification of Transcendental Expressions: Miscellaneous Techniques

In this section we review miscellaneous techniques for canonical simplification of transcendental expressions that have been developed before the technique based on the "structure theorems" (see next section) has been matured into its present state. There are three groups of methods falling into this category: *specific techniques* for relatively limited classes of transcendental expressions, *point-evaluation* (a probabilistic method), and general *reduction methods* (reducing the zero-equivalence problem for a given expression to that of simpler expressions).

Examples for methods in the first group are a zero-equivalence simplifier for the "*rational exponential expressions*" (rational numbers, i, π, $x_1, \ldots, x_n$, $+$, $-$, $\cdot$, $/$, exp) given in Brown [9] and a canonical simplifier for the "*exponential polynomials*" (rational numbers, x, $+$, $-$, $\cdot$, exp) given in Caviness [17], see also Moses [94]. These algorithms, essentially, consist in a systematic application of known identities for exp (as, for instance, $\exp(x \cdot y) = \exp(x) + \exp(y)$) and applying tests for (linear or algebraic) dependencies between subterms. The proof of the canonicality (normality) of these algorithms is based on number theoretic *conjectures* that are plausible although their proofs seem to be extremely difficult. (For instance, Brown uses the conjecture that if $\{q_1, \ldots, q_k, i\pi\}$ is linearly independent over the rational numbers, $\{e^{q_1}, \ldots, e^{q_k}, z, \pi\}$ is algebraically independent over the rational numbers.)

Point-evaluation (Martin [87], Oldehoeft [102], Fitch [40]), may be used for testing functional equivalence of (transcendental) expressions. Theoretically, the idea is simple: If s is not functionally equivalent to t (s, t transcendental expressions), then Fun($s - t$) has, at most, a countable number of solutions, while the real (or complex) numbers are uncountable. Thus, for a point z chosen at random, Fun($s - t$)(z) = 0 implies $s \sim t$ "with probability 1". Practically, because of rounding errors, overflow and underflow, floating-point arithmetic is not satisfactory for performing the evaluations necessary in this context. Finite field computations (Martin [87]), interval arithmetic (Fitch [40]) and computing with transcendental ground terms are the alternatives that have been proposed. The latter method, though seemingly natural, cannot be applied in general because very little is known about the algebraic dependence of transcendental ground terms (the *constant problem*) (see the survey by (Lang [68])).

As an example of a general reduction method we describe *Johnson's method* [61]. (A second reduction method, based on the notion of Budan-sequences, has been presented by Richardson [114]). Johnson's method is applicable to any class of transcendental expressions for which a general notion of "*eigenelements*" can be defined. The method presupposes that the zero-equivalence problem can be solved effectively for certain subclasses of the term class considered. Since this is not generally the case in practical examples, Johnson's method is an algorithm in a relativized sense only.

Let R and K ($K \subseteq R$) be classes of transcendental expressions such that $R/\sim$ is a ring without zero divisors and $K/\sim$ is a field, where $\sim$ denotes functional equivalence on R. Let $\phi: R \twoheadrightarrow R$ and $E: R \twoheadrightarrow K$ be computable functions such that E determines "*eigenvalues*" $E(e)$ for certain "*eigenelements*" e of R:

$e \in R$ is an *eigenelement* (w.r.t. ϕ) iff

$$\text{not}(e \sim 0) \text{ and } \phi(e) \sim E(e) \cdot e.$$

We assume:

(1) $\phi(s + t) \sim \phi(s) + \phi(t)$.
(2) $\phi(K) \subseteq K$, $\phi(1) \sim 0$.
(3) If e_1, e_2 are eigenelements then e_1 is invertible and e_1^{-1}, $e_1 \cdot e_2$ are eigenelements.

(4) "$t \sim 0$" is decidable for $t \in K$.

(5) "$t \sim 0$" is decidable for $t \in \ker(\phi) := \{t/\phi(t) \sim 0\}$.

From (1)−(3) one can easily prove that

(6) $\phi(0) \sim 0$,

(7) all elements in K are eigenelements,

(8) e is an eigenelement and not $(\phi(e) \sim 0) \Rightarrow \phi(e)$ is an eigenelement.

Algorithm (Test for zero-equivalence. Johnson 1971).

Input: eigenvalues $e_1, \ldots, e_n$ of an effective operator ϕ on R.

Question: Is $S := e_1 + \cdots + e_n \sim 0$?

1. If $n = 1$: Answer "Not($S \sim 0$)". Return.
2. $T := b_1 \cdot e_n^{-1} \cdot e_1 + \cdots + b_{n-1} \cdot e_n^{-1} \cdot e_{n-1}$,
     where $b_i := E(e_n^{-1} \cdot e_i)$ $(i := 1, \ldots, n-1)$.
     [verify $T \sim \phi(e_n^{-1} \cdot S)$; use (1), (2)].
3. If not $(T \sim 0)$: Answer "Not($S \sim 0$)". Return.
     [Use the fact that $\phi(0) \sim 0$.
     Note that "$T \sim 0$" may decided effectively:
     If all $b_i \sim 0$: Answer "$T \sim 0$".
       [Use (4); note that all $b_i \in K$.]
     Otherwise apply the algorithm recursively to T.
       [Note that all $b_i \cdot e_n^{-1} \cdot e_i$ with not $(b_i \sim 0)$ are eigenelements by (3),
       (7)]].
4. [Case: $T \sim 0$, in this case: $e_n^{-1} \cdot S \in \ker(\phi)$.]
   4.1. If not$(e_n^{-1} \cdot S) \sim 0$: Answer "Not($S \sim 0$)". Return.
       [Note that "$e_n^{-1} \cdot S \sim 0$" can be decided by (5)].
   4.2. Answer "$S \sim 0$". Return. ■

Examples for the application of this algorithm are given in Johnson [61]. For instance, ϕ may be the operation of formal differentiation on a set R of transcendental expressions. In this context, K can be chosen to be the set of rational expressions. The computation of eigenvalues may be effected by using the following rules: $E(r) = 0$, $E(u \cdot v) = E(u) + E(v)$, $E(u/v) = E(u) - E(v)$, $E(u^r) = rE(u)$, $E(s) = s'/s$, $E(e^s) = s'$ ($r \ldots$ rational number; $u, v \ldots$ arbitrary expressions in R; $s \ldots$ rational expression).

7. Canonical Simplification of Transcendental Expressions: Structure Theorems

The most advanced and systematic method of simplifying transcendental expressions t is based on the so-called structure theorems. The basic idea is similar to the procedure followed in Section 4: in a first step a (transcendental) extension field K of $\mathbb{Q}$ is constructed such that t may be conceived as denoting an element in K. Then t is simplified using rational arithmetic. The structure theorems guide the extension procedure. This procedure is described in the chapter on computing in transcendental extensions.

8. Complete Reduction Systems, Critical Pairs, and Completion Algorithms

In classes T of linguistic objects with an equivalence relation $\sim$ one often can define in a natural way a "reduction relation" $\rightarrow$, which describes how "one can get, in one step, from an object s to a more simple but equivalent object t". The *iteration of this reduction* yields a *canonical simplifier* for $\sim$, if $\rightarrow$ can be handled algorithmically, the reflexive, symmetric and transitive closure of $\rightarrow$ is $\sim$, and $\rightarrow$ is *complete* in the sense that

> the iteration of the reduction process always terminates after finitely many steps at an irreducible object (*finite termination* property) and

> different reductions processes starting from the same object always terminate at the same irreducible object (*uniqueness*).

For concrete reduction relations $\rightarrow$, the proof of the finite termination property mostly needs particular techniques. The test for uniqueness can often be carried out in an algorithmic way by using two important ideas: *localization* (generally applicable) and the method of *critical pairs* (often applicable for reduction relations that are "generated" from finitely many reduction patterns). Finitely generated reduction relations that are not complete may be completed by adding reduction patterns that arise in the analysis of the critical pairs. This method of *completion* is the third basic idea, which together with localization and critical pair analysis constitutes a powerful and widely applicable methodology for constructing canonical simplifiers.

More formally, let $T \neq 0$ be an arbitrary set (of linguistic objects). In the context of this section an arbitrary binary relation $\rightarrow \subseteq T \times T$ will be called a *reduction* relation. The *inverse* relation, the *transitive closure*, the *reflexive-transitive* closure and the *reflexive-symmetric-transitive* closure of $\rightarrow$ will be denoted by $\leftarrow$, $\rightarrow^+$, $\rightarrow^*$, and $\leftrightarrow^*$, respectively. Also, $\rightarrow^0 = $ identity, $\rightarrow^{n+1} = \rightarrow \circ \rightarrow^n$. If $\rightarrow$ is clear from the context, by $\underline{x}$ we denote that $x \in T$ is in *normal form* with respect to $\rightarrow$ (i.e. there is no $y \in T$ such that $x \rightarrow y$). x and y have a *common successor* (in symbols, $x \downarrow y$) iff for some $z : x \rightarrow^* z \leftarrow^* y$. $\rightarrow$ is called *noetherian* ($\rightarrow$ has the finite termination property) iff there is no infinite chain of the form $x_1 \rightarrow x_2 \rightarrow x_3 \rightarrow \cdots$.

Now, let $\sim$ be an equivalence relation on T and $\rightarrow$ be a noetherian reduction relation on T such that $\leftrightarrow^* = \sim$. Suppose we have a computable "*selector*" function Sel : $T \rightarrow T$ such that $x \rightarrow \text{Sel}(x)$ for all $x \in T$ that are not in normal form. Consider the computable function S recursively defined by

$$S(x) := \textit{if } x \textit{ is in normal form then } x \textit{ else } S(\text{Sel}(x)).$$

Let us call an S of this kind a *normal-form algorithm* for $\rightarrow$. Our objective is to provide an algorithmic test for deciding whether S is a canonical simplifier. First, observe that S satisfies (SE), i.e. $S(x) \leftrightarrow^* x$. We even have

(SE')   for all $x \in T$: $x \rightarrow^* \underline{S(x)}$.

In the sequel we shall present a number of lemmas showing that the test for (SC), i.e. the property: $x \leftrightarrow^* y \Rightarrow S(x) = S(y)$, can be reduced successively to intuitively easier tests and finally, can be "*localized*". The method of *critical pairs*, then, will make the test effective. Let S be a normal-form algorithm for $\rightarrow$ ($\rightarrow$ noetherian!).

Lemma (Reduction of canonicality to the Church-Rosser property). *S is a canonical simplifier for $\leftrightarrow^*$ iff $\rightarrow$ has the Church-Rosser property.* ($\rightarrow$ *has the Church-Rosser property iff for all x, $y \in T$: $x \leftrightarrow^* y \Rightarrow x \downarrow y$*). ■

Proof. "$\Rightarrow$": Easy. "$\Leftarrow$": If $x \leftrightarrow^* y$, then for some z we have: $\underline{S(x)} \leftarrow^* x \rightarrow^* z \leftarrow^* y \rightarrow^* \underline{S(y)}$ (apply the Church-Rosser property). Again by the Church-Rosser property (applied to $S(x)$ and z), we have $S(x) = z$ and, analogously, $z = S(y)$. ■

Lemma (Reduction of the Church-Rosser property to confluence). *$\rightarrow$ has the Church-Rosser property iff $\rightarrow$ is confluent.* ($\rightarrow$ *is confluent iff for all x, y, $z \in T: x \leftarrow^* z \rightarrow^* y \Rightarrow x \downarrow y$*). ■

Proof. "$\Rightarrow$": Easy. "$\Leftarrow$": Since $\leftrightarrow^*$ is the union of all $\leftrightarrow^n$ we can use induction on n. $n = 0$: easy. If $x \leftrightarrow^{n+1} y$, then $x \rightarrow z \leftrightarrow^n y$ or $x \leftarrow z \leftrightarrow^n y$ for some z. In each case, by induction hypothesis: $z \rightarrow^* u \leftarrow^* y$ for some u. In the first case, $x \rightarrow^* u \leftarrow^* y$, hence, $x \downarrow y$. In the second case, $x \rightarrow^* v \leftarrow^* u$ for some v by confluence, hence, $x \rightarrow^* v \leftarrow^* u \leftarrow^* y$, i.e. $x \downarrow y$. ■

Lemma (Reduction of confluence to local confluence. Newman 1942). *$\rightarrow$ confluent iff $\rightarrow$ locally confluent.* ($\rightarrow$ *is locally confluent iff for all x, z, $y \in T: x \leftarrow z \rightarrow y \Rightarrow x \downarrow y$*.) ■

Proof: "$\Rightarrow$": Easy. "$\Leftarrow$": By "noetherian" induction. Let $z_0 \in T$ be arbitrary but fixed. Induction hypothesis: For all z with $z_0 \rightarrow^+ z$: for all x, y: ($x \leftarrow^* z \rightarrow^* y \Rightarrow x \downarrow y$). We show: For all x, y: ($x \leftarrow^* z_0 \rightarrow^* y \Rightarrow x \downarrow y$). Case $x = z_0$ and case $z_0 = y$: easy. In the other case: $x \leftarrow^* x_1 \leftarrow z_0 \rightarrow y_1 \rightarrow^* y$ for some x_1, y_1. Then, by local confluence and induction hypothesis, there exist u, v, w with the properties shown in Fig. 1. ■

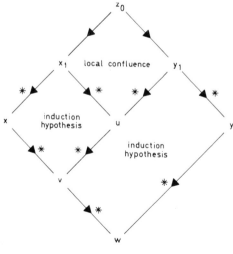

Fig. 1

The above lemmata can also be combined in the following way in order to yield a test for canonicality of S that involves S only:

Lemma (Local test for canonicality). *S is a canonical simplifier for $\leftrightarrow^*$ iff for all x, z, $y: (x \leftarrow z \to y \Rightarrow S(x) = S(y))$.* ∎

Still, this test for canonicality is not effective, because infinitely many x, y, z must be considered, in general. In the next section we will see how this test can be made effective in situations where $\to$ is generated from finitely many "reduction patterns" by operations as "substitution" and "replacement" (in a very general sense).

Historically, the central role of the Church-Rosser property for canonical simplification has been observed first in the development of λ-calculus, see Church, Rosser [20], Hindley [51], Hindley, Lercher, Seldin [52]. Newman's lemma as a general technique appeared first in Newman [100]. A complete yet concise presentation of the conceptual framework given above including a useful generalization for the case of reduction relations on quotient sets is given in Huet [55]. It is based on various earlier contributions, for instance Aho, Sethi, Ullman [1], Rosen [120], Lankford [70], Sethi [124], Staples [129].

It turns out that, implicitly, the idea of critical pair and completion is at the beginning of algorithmic mathematics: Euclid's algorithm and Gauss' algorithm may be viewed as critical pair/completion (cpc-) algorithms (see Section 11). The algorithm in Buchberger [12] for constructing canonical bases for polynomial ideals (see Section 11) seems to be the first one that explicitly stated the cpc-method, although several older algebraic algorithms (for instance, in group theory) have much of the flavor of cpc, for instance Dehn [28], Todd-Coxeter [136], Evans [33]. Also the resolution algorithm of Robinson [117] for automated theorem proving might be considered as a cpc-algorithm. The most general cpc-algorithm in the context of computer algebra is the Knuth-Bendix algorithm for rewrite rules (see Sections 9 and 10), whose forerunner, in fact, is Evans [33]. Later algorithms of the cpc type are the collection algorithm of McDonald [91], Newman [99] in computational group theory (see the respective chapter in this volume) and the algorithm of Bergman [5] for associative k-algebras. In Loos [81] some of these algorithms are analyzed under the cpc point of view and ideas for conceiving them as special cases of the Knuth-Bendix algorithms are presented.

9. The Knuth-Bendix Critical Pair Algorithm

The Knuth-Bendix critical pair algorithm is intended to yield a solution to the simplification problem for equivalence relations of the form $=_E$ (see Section 2) on sets of terms, where E is a set of equations in $\text{Term}(X, a)$. For an exact formulation of the algorithm some additional notions for describing the replacement of terms in terms are needed. We explain these notions in an example (a formal definition may be based on a formalization of the concept of "tree", see for instance O'Donnell [101]):

Example. The term $t := gfxygxy3x \, (a(f) = 2, a(g) = 3)$ has the tree representation shown in Fig. 2.

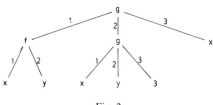

Fig. 2

One says that fxy *occurs* at the place (1) in t, $gxy3$ occurs at the place (2) in t, y occurs at the place (1, 2) in t etc., in symbols: $t/(1) = fxy$, $t/(2) = gxy3$, $t/(1, 2) = y$ etc. In addition, $t/\Lambda = t$ where Λ is the empty sequence of natural numbers. The set of *occurrences* ("addresses", "places") $\mathrm{Oc}(t)$ of t in the above example is $\{\Lambda, (1), (2), (3), (1, 1), (1, 2), (2, 1), (2, 2), (2, 3)\}$. ■

Let $\cdot$ denote concatenation on the set $\mathbb{N}^*$ of occurrences. The *prefix ordering* $\leqslant$ on $\mathbb{N}^*$ is defined by: $u \leqslant v$ iff $u \cdot w = v$ for some w. Furthermore, $v/u := w$ if $u \cdot w = v$, and $u \mid v$ (u and v are *disjoint*) iff neither $u \leqslant v$ nor $v \leqslant u$.

Finally, for terms s, t and occurrences u, $t[u \leftarrow s]$ is the term that derives from t if the term occurring at u in t is *replaced* by the term s. For instance, in the above example $t[(2) \leftarrow ffxx2] = gfxyffxx2x$.

Now, let E be a set of equations on $\mathrm{Term}(X, a)$. In a natural way, a reduction relation $\rightarrow_E$ on Term may be defined such that $\leftrightarrow^*_E = =_E$. Roughly, $s \rightarrow_E t$ iff t derives from s by applying one of the equations in E as "rewrite rule", i.e. in a directed way from left to right.

Definition. $s \rightarrow_E t$ (s *reduces* to t by E) iff there is a rule $(a, b) \in E$, a substitution σ and an occurrence $u \in \mathrm{Oc}(s)$ such that

$$s/u = \sigma(a) \qquad \text{and} \qquad t = s[u \leftarrow \sigma(b)]. \quad ■$$

Example. Let E be the following axiom system for group theory (in infix notation): (1) $1 \cdot x = x$, (2) $x^{-1} \cdot x = 1$, (3) $(x \cdot y) \cdot z = x \cdot (y \cdot z)$. Then, $(x^{-1} \cdot x) \cdot z \rightarrow_E 1 \cdot z \rightarrow_E z$. ■

In the sequel, we admit only those E, which satisfy: variable set of $b \subseteq$ variable set of a (for all $(a, b) \in E$). The next definition is crucial.

Definition. The terms p and q form a *critical pair* in E iff there are rules (a_1, b_1) and (a_2, b_2) in E and an occurrence u in $\mathrm{Oc}(a_1)$ such that

a_1/u is not a variable,
a_1/u and a_2 are unifiable,
$p = \sigma_1(a_1)[u \leftarrow \sigma_2(b_2)]$ and $q = \sigma_1(b_1)$,

where σ_1, σ_2 are substitutions such that

$\sigma_1(a_1/u) = \sigma_2(a_2)$ is a most general common instance of a_1/u and a_2 (with the property that no variable of $\sigma_1(a_1/u)$ occurs in a_1)

(i.e. p and q result from applying the rules (a_1, b_1) and (a_2, b_2) to a "most general match" of a_1 and a_2). ■

In the above definition, the notions of a "most general common instance" and of "unifiable terms" have been used. Let s, t_1, t_2 be terms. t_1 is an *instance of* t_2 iff $t_1 = \sigma(t_2)$ for some substitution σ. s is a *common instance* of t_1 and t_2 iff s is an instance of t_1 and t_2. t_1 and t_2 are *unifiable* iff t_1 and t_2 have a common instance. s is a *most general common instance* of t_1 and t_2 iff s is a common instance of t_1 and t_2 and every common instance of t_1 and t_2 is an instance of s. There is a straightforward algorithm ("*unification algorithm*", Robinson [117]) that decides for given t_1 and t_2 whether t_1 and t_2 are unifiable and, in the positive case, finds a most general common instance s of t_1 and t_2 (which is unique up to "permutations" of the variables) and substitutions σ_1 and σ_2 such that $s = \sigma_1(t_1) = \sigma_2(t_2)$. More sophisticated unification algorithms have been developed, for instance in Robinson [118], Paterson, Wegman [103], Huet [53], Martelli, Montanari [86].

Example. Consider rules (3) and (2) in the above example and the subterm $x \cdot y$ in (3), i.e. consider $a_1 = (x \cdot y) \cdot z$, $b_1 = x \cdot (y \cdot z)$, $a_2 = x^{-1} \cdot x$, $b_2 = 1$, and take $u = (1)$. $x^{-1} \cdot x$ is a most general instance of $a_1/u = x \cdot y$ and $a_2 = x^{-1} \cdot x$ (such that the condition on the variables is satisfied; take $\sigma_1(x) = x^{-1}$, $\sigma_1(y) = x$, $\sigma_2 = $ identity). Hence, $p = \sigma_1(a_1)[u \leftarrow \sigma_2(b_2)] = (x^{-1} \cdot x) \cdot z[(1) \leftarrow 1] = 1 \cdot z$ and $q = \sigma_1(b_1) = x^{-1} \cdot (x \cdot z)$ form a critical pair in E. ∎

Theorem (Reduction of local confluence to confluence of critical pairs. Knuth-Bendix 1967). $\rightarrow_E$ is *locally confluent iff for all critical pairs* (p, q) *of* E: $p \downarrow_E q$. ∎

If E is finite, one always can construct an algorithm Sel such that $t \rightarrow_E \mathrm{Sel}(t)$ if t is not in normal form. If $\rightarrow_E$ is noetherian, let S_E be the normal-form algorithm based on Sel.

Algorithm (Critical pair algorithm for rewrite rules, Knuth-Bendix 1967).

Input: E (a set of equations).

Question: Is the normal-form algorithm S_E a canonical simplifier for $=_E$?

$C := $ set of critical pairs of (E)
for all $(p, q) \in C$ *do*
    $(p_0, q_0) := (S_E(p), S_E(q))$
    if $p_0 \neq q_0$ *then* answer: "S_E is not canonical".
answer: "S_E is canonical". ∎

The correctness of the algorithm is an easy consequence of the above theorem and the lemmas in the preceding section.

Proof of the Theorem (sketch). "⇒": In the notation of the above definition, $p \leftarrow \sigma_1(a_1) \rightarrow q$ for critical pairs (p, q). (For simplicity, we write $\rightarrow$ instead of $\rightarrow_E$ etc.). Hence, $p \downarrow q$ by local confluence.

"⇐": Let t_1, s, t_2 be arbitrary terms and assume $t_1 \leftarrow s \rightarrow t_2$. We have to show $t_1 \downarrow t_2$, i.e. $t_1 \rightarrow^* w \leftarrow^* t_2$ for some w. By assumption, there are rules (a_1, b_1) and (a_2, b_2) in E, occurrences u_1, u_2 in $\mathrm{Oc}(s)$, and substitutions σ_1, σ_2 such that $s/u_1 = \sigma_1(a_1)$, $s/u_2 = \sigma_2(a_2)$, $t_1 = s[u_1 \leftarrow \sigma_1(b_1)]$, $t_2 = s[u_2 \leftarrow \sigma_2(b_2)]$. There are essentially the two cases shown in Fig. 3:

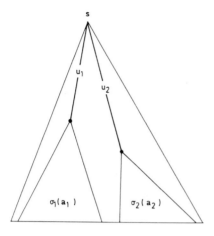

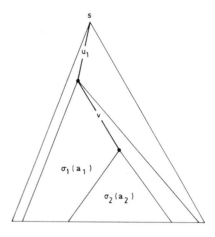

Case: $u_1 | u_2$ Case: $u_1 \cdot v = u_2$ for some v

Fig. 3

(The case $u_2 \cdot v = u_1$ is symmetric. Drawings in the above style may be very helpful for clarifying the subsequent arguments).

In the first case: $t_1/u_2 = s[u_1 \leftarrow \sigma_1(b_1)]/u_2 = s/u_2 = \sigma_2(a_2)$, $t_2/u_1 = \sigma_1(a_1)$. Define $w := t_1[u_2 \leftarrow \sigma_2(b_2)]$. Then $t_1 \rightarrow w$ and, also, $t_2 \rightarrow w$, because

$$w = s[u_1 \leftarrow \sigma_1(b_1)][u_2 \leftarrow \sigma(b_2)] = s[u_2 \leftarrow \sigma_2(b_2)][u_1 \leftarrow \sigma_1(b_1)] =$$
$$= t_2[u_1 \leftarrow \sigma_1(b_1)].$$

In the second case: $t_1 = s[u_1 \leftarrow \sigma_1(b_1)]$, $t_2 = s[u_1 \leftarrow \sigma_1(a_1)[v \leftarrow \sigma_2(b_2)]]$. If one can show that there exists w_0 such that $\sigma_1(b_1) \rightarrow^* w_0 \leftarrow^* \sigma_1(a_1)[v \leftarrow \sigma_2(b_2)]$, then also $t_1 \downarrow t_2$ by the *compatibility* property of $\rightarrow_E$ (see definition below).

There are the two subcases shown in Fig. 4.

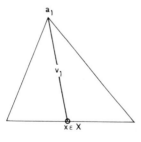

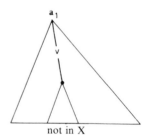

(for some $v_1 \leqslant v$)

Fig. 4

In the first subcase: $\sigma_2(a_2) = \sigma_1(x)/v_2$, where v_2 is such that $v_1 \cdot v_2 = v$. Define a new substitution σ_1' by $\sigma_1'(x) := \sigma_1(x)[v_2 \leftarrow \sigma_2(b_2)]$, $\sigma_1'(y) := \sigma_1(y)$ for $y \neq x$, and define $w_0 := \sigma_1'(b_1)$. Then $\sigma_1(b_1) \to^* \sigma_1'(b_1)$, because $\sigma_1(x) \to \sigma_1'(x)$. Also, $\sigma_1(a_1)[v \leftarrow \sigma_2(b_2)] = \sigma_1(a_1)[v_1 \leftarrow \sigma_1'(x)] \to^* \sigma_1'(a_1)$. Since $a_1 \to b_1$ we also have $\sigma_1'(a_1) \to \sigma_1'(b_1)$ ($\to_E$ is *stable*, see definition below). Hence, $\sigma_1(b_1) \to^* w_0 \leftarrow^* \sigma_1(a_1)[v \leftarrow \sigma_2(b_2)]$.

In the second subcase it is not difficult to show that the situation considered is an instance of a *critical pair*, i.e. there is a critical pair (p, q) of E and a substitution ρ such that $\sigma_1(a_1)[v \leftarrow \sigma_2(b_2)] = \rho(p)$ and $\sigma_1(b_1) = \rho(q)$. By assumption there is an r such that $p \to^* r \leftarrow^* q$. Define $w_0 := \rho(r)$. Again by the *stability* of $\to_E$ we have $\sigma_1(b_1) \to^* w_0 \leftarrow^* \sigma_1(a_1)[v \leftarrow \sigma_2(b_2)]$. ∎

For providing the details of the above proof a number of lemmas must be proven, which are pictorially evident (however, pictures may be misleading!) but tedious. (An exact recursive definition of the basic notions as, for instance, replacement, occurrence etc. would be necessary.) These lemmas have been tacitly used in the above sketch. One example of such a lemma is the "*commutativity* of replacement":

$$u_1 | u_2 \Rightarrow s[u_1 \leftarrow t_1][u_2 \leftarrow t_2] = s[u_2 \leftarrow t_2][u_1 \leftarrow t_1].$$

Two important lemmas, which have been mentioned explicitly in the above proof, are:

Lemma.

$\to_E$ is stable,   *i.e. for all terms s, t and substitutions σ:*

$$(s \to_E t \Rightarrow \sigma(s) \to_E \sigma(t)).$$

$\to_E$ is compatible,   *i.e. for all terms s, t_1, t_2 and $u \in \mathrm{Oc}(s)$:*

$$(t_1 \to_E t_2 \Rightarrow s[u \leftarrow t_1] \to_E s[u \leftarrow t_2]). \quad ∎$$

Example. In the above example of group theory the critical pair (p, q) resulting from rules (3) and (2) has no common successor. The only reduction possible for p is $p = 1 \cdot z \to z$ and $q = x^{-1} \cdot (x \cdot z)$ is already in normal form. Hence, $\to_E$ is not locally confluent and the normal-form algorithm S_E is not canonical.

Consider now the following axiom system E' for group theory (Knuth-Bendix [65]), which results from the above system by adding some identities that are theorems in group theory:

(1) $1 \cdot x = x$,   (2) $x^{-1} \cdot x = 1$,   (3) $(x \cdot y) \cdot z = x \cdot (y \cdot z)$,

(4) $1^{-1} = 1$,   (5) $x^{-1} \cdot (x \cdot y) = y$,   (6) $x \cdot 1 = x$,

(7) $(x^{-1})^{-1} = x$,   (8) $x \cdot x^{-1} = 1$,   (9) $x \cdot (x^{-1} \cdot y) = y$,

(10) $(x \cdot y)^{-1} = y^{-1} \cdot x^{-1}$.

It can be shown (Knuth-Bendix [65]) that $\to_{E'}$ is noetherian. Furthermore, by finitely many checks it can be seen that $S_{E'}(p) = S_{E'}(q)$ for all critical pairs of E'. In particular, the above critical pair has now the common successor z. Hence, by the theorem $\to_{E'}$ is (locally) confluent, $S_{E'}$ is a canonical simplifier for $\leftrightarrow^*_{E'} = \leftrightarrow^*_E = =_E$. This means that the decision about the validity of identities in group theory can be fully automatized. ∎

For applying the Knuth-Bendix algorithm to a set E of equations $\to_E$ must be proven noetherian. In general this is a non-trivial task that needs specific methods for each example. The uniform halting problem, i.e. the question whether there exists an infinite chain $t \to_E t_1 \to_E t_2 \to_E \cdots (t$ and E free parameters) is undecidable, see Huet, Lankford [57]). On the other hand, the uniform halting problem for rewrite system E that consist entirely of ground terms is decidable, see Huet, Lankford [57]. A number of important techniques have been developed for proving $\to_E$ noetherian, for instance Knuth, Bendix [65], Manna, Ness [85], Lankford [69, 70], Plaisted [106, 107], Dershowitz [29, 30], Dershowitz, Manna [31], see the survey in Huet, Oppen [58]. Basically, these techniques proceed from a well-founded partial ordering $>$ on the set of ground terms. It then is shown that $s \to_E t \Rightarrow s > t$ (for ground terms s, t). This suffices for proving $\to_E$ noetherian because it is known that $\to_E$ is noetherian iff there is no ground term having an infinite sequence of reductions.

Important modifications of the Knuth-Bendix (critical pair and completion algorithm) have been given in Plotkin [108], Lankford, Ballantyne [73, 74, 75], Huet [55], and Peterson, Stickel [105]. Mainly, these modifications consist in considering confluence modulo equivalence relations generated by standard axioms (as, for instance, commutativity), which do not naturally suggest a distinction between left-hand and right-hand side. Unification modulo such sets of axioms plays an essential role in this context, see the surveys of Raulefs, Siekmann, Szabo, Unvericht [111] and Huet, Oppen [58].

10. The Knuth-Bendix Completion Algorithm

If the Knuth-Bendix critical pair algorithm for a set E of equations shows that the reduction of a critical pair (p, q) yields $S(p) \neq S(q)$ then it suggests itself to try to augment E by the rule $(S(p), S(q))$ (or $(S(q), S(p)))$ in order to achieve completeness: in fact, the reflexive-symmetric-transitive closure of $\to_E$ is not changed by this adjunction, whereas $\to_E$ itself is properly extended. This process may be iterated until, hopefully, "saturation" will be reached, i.e. all critical pairs have a unique normal form. In this case the final set of equations still generates the same equivalence relation $=_E$ as the original one, but its reduction relation $\to_E$ is unique, i.e. the associated normal form algorithm is a canonical simplifier for $=_E$ (and hence, $=_E$ is decidable). Of course, before extending E one must check whether $\to_E$ remains noetherian by this extension and in order to meet this condition one must choose between the two possibilities $(S(p), S(q))$ and $(S(q), S(p))$. It may also happen that the preservation of the finite termination property cannot be proven for either possibility. In this case the completion process cannot be continued reasonably. We present the rough structure of this procedure:

Algorithm (Completion algorithm for rewrite rules. Knuth-Bendix 1967).

Given: E, a finite set of equations such that $\to_E$ is noetherian.
Find: F, a finite set of equations such that
$$\leftrightarrow^*_E = \leftrightarrow^*_F \text{ and}$$
$\to_F$ is (locally) confluent ($\Leftrightarrow S_F$ is a canonical simplifier).

$F := E$
$C :=$ set of critical pairs of (F)

while $C \neq 0$ *do*
      $(p, q) :=$ an element in (C)
      $(p_0, q_0) := (S_F(p), S_F(q))$
      if $p_0 \neq q_0$ *then*
          Analyze (p_0, q_0)
          $C := C \cup$ set of new critical pairs $((p_0, q_0), F)$
            $F := F \cup \{(p_0, q_0)\}$
      $C := C - \{(p, q)\}$

stop with success. ■

The subroutines "set of critical pairs of" and "an element in" seem to be self-explanatory. The subroutine "set of new critical pairs" computes the critical pairs deriving from the new rule (p_0, q_0) and the rules in F. The subroutine "Analyze" determines whether $\rightarrow_E$ remains noetherian when (p_0, q_0) or (q_0, p_0) is added to E. In the first case (p_0, q_0) is left unchanged, in the second case the roles of p_0 and q_0 are interchanged. If none of the two alternatives holds, the subroutine "Analyze" executes a *stop* with failure. There are three possibilities: 1. The algorithm stops with success. In this case the final F meets the specifications stated in the heading of the algorithm. 2. The algorithm stops with failure. In this case nothing interesting can be said. 3. The algorithm never stops: In this case the algorithm is at least a semidecision procedure for $=_E$. It can be shown (Huet [54]) that $s =_E t$ can be semidecided by reducing s and t with respect to the steadily increasing set of equations produced by the algorithm.

The above crude form of the algorithm can be organizationally refined in many ways. For instance, the two sides of an equation in F can be kept in reduced form relative to the other equations in F. Furthermore, the critical pairs can be generated in an "incremental way" (Huet [54]). The sequence of critical pairs chosen by the procedure "an element in" may have a crucial influence on the efficiency of the algorithm. These questions are subtle.

Implementations of the Knuth-Bendix completion algorithm are reported in Knuth, Bendix [65], Hullot [59], Richter, Kemmenich [115], Loos [80], Peterson, Stickel [105]. Investigations of the *complexity* of the algorithm are difficult. Necessarily they are confined to special applications of the algorithm. A survey on known complexity results is given in Lankford [71]. An analysis of various subalgorithms used in the Knuth-Bendix algorithm may be found in Loos [80].

An impressive variety of *axiom systems* have been successfully completed by the Knuth-Bendix algorithm (Knuth, Bendix [65], Lankford, Ballantyne [73, 74, 75], Stickel, Peterson [130], Ballantyne, Lankford [4], Huet, Hullot [56], Hullot [59], Bücken [14], Richter, Kemmenich [115]), including axiom systems for groups (see example in the last section), central groupoids, loops, (l, r) systems, commutative rings with 1, R-modules, fragments of number theory, distributive lattices, non-deterministic machines, "robot" theory, special finitely presented groups and commutative semigroups. Recently, the Knuth-Bendix completion algorithm is extensively used for completing the defining equations of *abstract data types*, for instance in the AFFIRM system (Gerhart et al. [42], Musser [97]). By the theorem on canonical simplification and computing in algebraic domains (see Section 1) completed axiom systems allow to effectively compute in the "direct implementations of the abstract data types", i.e. in the term algebra modulo the equivalence generated by the equations (Musser [97], Lichtenberger [78], Kapur, Musser, Stepanov [62]). Examples of complete equational specifications of abstract data types are given in Musser [97, 98].

Recently (Courcelle [24], Musser [98], Goguen [43], Huet, Hullot [56]), a new type of application of the completion process has been proposed: completion as a substitute for induction, see also Lankford [72]. By this method, the validity of equations e in the "initial model" of a set E of equations (Goguen et al. [15, 44]) may be proven by testing whether the Knuth-Bendix completion algorithm applied to $E \cup \{e\}$ stops without generating any inconsistency.

11. A Completion Algorithm for Polynomial Ideal Bases

Let E be a (finite) set of *polynomials* in $K[x_1, \ldots, x_n]$, K a field and let $<_T$ be a linear *ordering of the monomials*, which is "*admissible*" in the sense that $\sigma_1 <_T \sigma_2 \Rightarrow \tau \cdot \sigma_1 <_T \tau \cdot \sigma_2$ and $1 \leqslant_T \sigma$ (τ, σ_1, σ_2 monomials). We study the problem of constructing a canonical simplifier for the *congruence relation* $\equiv_E$ modulo the polynomial ideal Ideal(E) generated by the polynomials in E. It is easy (although non-trivial) to show that the following reduction relation $\to_E$ is such that $\leftrightarrow^*_E = \equiv_E$:

Definition. $s \to_E t$ (s *reduces to* t modulo E) iff there is a polynomial (a, b) in E, a monomial σ and a monomial u occurring in s such that

$$u = \sigma \cdot a \qquad \text{and} \qquad t = s[u \leftarrow \sigma \cdot b]. \quad \blacksquare$$

The polynomials of E are thought to be represented in E in the form (a, b), where a is the leading monomial in the polynomial with respect to $<_T$ with its leading coefficient normed to 1 and b is the rest of the polynomial. Structurally, this definition is very similar to the definition of $\to_E$ in the preceding section. The multiplication with a monomial corresponds to the application of a substitution and the operation of replacement in the present context is defined as follows:

$s[u \leftarrow t] :=$ the polynomial that results from the polynomial s by replacing the monomial u by the polynomial t.

Example.

$$E := \{(x^2 y, \ -3y^2 + 2y), (y^2, \ -2x)\}.$$

$$s := x^4 - 2x^2 y^2 + 5y^3 + x \to_E x^4 - 11y^3 - 4y^2 + x = :t,$$

because the monomial $x^2 y^2$ occurring in s is a multiple $u \cdot x^2 y$ of the left-hand side of the first polynomial in E, where u is the monomial y. t results from s by replacing $x^2 y^2$ by $u \cdot (-3y^2 + 2y)$. $\blacksquare$

It is not difficult to show that $\to_E$ is noetherian for arbitrary E. It is clear that there exists a selector function Sel_E and, based on Sel_E, a normal form algorithm S_E

for $\to_E$. Again, the following notion is basic for the approach to constructive ideal theory given here.

Definition. The polynomials p and q form a *critical pair* of E iff there are polynomials (a_1, b_1) and (a_2, b_2) in E such that

$$p = \sigma_1(b_1) \qquad \text{and} \qquad q = \sigma_2(b_2)$$

where σ_1, σ_2 are monomials such that

$$\sigma_1 \cdot a_1 = \sigma_2 \cdot a_2 \text{ is the least common multiple of } a_1 \text{ and } a_2. \quad \blacksquare$$

Example.

$$E := \{(x^3yz, xz^2), (xy^2z, xyz), (x^2y^2, z^2)\}.$$

x^3y^2z is a least common multiple of the left-hand sides of the first two polynomials in E. $\sigma_1 = y$, $\sigma_2 = x^2$. $p = xyz^2$, $q = x^3yz$. $\quad \blacksquare$

Again, one can show that the consideration of the critical pairs is sufficient for deciding the (local) confluence of $\to_E$:

Theorem (Critical pair algorithm for polynomial ideals. Buchberger 1965). S_E *is a canonical simplifier for* $\to_E$ *iff*

$$\text{for all critical pairs of } (p, q) \text{ of } E: S_E(p) = S_E(q). \quad \blacksquare$$

Proof. Again, the proof of this theorem may be organized in such a way that, essentially it consists in the proof of: $\to_E$ is locally confluent iff for all critical pairs (p, q) of E: $p \downarrow q$. At present, no proof of this property is known which would proceed by a specialization of the Knuth-Bendix proof. A special proof is necessary, see Buchberger [12, 13], Bachmair, Buchberger [3]. The main reason for this seems to be the fact that $\to_E$ is not compatible, but only "semi-compatible":

$\to_E$ is *stable*,   i.e. for all polynomials s, t and monomials σ:

$$(s \to_E t \Rightarrow \sigma \cdot s \to_E \sigma \cdot t).$$

$\to_E$ is *semi-compatible*,   i.e. for all polynomials s, t_1, t_2 and monomials

u occurring in s:

$$(t_1 \to_E t_2 \Rightarrow s[u \leftarrow t_1] \downarrow_E s[u \leftarrow t_2]). \quad \blacksquare$$

Example. In the above example $S_E(p) = \underline{xyz^2}$, $S_E(q) = \underline{xz^2}$. Hence, by the theorem S_E is not canonical. $\quad \blacksquare$

Again the critical pair algorithm can naturally be modified to obtain the following completion algorithm:

Algorithm (Completion algorithm for polynomial ideal bases. Buchberger 1965).

Given: E, a finite set of polynomials in $K[x_1, \ldots, x_n]$.
Find: F, a finite set of polynomials in $K[x_1, \ldots, x_n]$ such that
$\qquad \equiv_E = \equiv_F$ and
$\qquad \to_F$ is (locally) confluent ($\Leftrightarrow S_F$ is a canonical simplifier)
$\qquad$ (An F with this property is called *Gröbner* basis).

The algorithm is formally identical to the Knuth-Bendix completion algorithm, but with different meaning associated to the subroutines. In particular the subroutine "Analyze" has no *stop* with failure. Instead "Analyze" performs $(p_0, q_0) :=$ leading monomial and rest of the polynomial $p_0 - q_0$ (after normalizing the leading coefficient to 1). ∎

The algorithm stops for arbitrary input sets E, see Buchberger [12], and Bergman [5], p. 208. Again, a number of organizational improvements are possible for the algorithm, the most important being the one developed in Buchberger [13], which shows that, instead of critical pairs, "chains" of critical pairs may be considered. This results in a drastic improvement in the speed of the algorithm. This approach might also prove fruitful for other instances of cpc algorithms. Research in this direction has now been undertaken.

In the special case of $E \subseteq K[x]$ the above algorithm specializes to Euclid's algorithm. In the special case of a set of linear polynomials in $K[x_1, \ldots, x_n]$ the algorithm specializes to Gauss' algorithm. In the special case of having only monomials on the right-hand sides of the polynomials in E, E are the defining relations of a finitely generated commutative semigroup and the algorithm is a decision procedure for the uniform word problem for commutative semigroups. It has been shown by Cardoza, Lipton, Meyer [16], and Mayr, Meyer [90] that this problem is complete in exponential space under log-space transformability. Hence, also the complexity of the above algorithm necessarily will be high. Nevertheless, in practical examples it has proven feasible. An explicit complexity bound for the case $n = 2$ has been derived in Buchberger [13]. Various implementations of the algorithm have been reported, see Buchberger [12], Schrader [122], Lauer [76], Spear [128], Trinks [137], Zacharias [141], Schaller [121], Winkler et al. [139], including generalizations for polynomials over rings and various applications (effective computation in residue rings modulo polynomial ideals, computation of elimination ideals and solution of sets of algebraic equations, effective Lasker-Noether decomposition of polynomial ideals, effective computation of the Hilbert function and the sycygies of polynomial ideals, interpolation formulae for numerical cubature (Möller [92])). A generalization of the algorithm for associative algebras with many interesting applications (for instance for Lie-algebras) over commutative rings with 1 has independently been derived in Bergman [5].

For improving the computational feasibility of the algorithm the application of the techniques derived for Euclid's algorithm (Collins [23], Brown [10]; see the chapter on polynomial remainder sequences) should be investigated. A comparison of the notion of resultant, polynomial remainder sequence and reduction $\rightarrow_E$ is necessary for this purpose. The first observations in this direction appear in Schaller [121], Pohst, Yun [109]. Also a comparison with other approaches for deriving "canonical bases" for polynomial ideals and constructive methods in ring theory (Hermann [50], Szekeres [134], Kleiman [64], Seidenberg [123], Shtokhamer [125], Trotter [138], Ayub [2]), seems to be promising.

References

[1] Aho, A., Sethi, R., Ullman, J.: Code Optimization and Finite Church-Rosser Systems. Proc. of Courant Comput. Sci. Symp. 5 (Rustin, R., ed.). Englewood Cliffs, N. J.: Prentice-Hall 1972.

[2] Ayoub, C. W.: On Constructing Bases for Ideals in Polynomial Rings over the Integers. Pennsylvania State University, Univ. Park: Dept. Math. Rep. **8184** (1981).

[3] Bachmair, L., Buchberger, B.: A Simplified Proof of the Characterization Theorem for Gröbner Bases. ACM SIGSAM Bull. **14**/4, 29 – 34 (1980).

[4] Ballantyne, A. M., Lankford, D. S.: New Decision Algorithms for Finitely Presented Commutative Semigroups. Comp. Math. Appl. **7**, 159 – 165 (1981).

[5] Bergman, G. M.: The Diamond Lemma for Ring Theory. Adv. Math. **29**, 178 – 218 (1978).

[6] Birkhoff, G.: On the Structure of Abstract Algebras. Proc. Cambridge Phil. Soc. **31**, 433 – 454 (1935).

[7] Birkhoff, G., Lipson, J. D.: Heterogeneous Algebras. J. Comb. Theory **8**, 115 – 133 (1970).

[8] Boyer, R. S., Moore, J. S.: A Computational Logic. New York-London: Academic Press 1979.

[9] Brown, W. S.: Rational Exponential Expressions and a Conjecture Concerning π and e. Am. Math. Mon. **76**, 28 – 34 (1969).

[10] Brown, W. S.: On Euclid's Algorithm and the Computation on Polynomial Greatest Common Divisor. J. ACM **18**/4, 478 – 504 (1971).

[11] Brown, W. S.: On Computing with Factored Rational Expressions. EUROSAM **1974**, 26 – 34.

[12] Buchberger, B.: An Algorithm for Finding a Basis for the Residue Class Ring of a Zero-Dimensional Polynomial Ideal (German). Ph.D. Thesis, Math. Inst., Univ. of Innsbruck, Austria, 1965, and Aequationes Math. **4**/3, 374 – 383 (1970).

[13] Buchberger, B.: A Criterion for Detecting Unnecessary Reductions in the Construction of Gröbner Bases. EUROSAM **1979**, Lecture Notes in Computer Science, Vol. 72, pp. 3 – 21. Berlin-Heidelberg-New York: Springer 1979.

[14] Bücken, H.: Reduktionssysteme und Wortproblem. Rhein.-Westf. Tech. Hochschule, Aachen: Inst. für Informatik, Rep. **3**, 1979.

[15] Burstall, R. M., Goguen, J. A.: Putting Theories Together to Make Specifications. Proc. 5th Internat. Joint Conf. on Artificial Intelligence **1977**, 1045 – 1058.

[16] Cardoza, E., Lipton, R., Meyer, A. R.: Exponential Space Complete Problems for Petri Nets and Commutative Semigroups. Conf. Record of the 8th Annual ACM Symp. on Theory of Computing, 50 – 54 (1976).

[17] Caviness, B. F.: On Canonical Forms and Simplification, Ph.D. Diss., Pittsburgh, Carnegie-Mellon University, 1967, and J. ACM **17**/2, 385 – 396 (1970).

[18] Caviness, B. F., Fateman, R.: Simplification of Radical Expressions. SYMSAC **1976**, 329 – 338.

[19] Caviness, B. F., Pollack, P. L., Rubald, C. M.: An Existence Lemma for Canonical Forms in Symbolic Mathematics. Inf. Process. Lett. **1**, 45 – 46 (1971).

[20] Church, A., Rosser, J. B.: Some Properties of Conversion. Trans. AMS **39**, 472 – 482 (1936).

[21] Cohn, P. M.: Universal Algebra. New York: Harper and Row 1965.

[22] Collins, G. E.: The SAC-1 Polynomial System. Univ. of Wisconsin, Madison: Comput. Sci. Dept., Tech. Rep. **115**, 1968.

[23] Collins, G. E.: The Calculation of Multivariate Polynomial Resultants. J. ACM **18**/4, 515 – 532 (1971).

[24] Courcelle, B.: On Recursive Equations Having a Unique Solution. IRIA – Laboria Rep. **285** (1976).

[25] Davenport, J.: On the Integration of Algebraic Functions. Univ. of Cambridge: Ph.D. Thesis. (Lecture Notes in Computer Science, Vol. 102.) Berlin-Heidelberg-New York: Springer 1981.

[26] Davis, M.: Hilbert's Tenth Problem is Unsolvable. Am. Math. Mon. **80**/3, 233 – 269 (1973).

[27] Davis, M., Putnam, H., Robinson, J.: The Decision Problem for Exponential Diophantine Equations. Ann. Math. **74**, 425 – 436 (1961).

[28] Dehn, M.: Über unendliche diskontinuierliche Gruppen. Math. Ann. **71**, 116 – 144 (1911).

[29] Dershowitz, N.: A Note on Simplification Orderings. Inf. Process. Lett. **9**/5, 212 – 215 (1979).

[30] Dershowitz, N.: Orderings for Term-Rewriting Systems. Proc. 20th Symp. on Foundations of Comp. Sci. **1979**, 123 – 131.

[31] Dershowitz, N., Manna, Z.: Proving Termination with Multiset Ordering. Commun. ACM **22**, 465 – 476 (1979).

[32] Epstein, H. I.: Using Basis Computation to Determine Pseudo-Multiplicative Independence. SYMSAC **1976**, 229 – 237.

[33] Evans, T.: The Word Problem for Abstract Algebras. J. London Math. Soc. **26**, 64 – 71 (1951).

[34] Fateman, R. J.: The User-Level Semantic Pattern Matching Capability in MACSYMA. Proc. SYMSAM **1971**, 311 – 323.

[35] Fateman, R. J.: Essays in Algebraic Simplification. Thesis, MIT, Project MAC 1972.

[36] Fateman, R. J.: MACSYMA's General Simplifier: Philosophy and Operation. MACSYMA **1979**, 563 – 582.

[37] Fateman, R. J.: Symbolic and Algebraic Computer Programming Systems. ACM SIGSAM Bull. **15**/1, 21 – 32 (1981).

[38] Fenichel, J.: An One-Line System for Algebraic Manipulation. Harvard Univ., Cambridge, Mass.: Ph.D. Thesis 1966.

[39] Fitch, J. P.: An Algebraic Manipulator. Ph.D. Thesis, Univ. of Cambridge, 1971.

[40] Fitch, J. P.: On Algebraic Simplification. Comput. J. **16**/1, 23 – 27 (1973).

[41] Foderaro, J. K.: Typesetting MACSYMA Equations. MACSYMA **1979**, 345 – 361.

[42] Gerhart, S. L., Musser, D. R., Thompson, D. H., Baker, D. A., Bates, R. L., Erickson, R. W., London, R. L., Taylor, D. G., Wile, D. S.: An Overview of AFFIRM: A Specification and Verification System. In: Information Processing 80 (Lavington, S. H., ed.), pp. 343 – 347. Amsterdam: North-Holland 1980.

[43] Goguen, J. A.: How to Prove Algebraic Inductive Hypotheses without Induction, with Applications to the Correctness of Data Type Implementations. Proc. 5th Conf. on Aut. Deduction. Lecture Notes in Computer Science, Vol. 87, pp. 356 – 373. Berlin-Heidelberg-New York: Springer 1980.

[44] Goguen, J. A., Thatcher, J. W., Wagner, E. G., Wright, J. B.: Abstract Data Types as Initial Algebras and Correctness of Data Representations. Proc. Conf. on Comput. Graphics, Pattern Recognition and Data Structure **1975**, 89 – 93, Beverly Hills, California.

[45] Griss, M. L: The Definition and Use of Datastructures in REDUCE. SYMSAC **1976**, 53 – 59.

[46] Hall, A. D.: Factored Rational Expressions in ALTRAN. EUROSAM **1974**, 35 – 45.

[47] Hearn, A. C.: A New REDUCE Model for Algebraic Simplification. SYMSAC **1976**, 46 – 52.

[48] Hearn, A. C.: Symbolic Computation: Past, Present, Future. Dept. of Comput. Sci., Univ. Utah, Salt Lake City 1978.

[49] Henrici, P.: A Subroutine for Computations with Rational Numbers. J. ACM **3**/1, 6 – 9 (1956).

[50] Hermann, G.: Die Frage der endlich vielen Schritte in der Theorie der Polynomideale. Math. Ann. **95**, 736 – 788 (1926).

[51] Hindley, R.: An Abstract Form of the Church-Rosser Theorem II: Applications. J. Symb. Logic **39**/1, 1 – 21 (1974).

[52] Hindley, R., Lercher, B., Seldin, J. P.: Introduction to Combinatory Logic. Cambridge: Camb. Univ. Press 1972.

[53] Huet, G. P.: Résolution d'équations dans des languages d'ordre 1, 2, ..., ω. Ph.D. Thesis, University of Paris VII, 1976.

[54] Huet, G.: A Complete Proof of the Knuth-Bendix Competion Algorithm. IRIA: Report (1979).

[55] Huet, G. P.: Confluent Reductions: Abstract Properties and Applications to Term Rewriting Systems. 18th IEEE Symp. on Foundat. of Comput. Sci., 30 – 45 (1977) and J. ACM **27**/4, 797 – 821 (1980).

[56] Huet, G. P., Hullot, J. M.: Proofs by Induction in Equational Theories with Constructors. 21st IEEE Symp. on Foundations of Comput. Sci. **1980**, 96 – 107.

[57] Huet, G., Lankford, D. S.: On the Uniform Halting Problem for Term Rewriting Systems. IRIA: Rapp. Laboria **283**, 1978.

[58] Huet, G. P., Oppen, D. C.: Equations and Rewrite Rules: A Survey. Techn. Rep. CSL-**111**, SRI International, Stanford, 1980.

[59] Hullot, J. M.: A Catalogue of Canonical Term Rewriting Systems. Stanford: SRI International Tech. Rep. CSL-**113**, 1980.

[60] Jenks, R. D.: A Pattern Compiler. SYMSAC **1976**, 60 – 65.

[61] Johnson, S. C.: On the Problem of Recognizing Zero. J. ACM **18**/4, 559 – 565 (1971).

[62] Kapur, D., Musser, D. R., Stepanov, A. A.: Operators and Algebraic Structures. Schenectady, N.Y.: General Electric Automation and Control Laboratory, Report 81CRD **114** (1981).

[63] King, J. C., Floyd, R. W.: An Interpretation Oriented Theorem Prover Over the Integers. J. Comput. Syst. Sci. **6**/4, 305 – 323 (1972).

[64] Kleiman, S. L.: Computing with Rational Expressions in Several Algebraically Dependent Variables. Columbia Univ., New York: Computing Science Techn. Report **42**, 1966.

[65] Knuth, D. E., Bendix, P. B.: Simple Word Problems in Universal Algebras. OXFORD 67, 263 – 298.

[66] Korsvold, K.: An On-Line Algebraic Simplify Program. Stanford Univ.: Art. Int. Project Memorandum 37, 1965, and Comm. ACM **9**, 553 (1966).

[67] Lafferty, E. L.: Hypergeometric Function Reduction – an Adventure in Pattern Matching. MACSYMA **1979**, 465 – 481.

[68] Lang, S.: Transcendental Numbers and Diophantine Approximations. Bull. AMS **77**/5, 635 – 677 (1971).

[69] Lankford, D. S.: Canonical Algebraic Simplification. Univ. of Texas, Austin: Dept. Math. Comput. Sci., Rep. ATP-**25**, 1975.

[70] Lankford, D. S.: Canonical Inference. Univ. of Texas, Austin: Dept. Math. Comput. Sci., Rep. ATP-**32**, 1975.

[71] Lankford, D. S.: Research in Applied Equational Logic. Louisiana Tech. Univ., Ruston: Math. Dept., MTP-**15**, 1980.

[72] Lankford, D. S.: A Simple Explanation of Inductionless Induction. Louisiana Tech. Univ., Ruston: Math. Dept., MTP-**14**, 1981.

[73] Lankford, D. S., Ballantyne, A. M.: Decision Procedures for Simple Equational Theories with Commutative Axioms: Complete Sets of Commutative Reductions. Univ. of Texas, Austin: Dept. Math. Comput. Sci., Rep. ATP-**35**, 1977.

[74] Lankford, D. S., Ballantyne, A. M.: Decision Procedures for Simple Equational Theories with Permutative Axioms: Complete Sets of Permutative Reductions. Univ. of Texas, Austin: Dept. Math. Comput. Sci., Rep. ATP-**37**, 1977.

[75] Lankford, D. S., Ballantyne, A. M.: Decision Procedures for Simple Equational Theories with Commutative-Associative Axioms: Complete Sets of Commutative-Associative Reductions. Univ. of Texas, Austin: Dept. Math. Comput. Sci., Rep. ATP-**39**, 1977.

[76] Lauer, M.: Canonical Representatives for Residue Classes of a Polynomial Ideal. SYMSAC **1976**, 339 – 345.

[77] Lausch, H., Nöbauer, W.: Algebra of Polynomials. Amsterdam: North-Holland 1973.

[78] Lichtenberger, F.: PL/ADT: A System for Using Algebraically Specified Abstract Data Types in PL/I (German). Ph.D. Thesis, Math. Inst., Univ. of Linz, Austria, 1980.

[79] Loos, R.: Toward a Formal Implementation of Computer Algebra. EUROSAM **1974**, 9 – 16.

[80] Loos, R.: Algorithmen und Datenstrukturen I: Abstrakte Datentypen. Univ. Karlsruhe, Federal Republic of Germany: Lecture Notes, 1980.

[81] Loos, R.: Term Reduction Systems and Algebraic Algorithms. Proc. 5th GI Workshop on Artif. Intell., Bad Honnef 1981, Informatik Fachberichte **47**, 214 – 234 (1981).

[82] Loveland, D. W., Shostak, R. E.: Simplifying Interpreted Formulas. 5th Conf. on Automated Deduction, Les Arces, France, 1980. Lecture Notes in Computer Science, Vol. 87, pp. 97 – 109. Berlin-Heidelberg-New York: Springer 1980.

[83] Luckham, D. C., German, S. M., Henke, F. W., Karp, R. A., Milne, P. W., Oppen, D. C., Polak, W., Scherlis, W. L.: Stanford PASCAL Verifier User Manual. Stanford Univ.: Comput. Sci. Dept., Rep. STAN-CS-79-**731**, 1979.

[84] Manin, Y. I.: A Course in Mathematical Logic. Berlin-Heidelberg-New York: Springer 1977.

[85] Manna, Z., Ness, S.: On the Termination of Markov Algorithms. Third Hawaii International Conference on System Sciences **1970**, 789 – 792.

[86] Martelli, A., Montanari, U.: An Efficient Unification Algorithm. Unpublished Report (1979).

[87] Martin, W. A.: Computer Input/Output of Mathematical Expressions. SYMSAM **1971**, 78 – 89.

[88] Martin, W. A.: Determining the Equivalence of Algebraic Expressions by Hash Coding. SYMSAM **1971**, 305 – 310.

[89] Matiyasevic, J.: Diophantine Representation of Recursively Enumerable Predicates. Proc. Second Scandinavian Logic Symp. Amsterdam: North-Holland 1970.

[90] Mayr, E. W., Meyer, A. R.: The Complexity of the Word Problems for Commutative Semigroups and Polynomial Ideals. M.I.T.: Lab. Comput. Sci. Rep. LCS/TM-199 (1981).

[91] McDonald, I. D.: A Computer Application to Finite p-Groups. J. Aust. Math. Soc. **17**, 102 – 112 (1974).

[92] Möller, H. M.: Mehrdimensionale Hermite-Interpolation und numerische Integration. Math. Z. **148**, 107 – 118 (1976).

[93] Moses, J.: Symbolic Integration. Ph.D. Thesis, Cambridge, Mass., M.I.T., Math. Dept., Rep. MAC-**47**, 1967.

[94] Moses, J.: Algebraic Simplification: A Guide for the Perplexed. SYMSAM **1971**, 282 – 304 and Commun. ACM **14**, 527 – 537 (1971).

[95] Musser, D. R.: Algorithms for Polynomial Factorization. Univ. of Wisconsin, Madison: Ph.D. Thesis, Techn. Rep. **134**, 1971.

[96] Musser, D. R.: A Data Type Verification System Based on Rewrite Rules. USC Information Science Institute, Marina del Rey, Calif.: 1977.

[97] Musser, D. R.: Convergent Sets of Rewrite Rules for Abstract Data Types. Information Sciences Institute: Report 1978.

[98] Musser, D. R.: On Proving Inductive Properties of Abstract Data Types. Seventh ACM Symp. on Principles of Programming Languages **1980**, 154 – 162.

[99] Newman, M. F.: Calculating Presentations for Certain Kinds of Quotient Groups. SYMSAC **1976**, 2 – 8.

[100] Newman, M. H. A.: On Theories with a Combinatorial Definition of "Equivalence". Ann. Math. **43**/2, 223 – 243 (1942).

[101] O'Donnell, M.: Computing in Systems Described by Equations. Lecture Notes in Computer Science, Vol. 58. Berlin-Heidelberg-New York: Springer 1977.

[102] Oldehoeft, A.: Analysis of Constructed Mathematical Responses by Numeric Tests for Equivalence. Proc. ACM 24th Nat. Conf., 117 – 124 (1969).

[103] Paterson, M. S., Wegmann, M. N.: Linear Unification. J. Comput. Syst. Sci. **16**, 158 – 167 (1978).

[104] Pavelle, R., Rothstein, M., Fitch, J.: Computer Algebra. Preprint, Scientific American 1981.

[105] Peterson, G. E., Stickel, M. E.: Complete Set of Reductions for Some Equational Theories. J. ACM **28**/2, 233 – 264 (1981).

[106] Plaisted, D.: Well-Founded Orderings for Proving Termination of Systems of Rewrite Rules. Univ. of Illinois, Urbana-Champaign: Rep. 78-**932**, 1978.

[107] Plaisted, D.: A Recursively Defined Ordering for Proving Termination of Term Rewriting Systems. Univ. of Illinois, Urbana-Champaign: Rep. 78-**943**, 1978.

[108] Plotkin, G.: Building-In Equational Theories. Machine-Intelligence **7**, 73 – 90 (1972).

[109] Pohst, M., Yun, D. Y. Y.: On Solving Systems of Algebraic Equations Via Ideal Bases and Elimination Theory. SYMSAC **1981**, 206 – 211.

[110] Polak, W.: Program Verification at Stanford: Past, Present, Future. Stanford University: Comput. Syst. Lab., Rep. 1981.

[111] Raulefs, P., Siekmann, J., Szabó, P., Unvericht, E.: A Short Survey on the State of the Art in Matching and Unification Problems. SIGSAM Bull. **13**/2, 14 – 20 (1979).

[112] Reynolds, J. C.: Reasoning About Arrays. Commun. ACM **22**/5, 290 – 299 (1979).

[113] Richardson, D.: Some Unsolvable Problems Involving Elementary Functions of a Real Variable. J. Symb. Logic **33**, 511 – 520 (1968).

[114] Richardson, D.: Solution of the Identity Problem for Integral Exponential Functions. Math. Logik Grundlagen Math. **15**, 333 – 340 (1969).

[115] Richter, M. M., Kemmenich, S.: Reduktionssysteme und Entscheidungsverfahren. Rhein.-Westf. Tech. Hochschule, Aachen: Inst. f. Informatik, Rep. **4**, 1980.

[116] Risch, R. H.: The Problem of Integration in Finite Terms. Trans. AMS **139**, 167 – 189 (1969).

[117] Robinson, J. A.: A Machine-Oriented Logic Based on the Resolution Principle. J. ACM **12**, 23 – 41 (1965).

[118] Robinson, J. A.: Computational Logic: The Unification Computation. Machine Intelligence **6**, 63 – 72. New York: Edinb. Univ. Press 1971.

[119] Rogers, H.: Theory of Recursive Functions and Effective Computability. New York: McGraw-Hill 1967.

[120] Rosen, B. K.: Tree-Manipulating Systems and Church-Rosser Theorems. J. ACM **20**/1, 160 – 187 (1973).

[121] Schaller, S.: Algorithmic Aspects of Polynomial Residue Class Rings. Ph.D. Thesis, Comput. Sci. Tech., University of Wisconsin, Madison, Rep. **370**, 1979.

[122] Schrader, R.: Zur konstruktiven Idealtheorie. Dipl. Thesis, Math. Inst., Univ. of Karlsruhe, Federal Republic of Germany 1976.

[123] Seidenberg, A.: Constructions in Algebra. Trans. AMS **197**, 273 – 313 (1974).

[124] Sethi, R.: Testing for the Church-Rosser Property. J. ACM **21**/4, 671 – 679 (1974), **22**/3, 424 (1975).

[125] Shtokhamer, R.: Simple Ideal Theory: Some Applications to Algebraic Simplification. Univ. of Utah, Salt Lake City: Tech. Rep. UCP-**36**, 1975.

[126] Shtokhamer, R.: A Canonical Form of Polynomials in the Presence of Side Relations. Technion, Haifa: Phys. Dept., Rep. **25**, 1976.

[127] Shtokhamer, R.: Attempts in Local Simplification of Non-Nested Radicals. SIGSAM Bull. **11**/1, 20 – 21 (1977).

[128] Spear, D.: A Constructive Approach to Commutative Ring Theory. MACSYMA **1977**, 369 – 376.

[129] Staples, J.: Church-Rosser Theorems for Replacement Systems. Algebra and Logic (Crossley, J., ed.), Lecture Notes in Mathematics, Vol. 450, pp. 291 – 307. Berlin-Heidelberg-New York: Springer 1975.

[130] Stickel, M. E., Peterson, G. E.: Complete Sets of Reductions for Equational Theories with Complete Unification Algorithm. Univ. of Arizona: Dept. Comp. Sci. 1977.

[131] Stoutemyer, D. R.: Qualitative Analysis of Mathematical Expressions Using Computer Symbolic Mathematics. SYMSAC **1976**, 97 – 104.

[132] Stoutemyer, D. R.: Automatic Simplification for the Absolute Value Function and its Relatives. ACM SIGSAM Bull. **10**/4, 48 – 49 (1976).

[133] Suzuki, N., Jefferson, D.: Verification Decidability of Presburger Array Programs. J. ACM **27**/1, 191 – 205 (1980).

[134] Szekeres, G.: A Canonical Basis for the Ideals of a Polynomial Domain. Am. Math. Mon. **59**/6, 379 – 386 (1952).

[135] Tobey, R. G.: Experience with FORMAC Algorithm Design. Commun. ACM **9**, 589 – 597 (1966).

[136] Todd, J. A., Coxeter, H. S. M.: A Practical Method for Enumerating Cosets of a Finite Abstract Group. Proc. Edinb. Math. Soc. (2) **5**, 26 – 34 (1936).

[137] Trinks, W.: Über B. Buchbergers Verfahren, Systeme algebraischer Gleichungen zu lösen. J. Number Theory **10**/4, 475 – 488 (1978).

[138] Trotter, P. G.: Ideals in $Z[x, y]$. Acta Math. Acad. Sci. Hungar. **32** (1 – 2), 63 – 73 (1978).

[139] Winkler, F., Buchberger, B., Lichtenberger, F., Rolletschek, H.: An Algorithm for Constructing Canonical Bases (Gröbner-Bases) for Polynomial Ideals. (Submitted.)

[140] Yun, D. Y. Y., Stoutemyer, R. D.: Symbolic Mathematical Computation. Encyclopedia of Computer Science and Technology (Belzer, J., Holzman, A. G., Kent, A., eds.), Vol. 15, pp. 235 – 310. New York-Basel: Marcel Dekker 1980.

[141] Zacharias, G.: Generalized Gröbner Bases in Commutative Polynomial Rings. Bachelor Thesis, Dept. Comput. Sci., M.I.T., 1978.

[142] Zippel, R. E. B.: Simplification of Radicals with Applications to Solving Polynomial Equations. Cambridge, Mass.: M.I.T., Dept. of Electrical Engineering and Computer Science, Master's Thesis, 1977.

Prof. Dr. B. Buchberger
Institut für Mathematik
Johannes-Kepler-Universität Linz
Altenbergerstrasse 69
A-4040 Linz
Austria

Prof. Dr. R. Loos
Institut für Informatik I
Universität Karlsruhe
Zirkel 2
D-7500 Karlsruhe
Federal Republic of Germany

Computing with Groups and Their Character Tables*

J. **Neubüser,** Aachen

Abstract

In this survey an attempt is made to give some impression of the capabilities of currently available programs for computations with finitely generated groups and their representations.

Introduction

Group theory started from one of the classical problems of algebra, the question whether polynomial equations can be solved by radicals or not. However, like other parts of "modern" algebra, it has developed into an "abstract" structure theory with questions of its own. The algorithms of computational group theory were primarily designed to answer such questions about individual groups, most of them start from and use elementary, but typical group theoretic notions and facts. So except for the joint use of standard implementation techniques, there are rather few connections (the matrix treatment of abelian groups, section 1.1; representation of cyclotomic integers for characters, section 7) between computational group theory and other parts of computer algebra.

It is only recently that in a few instances common background structures of algorithms for groups and in more classical algebraic areas attracted attention, such as the use of the Church-Rosser property and the idea of the Knuth-Bendix algorithm [39] (see also the chapter on algebraic simplification in this volume). Some recent papers may indicate that group theoretic algorithms provide interesting examples for complexity theory; however so far the latter has been of no help in improving or illuminating the methods used to investigate groups with the help of computers.

Because of its relative isolation from the rest of computer algebra a full description of data structures and algorithms of computational group theory can unfortunately not build on the contents of the other articles in this volume and so is unfeasible in the limited space available. So it seemed best to give at least an idea of *what* presently can be done in computational group theory by just reporting what kind of output can be obtained from a given input. For general ideas and their development the reader is e.g. referred to the surveys [8, 42, 54, 55], for more details on coset table methods to [45]. In particular, no explanation of group

---

* This paper is a modified version of the paper presented at the 10th International Colloquium on "Group-Theoretical Methods in Physics" and published in Physica A **114A**, 493 – 506 (1982).

theoretic terms or facts can be given here; background about the general theory of finite groups can be found in e.g. [32], about presentations in [40, 34], about characters and representations in [33]. References to papers describing the method used will be given for each implementation separately. No claim of completeness is made for these; rather than the papers first dealing with a method, more recent ones with further references are quoted.

Guide to the Lists

All programs listed investigate individual groups, determined either concretely (e.g. as a particular group of permutations or matrices) or up to isomorphism (e.g. when the group is given by a presentation). The group to be investigated must be given to the program in a specified form (cf. [55] for a closer discussion of this point), usually in terms of generators or, in the case of the character table programs, by some faithful characters.

We shall list for each program

(i) under "Input" the data that have to be given to the program,

(ii) under "Output" some typical information that the program produces – it should be understood that often further output can be obtained,

(iii) under "Method" some indication of the kind of algorithms used, in particular whether the program must be handled interactively,

(iv) under "Examples" some data about the performance of the programs, usually for routine runs on an average main-frame computer, sometimes also for extreme cases obtainable only by special implementations and/or computers. It should be noted that in computational group theory the feasibility of a computation usually depends on a large number of parameters characterizing the "complexity" (this term being used as in common language, not in its technical meaning in computer science) of a group, rather than on a single parameter like the order of the group. So, for instance, while a certain program may just about be able to determine the subgroup lattice of the alternating group A_8 of order 20160, it would be hopelessly defeated by the elementary abelian group of order 256.

With some arbitrariness the list is subdivided into 8 sections. The first four sections are characterized by the kind of input required: "Presentations", "AG-systems", "Permutations", and "Matrices". The fifth, "Details for small groups", summarizes programs which do not depend much on the kind of input. Section 6 contains some information about the availability of group theoretic programs. In the last two sections "Characters" and "Representations", we turn to ordinary and, very briefly, to modular representation theory of finite groups.

List of Programs

1. Presentations

We shall start by listing programs that work from a finite presentation [40, 34]:

$$(*) \qquad G = \langle g_1, \ldots, g_k | r_1(g_1, \ldots, g_k) = 1, \ldots, r_l(g_1, \ldots, g_k) = 1 \rangle.$$

1.1. Input: (∗) only.

Output: Maximal abelian factor group G/G' as direct product of cyclic groups, finite and infinite.

Method: Smith normal form of relation matrix [52, 30].

Examples: $k = 20$, $l = 38$ (routine case, several methods); $k = 191$, $l = 380$ (extreme case, modular arithmetic) [30].

1.2. Input: (∗) plus subgroup $U = \langle u_1(g_1, \ldots, g_k), \ldots, u_m(g_1, \ldots, g_k) \rangle$ of G.

Output: Index $G:U$; coset representatives of U in G; permutation representation of G on cosets of U, usable as input to 3.1; with $U = \{id\}$ (id being the unit element of G) a regular representation of G, usable as input to 5.1 (Cayley graph); normalizer of U in G; etc.

Method: Todd-Coxeter coset enumeration (see ref. [15, 45] and many papers quoted there).

Warning: The enumeration terminates eventually whenever $G:U$ is finite, but no a priori bounds for termination can be given.

Examples: Weyl group B_6, $U = \{id\}$, B_6: $\{id\} = 46080$ [15]; Index up to 700000 possible (G. Havas, private communication, June 1980).

1.3. Input: (∗) plus a bound $b \in \mathbb{N}$.

Output: All classes of conjugate subgroups $U \leqslant G$ with index $G:U \leqslant b$.

Method: Tree searching using coset tables [53, 17, 45].

Example: $G = \langle x, y, t \,|\, x^2 = y^3 = (xy)^7 = t^2 = (xt)^2 = (yt)^2 = 1 \rangle$ has 2042 classes of conjugate subgroups U of index $G:U \leqslant 97$.

1.4. Input: (∗) plus coset representatives of a subgroup $U \leqslant G$ (if $U \unlhd G$, a presentation of G/U can be used instead).

Output: A presentation of U.

Method: Either (i) Schreier-Reidemeister [26] or (ii) modified Todd-Coxeter [3a, 35, 41, 45].

Example: (i) Subgroup of index 152 (and order at least 5^{741}) of the Fibonacci group $F(2, 9)$ [29].

1.5. Input: (∗) plus a prime number p.

Output: Order of the maximal p-factor group $G/O^p(G)$; rank of the "p-multiplier" $M_p(G/O^p(G))$; lower exponent-p-central chain of G; "power-commutator presentation" for factor groups of $G/O^p(G)$, usable as input to 2.1 and 2.2.

Method: Nilpotent quotient algorithm [47, 27] using collection methods [28].

Examples: Burnside groups: $|B(4, 4)| = 2^{422}$; $|B(3, 5; 9)| = 5^{916}$; the rank of $M_2(B(4, 4))$ is 1055 [27].

2. AG-Systems

The power-commutator presentation delivered by the nilpotent quotient algorithm (1.5) is a special case of a wider class of the so-called power-commutation presentations [47] (or AG-systems [20]), which in the case of a finite group G is a presentation of the form

$$G = \langle g_1, \ldots, g_k | g_i^{s(i)} = w_i, 1 \leqslant i \leqslant k, g_j g_i = g_i w_{ij}, 1 \leqslant i < j \leqslant k \rangle,$$

where $s(i) \in \mathbb{N}$, and w_i, w_{ij} are words in $g_{i+1}, \ldots, g_k$. Power-commutation presentations and even better power-commutator presentations (or psg-systems [22]) are used in fairly efficient programs for the investigation of finite soluble and finite p-groups, respectively, which we list next:

2.1. Input: Power-commutation presentation of a finite soluble group G.

Output: Derived series; lower central series; Sylow subgroups; $\langle U, V \rangle$ and $U \cap V$ for given $U, V \leqslant G$; $\langle U^g | g \in G \rangle$, $\bigcap_{g \in G} U^g$, normalizer $N_G(U)$, all U^g, for given $U \leqslant G$.

Method: Interactively controlled handling of AG-systems for subgroups of G, collection algorithms [20; E. Oppelt, Diplomarbeit, Aachen 1981, unpublished].

Example: Derived series of a group of order 2^{34}.

2.2. Input: Power-commutator presentation of a p-group G.

Output: As in 2.1 plus: Conjugacy classes and centralizers of elements; lattice of normal subgroups; maximal abelian subgroups of G; lattice of all subgroups ordered into classes of conjugate subgroups or prescribed sections of this lattice; centralizer $C_G(U)$, maximal subgroups of U for prescribed $U \leqslant G$.

Method: As in 2.1 [22; F. L. Chomse, U. Paul, Diplomarbeiten, Aachen 1981/82, unpublished].

Examples: Determination of all (24) conjugacy classes of length 2^{34} in a group of order 2^{41} [22]; all (2082) normal subgroups of $Z_{32} \times (Z_{32}/Z_2)$, all (1504) classes of supplements of $\mathbb{Z}_5^4$ in the natural split extension of $\mathbb{Z}_5^4$ by $\mathrm{Syl}_5(GL(4, 5))$.

3. Permutations

While all programs described so far started directly or indirectly from a presentation, we now turn to the cases in which a group is given by a generating set of "concrete" elements such as permutations or matrices, i.e. in which a group is given as a subgroup of a well known group such as a symmetric group or some linear group. In this section we discuss algorithms for the investigation of permutation groups.

3.1. Input: A set $\{g_1, \ldots, g_k\}$ of permutations of degree n.

Output: Orbits of $\langle g_1, \ldots, g_k \rangle = G$; blocks (sets of imprimitivity) of G; representation of G by block permutations.

Method: Elementary orbit algorithm; Atkinson's algorithm for blocks [3].

Example: Blocks of a group of degree 998 [3].

For the rest of the programs discussed in this section the notions of a base and a strong generating set of a permutation group, introduced by C. Sims [50] are fundamental: Let G be a permutation group on a finite set Ω, and

$$G_{\alpha_1,\ldots,\alpha_r} = \{g \in G \mid g(\alpha_i) = \alpha_i, \ 1 \leqslant i \leqslant r\}.$$

A sequence $(\beta_1, \ldots, \beta_s)$ with $\beta_i \in \Omega$ is a base for G, if $G_{\beta_1,\ldots,\beta_s} = \{\text{id}\}$.

A subset $S \subseteq G$ is a strong generating set for G with respect to the base $(\beta_1, \ldots, \beta_s)$ if

$$G_{\beta_1,\ldots,\beta_{i-1}} = \langle S \cap G_{\beta_1,\ldots,\beta_{i-1}} \rangle \qquad \text{for all } i, \qquad 1 \leqslant i \leqslant s.$$

It is essential for all further programs in this section that a base and strong generating set can be computed from an arbitrary generating set of a permutation group:

3.2. Input: A set $\{g_1, \ldots, g_k\}$ of permutations of degree n.

Output: A base $B = (\beta_1, \ldots, \beta_s)$ for $G = \langle g_1, \ldots, g_k \rangle$; a strong generating set of G with respect to B; index and coset representatives of $G_{\beta_1,\ldots,\beta_i}$ in $G_{\beta_1,\ldots,\beta_{i-1}}, 1 \leqslant i \leqslant s$; order of G; a presentation of G.

Method: Schreier-Sims method (SS), Schreier-Todd-Coxeter-Sims method (STCS) [36].

Examples: (SS): degree 125, order $2^{58} \cdot 3^{10} \cdot 5^3 \cdot 7^2$; (STCS): Held's group, degree 2048, order 4 030 387 200: base, strong generating set, presentation; degrees up to and over 10 000 possible [36, 37].

3.3. Input: Strong generating set S of a permutation group G.

Output: Conjugacy classes and centralizers of elements; Sylow subgroups; derived series; descending central series; center; $\langle U^g \mid g \in G \rangle$, normalizer $N_G(U)$, centralizer $C_G(U)$ for given $U \leqslant G$; test for isomorphism to specified groups; $U \cap V$, $\langle U, V \rangle$, test for conjugacy for given $U, V \leqslant G$.

Method: Use of base and strong generating set, base-change algorithm, backtrack search [50, 51, 13, 7].

Example: Detailed analysis of the structure of Held's group [7].

4. Matrices

In section 3 we dealt specifically with programs making strong use of the fact that permutation groups are handled, and providing powerful means for their investigation. The basic Schreier-Sims and Schreier-Todd-Coxeter-Sims methods for the determination of base and strong generating set have been modified for matrix groups of finite dimension, in particular over finite fields by Cannon and Butler [6, 7] (see also forthcoming papers of Butler and Cannon). Almost all entries of 3.2 and 3.3 can be carried over, but it has to be understood that applications are more restricted and costly than with permutation groups because for matrix groups it is more difficult to find stabilizer chains with comparatively small indices and to test membership to a coset of such a stabilizer [6].

Examples: $Sp_{10}(2)$, order $2^{25} \cdot 3^6 \cdot 5^2 \cdot 7 \cdot 11 \cdot 17 \cdot 31$: base, strong generating set, order [6]; $Syl_2(GL(5, 2))$, order 2^{10}: upper and lower central series; $Sp_6(4)$, order $2^{18} \cdot 3^4 \cdot 5^3 \cdot 7 \cdot 13 \cdot 17$: centralizer of given element [7].

5. Details for Small Groups

In this section we collect a number of programs which assume that generating elements for a group are given in some way that allows multiplication, inversion, and test for equality of elements, but do not essentially depend on the special nature of these elements, so that the list of input types given below is open-ended and could be extended rather easily by implementing further element handling routines, e.g. for matrices over rings of algebraic integers. The common feature of the programs is that they produce rather detailed information about a group by processing usually many and extensive stacks describing structural properties of it. Their applicability is therefore usually limited by the store needed to hold this information.

5.1. Input: Generating set of a group G in some form allowing multiplication, inversion and comparison of elements, in particular: permutations; matrices over $\mathbb{Z}$, $GF(p^x)$; power-commutation presentation; Cayley graph.

Output: a) a presentation of G;

b) conjugacy classes of G;

c) lattice of normal subgroups of G;

d) lattice of all subgroups of G with many options: sorting into conjugacy classes; normalizer $N_G(U)$, centralizer $C_G(U)$, maximal subgroups of prescribed subgroups U; determination of abstract (e.g. abelian, nilpotent, soluble, perfect, simple) and representation dependent (e.g. transitive, primitive) properties of prescribed subgroups U;

e) automorphism group of G;

f) character table of G.

Method: a) exhaustive search in Schreier coset graph [10, 45],

b) orbit algorithm applied to conjugation [9];

c) starting from normal subgroups generated by a single conjugacy class of elements, lattice operations are used [9];

d) cyclic extension method [43], using the knowledge of the perfect groups of order $\leqslant 10^4$ [49]; interactive variations, allowing to build partial lattices adding cyclic extensions, normalizers, centralizers, intersections, joins, Sylow subgroups, perfect subgroups etc;

e) construction of base and strong generating set for $Aut(G)$ operating on G [H. Robertz, Diplomarbeit, Aachen 1976, unpublished];

f) J. D. Dixon's method, starting from class multiplication coefficients [18].

Examples: a) presentation of the Mathieu group M_{12} of order 95040 [10];

c) normal subgroups of $Syl_2(GL(5, 2))$;

b), d) alternating group A_8, order 20160;

e) dihedral group D of order 48:

$$|\text{Aut}(D)| = 2^6 \cdot 3, \qquad |\text{Aut}(\text{Aut}(D))| = 2^8 \cdot 3,$$

$$|\text{Aut}(\text{Aut}(\text{Aut}(D)))| = 2^{21} \cdot 3;$$

f) A_8, order 20160, 14 classes; split extension of $\mathbb{Z}_2^5$ by $D_8 \times D_{12}$, the direct product of the dihedral groups of orders 8 and 12, respectively; this has order 1536, 80 classes.

6. Availability of Group Theoretical Programs

Most of the earlier implementations, written in machine code for reasons of efficiency on the then much smaller and slower computers, became obsolete with the computers on which they were implemented. Machine-independent languages have improved the situation, but the necessity of handling unusual data-types has often led to implementations which, at least in parts, circumvent standards of languages, thus making transfer to different machines more difficult. We shall distinguish two kinds of implementations available.

First, there are programs implementing essentially one algorithm. Often being written with application to concrete problems in mind, they are usually tuned for high efficiency while often not too much attention is paid to user comfort or compatibility of input or output to those of other programs. Cannon who reviews the state of the art in computational group theory under the aspect of this section in [14] uses the term "autonomous programs" for these. The algorithm most often implemented as autonomous program is certainly the Todd-Coxeter method (1.2). A very versatile implementation of it is the one by Alford and Havas of the Australian National University, Canberra, a predecessor of which is described in [15]. References to further autonomous programs can be found in [14].

At the other end of the scale is the attempt to unify many programs into a big system with common data structures and a high-level user oriented handling language. By far the biggest enterprise of this kind is the system designed by J. Cannon (Sydney) which is often referred to as the CAYLEY system by the name originally given to its user language [11, 12, 14]. The system contains most of the programs listed in sections 1, 3, 4, 5, and some of 2. It can handle several groups simultaneously and allows to set up and use mappings between them, as well as to define "standard" examples like cyclic, dihedral or symmetric groups. The system is equipped with a library of input data for many permutation and matrix groups. Portability of the CAYLEY system is facilitated by a special storage allocation program, the STACKHANDLER, which is also used e.g. by the CAS system mentioned below. While still being developed earlier versions of it are operational in several places including Aachen (RWTH), Canberra (ANU), London (QMC), Melbourne, Montreal (Concordia), Sydney, and Zürich (ETH).

Having started as an offspring of a predecessor of the CAYLEY system, the CAMAC system [38] has since developed independently, containing e.g. programs for handling presentations (1.2) and permutation groups (section 3), CAMAC is particularly aimed at applications in fields like coding theory. It is located at the University of Illinois at Chicage Circle.

More recently programs for handling soluble and in particular *p*-groups described in section 2 are being combined into a system, SOGOS, at the RWTH Aachen. The CAS system for working with characters (to be described in section 7) has also been implemented there. CAS is presently also operational in Bayreuth, Sydney, Amsterdam (MC) and London (QMC).

7. Characters

As Dixon's method for the direct determination of the character table of a finite group (5.1f) works from the class multiplication coefficients, its applicability is not only limited by the size of the character table itself but also, since the classes must be known fairly explicitly, by the order of the group. On the other hand simple groups may have huge orders but comparatively very few classes. So, e.g., the so-called Monster, a sporadic simple group, has order $2^{46} \cdot 3^{20} \cdot 5^9 \cdot 7^6 \cdot 11^2 \cdot 13^3 \cdot 17 \cdot 19 \cdot 23 \cdot 29 \cdot 31 \cdot 41 \cdot 47 \cdot 59 \cdot 71$, but "only" 194 classes. Character tables of such groups have been determined "by hand", sometimes with assistance of computer programs that can perform certain operations with characters or character values. Such systems of handling routines have been implemented by several people, but except for [31] and a few theses I do not know of publications on such programs. Therefore in what follows I shall refer explicitly to the CAS system for generating and handling character tables and related objects which has been implemented in Aachen during the last few years [1, 2]. In designing CAS we have in particular built on the experience that Livingstone and Wynn gathered with their system in Birmingham and Gabrysch with his in Bielefeld.

A special feature of the CAS system, to our knowledge not present in its predecessors, is that it handles irrational character values in exact algebraic form. Since a character value $\chi(g)$ is a sum of roots of unity, whose orders divide the order $|g|$ of the group element g, calculations are done in rings $\mathbb{Z}(\zeta)$, where ζ is a primitive $|g|$th root of unity, and such a ring is represented as $\mathbb{Z}[x]/\varphi(x)$, where φ is the $|g|$th cyclotomic polynomial. Calculation with characters of big groups requires moreover that even in the representation of $\mathbb{Z}[x]$ arithmetic can be done with arbitrarily high multiple precision.

CAS is designed towards interactive handling, hence there are no definite requirements for the input.

The routines of CAS that under the control of a problem-oriented user language are provided for interactive use can be sorted into a few main categories:

7.1 Table Commands

These enable one to define, copy and erase character tables; to sort characters and/or classes according to defined rules or the user's wish; to supplement power maps from partial knowledge of a table; to set up and handle fusion maps between the classes of different groups; etc.

7.2 Commands Which Generate Characters

One can get: the tensor product of given characters; the conjugate character of a given character under a given Galois automorphism of a cyclotomic field or under an automorphism of the group; the symmetrized powers of a given character using "linear" or "orthogonal" symmetrization [24] (prerequisite: respective power map).

Known irreducible characters can be stripped off from a compound character; characters of a subgroup U of G can be induced up to G (prerequisite: fusion of the classes of U into the classes of G); characters of a group H containing G can be restricted to G (prerequisite: fusion of the classes of G into the classes of H); characters of a factor group G/N of G can be extended to characters of G. If possible, the splitting of several compound characters of G into irreducibles can be obtained from the triangle of scalar products of the compound characters.

7.3 Commands Which Test Characters

These include tests for the first and second orthogonality relations; test of p-block orthogonality; test of decomposition of tensor products and symmetrized powers into irreducibles. Tests can be performed with incompletely known characters or class columns in order to obtain missing values.

7.4 Commands Which Generate Character Tables

From the character tables of groups G and H the table of the direct product $G \times H$ can be obtained while from the table of G and a normal subgroup N of G the table of G/N can be obtained. There are programs generating the tables of cyclic groups as well as of PSL$(2, q)$, SL$(2, q)$, PGL$(2, q)$, GL$(2, q)$, PSL$(3, q)$, SL$(3, q)$, PSU$(3, q^2)$, and SU$(3, q^2)$ for prescribed values of q from generic formulae.

7.5 Commands Which Extract Information from Character Tables

Of group theoretic information e.g. the class multiplication coefficients, the kernels of characters, and the Schur-Frobenius indicators of characters can be computed. The above-mentioned exact algebraic description of irrational character values allows some steps towards modular representation theory. So the central characters can be formed and used to sort the characters into p-blocks, determine the defect of these, etc. Further it is possible to obtain coefficients of Molien series [56]. Finally a hierarchy of necessary condition tests for a character to be a permutation character is provided.

7.6 Examples

CAS has been used e.g. to determine the character table of $3 \cdot U(6, 2) \cdot 3$ of order $2^{15} \cdot 3^8 \cdot 5 \cdot 7 \cdot 11$ having 324 classes. It can handle character tables of up to about 1000 classes, there are no built-in limits to degrees of characters or the group order.

7.7 Library

The CAS system is supplemented by a library of character tables (presently containing e.g. all the sporadic simple groups) which incorporates tables given to us by many colleagues, in particular J. Conway and his collaborators of the Cambridge (U.K.) Group Atlas project.

8. Representations

8.1 Ordinary ($\mathbb{C}$-)Representations

All methods for the determination of the irreducible $\mathbb{C}$-representations of a group that have been proposed so far seem to be applicable only to fairly small groups or to fairly small degrees. S. Flodmark and P.-O. Jansson [23] describe the method of a working program; performance statistics are not yet given but applicability is probably restricted to groups of order a few hundred. Other methods have been proposed by J. D. Dixon [19] and by J. Gabriel in a series of papers [25]. I do not know of implementations of these. A program by C. Brott [5] has meanwhile become obsolete. In particular in view of possible applications, e.g. in physics and chemistry, it seems desirable to follow up proposals and to implement a unified system possibly also using information from character programs.

8.2 Modular (GF(p^{α})-)Representations

It may be surprising, at first sight, that while present programs for the determination of ordinary representations are limited to small groups, during the last years much more powerful interactive methods for finding irreducible representations over finite fields have been developed by R. Parker at Cambridge (U.K.). Like the interactive character programs they start from some known modular representation (given by matrices representing generators) and obtain further representations as tensor products, symmetric and antisymmetric squares, etc. At the heart of the matter are highly efficient trial-and-error methods for splitting off constituents or proving irreducibility and for testing equivalence. The probability of success is high for the field of two elements and drops for bigger fields. The programs have been used e.g. in the existence proof of the simple group J_4 of order $2^{21} \cdot 3^3 \cdot 5 \cdot 7 \cdot 11^3 \cdot 23 \cdot 29 \cdot 31 \cdot 37 \cdot 43$, a GF(2)-representation of degree 4995 of which had to be split off from a representation of degree 6216. Representations of degree up to about 10 000 can be handled [57]. No full account of these programs has yet been published.

Acknowledgement

I am indepted to several colleagues for information and in particular to Dr. V. Felsch, from whose bibliography on the use of computers in group theory [21] the references are taken and who helped in sorting out the examples.

References

[1] Aasman, M., Janißen, W., Lammers, H., Neubüser, J., Pahlings, H., Plesken, W.: The CAS System, User Manual (provisional version). RWTH Aachen: Lehrstuhl D f. Math., Mimeographed Rep., 83 p., 1981.
[2] Aasman, M., Janißen, W., Lammers, H., Neubüser, J., Pahlings, H., Plesken, W.: Das CAS System, Handbuch (vorläufige Version). RWTH Aachen: Lehrstuhl D f. Math., Mimeographed Rep., 555 p., 1981.
[3] Atkinson, M. D.: An Algorithm for Finding the Blocks of a Permutation Group. Math. Comput. **29**, 911–913 (1975).

[3a] Beetham, M. J., Campbell, C. M.: A Note on the Todd-Coxeter Coset Enumeration Algorithm. Proc. Edinburgh Math. Soc. (2) **20**, 73 – 79 (1976).

[4] Birkhoff, G., Hall, M., jr. (eds.): Computers in Algebra and Number Theory. (SIAM-AMS Proc. **4**.) Providence, R. I.: Amer. Math. Soc. 1971.

[5] Brott, C., Neubüser, J.: A Programme for the Calculation of Characters and Representations of Finite Groups. In: Proc. of the Conference on Computational Problems in Abstract Algebra, Oxford, 1967 (Leech, J., ed.), pp. 101 – 110. Pergamon Press 1970.

[6] Butler, G.: The Schreier Algorithm for Matrix Groups. SYMSAC **1976**, 167 – 170.

[7] Butler, G.: Computational Approaches to Certain Problems in the Theory of Finite Groups. Ph.D. Thesis, Univ. of Sydney, 316 p., 2 appendices on microfiche, 1979.

[8] Cannon, J. J.: Computers in Group Theory: A Survey. Commun. ACM **12**, 3 – 12 (1969).

[9] Cannon, J. J.: Computing Local Structure of Large Finite Groups. In [4], 161 – 176.

[10] Cannon, J. J.: Construction of Defining Relators for Finite Groups. Discrete Math. **5**, 105 – 129 (1973).

[11] Cannon, J. J.: A General Purpose Group Theory Program. In [46], 204 – 217.

[12] Cannon, J. J.: A Draft Description of the Group Theory Language Cayley. SYMSAC **1976**, 66 – 84.

[13] Cannon, J. J.: Effective Procedures for the Recognition of Primitive Groups. In [16], 487 – 493.

[14] Cannon, J. J.: Software Tools for Group Theory. In [16], 495 – 502.

[15] Cannon, J. J., Dimino, L. A., Havas, G., Watson, J. M.: Implementation and Analysis of the Todd-Coxeter Algorithm. Math. Comput. **27**, 463 – 490 (1973).

[16] Cooperstein, B., Mason, G. (eds.): The Santa Cruz Conference on Finite Groups. (Proc. Sympos. Pure Math. **37**.) Providence, R. I.: Amer. Math. Soc. 1980.

[17] Dietze, A., Schaps, M.: Determining Subgroups of Given Finite Index in a Finitely Presented Group. Can. J. Math. **26**, 769 – 782 (1974).

[18] Dixon, J. D.: High Speed Computation of Group Characters. Numer. Math. **10**, 446 – 450 (1967).

[19] Dixon, J. D.: Computing Irreducible Representations of Groups. Math. Comput. **24**, 707 – 712 (1970).

[20] Felsch, V.: A Machine Independent Implementation of a Collection Algorithm for the Multiplication of Group Elements. SYMSAC **1976**, 159 – 166.

[21] Felsch, V.: A Bibliography on the Use of Computers in Group Theory and Related Topics: Algorithms, Implementations, and Applications. Kept current and obtainable from Lehrstuhl D für Mathematik, RWTH Aachen, D-5100 Aachen, Federal Republic of Germany.

[22] Felsch, V., Neubüser, J.: An Algorithm for the Computation of Conjugacy Classes and Centralizers in p-Groups. EUROSAM **1979**, 452 – 465.

[23] Flodmark, S., Jansson, P.-O.: Irreducible Representations of Finite Groups (Proc. 10th Internat. Coll. on Group Theoret. Methods in Phys.). Physica **114 A**, 485 – 492 (1982).

[24] Frame, J. S.: Recursive Computation of Tensor Power Components. Bayreuther Math. Schriften Heft 10, 153 – 159 (1982).

[25] Gabriel, J. R.: Numerical Methods for Reduction of Group Representations. SYMSAM **1971**, 180 – 182.

[26] Havas, G.: A Reidemeister-Schreier Program. In [46], 347 – 356.

[27] Havas, G., Newman, M. F.: Application of Computers to Questions like Those of Burnside. In: Burnside Groups Proc., Bielefeld, Germany: 1977 (Mennicke, J. L., ed.). Lecture Notes in Mathematics, Vol. 806, pp. 211 – 230. Berlin-Heidelberg-New York: Springer 1980.

[28] Havas, G., Nicholson, T.: Collection. SYMSAC **1976**, 9 – 14.

[29] Havas, G., Richardson, J. S., Sterling, L. S.: The Last of the Fibonacci Groups. Proc. Roy. Soc. (Edinburgh) **A83**, 199 – 203 (1979).

[30] Havas, G., Sterling, L. S.: Integer Matrices and Abelian Groups. EUROSAM **1979**, 431 – 451.

[31] Hunt, D. C.: A Computer-Based Atlas of Finite Simple Groups. In [16], 507 – 510.

[32] Huppert, B.: Endliche Gruppen I. Berlin-Heidelberg-New York: Springer 1967.

[33] Isaacs, I. M.: Character Theory of Finite Groups. New York-San Francisco-London: Academic Press 1976.

[34] Johnson, D. L.: Topics in the Theory of Group Presentations. London Math. Soc. Lect. Note Series, Vol. 42. Cambridge: Cambridge University Press 1980.

[35] Leech, J.: Computer Proof of Relations in Groups. In: Topics in Group Theory and Comput. (Proc., Galway, 1973), (Curran, M. P. J., ed.), pp. 38–61. London-New York-San Francisco: Academic Press 1977.

[36] Leon, J. S.: On an Algorithm for Finding a Base and a Strong Generating Set for a Group Given by Generating Permutations. Math. Comput. **35**, 941–974 (1980).

[37] Leon, J. S.: Finding the Order of a Permutation Group. In [16], 511–517.

[38] Leon, J. S., Pless, V.: CAMAC 1979. EUROSAM **1979**, 249–257.

[39] Loos, R.: Term Reduction Systems and Algebraic Algorithms. Informatik Fachberichte **47**, 214–234 (1981).

[40] Magnus, W., Karrass, A., Solitar, D.: Combinatorial Group Theory: Presentations of Groups in Terms of Generators and Relations. New York-London-Sydney: Interscience 1966.

[41] McLain, D. H.: An Algorithm for Determining Defining Relations of a Subgroup. Glasg. Math. J. **18**, 51–56 (1977).

[42] Neubüser, J.: Investigations of Groups on Computers. In: Proc. of the Conference on Computational Problems in Abstract Algebra, Oxford, 1967 (Leech, J., ed.), pp. 1–19. Pergamon Press 1970.

[43] Neubüser, J.: Computing Moderately Large Groups: Some Methods and Applications. In [4], 183–190.

[44] Neubüser, J.: Some Computational Methods in Group Theory. In: 3rd Internat. Coll. Adv. Comput. Methods in Theoret. Phys., B-II-1 – B-II-35. Marseille: Centre de Physique Théorique CNRS 1973.

[45] Neubüser, J.: An Elementary Introduction to Coset Table Methods in Computational Group Theory. Groups – St. Andrews 1981 (Campbell, C. M., Robertson, E. F., eds.). Cambridge Univ. Press 1982.

[46] Newman, M. F. (ed.): Proc. 2nd Internat. Conf. on the Theory of Groups (Canberra: Austral. Nat. Univ. 1973). Lecture Notes in Mathematics, Vol. 372. Berlin-Heidelberg-New York: Springer 1974.

[47] Newman, M. F.: Calculating Presentations for Certain Kinds of Quotient Groups. SYMSAC **1976**, 2–8.

[48] Newman, M. F. (ed.): Topics in Algebra. (Canberra: Proc. 1978). Lecture Notes in Mathematics, Vol. 697. Berlin-Heidelberg-New York: Springer 1978.

[49] Sandlöbes, G.: Perfect Groups of Order Less than 10^4. Commun. Algebra **9**, 477–490 (1981).

[50] Sims, C. C.: Determining the Conjugacy Classes of a Permutation Group. In [4], 191–195.

[51] Sims, C. C.: Computation with Permutation Groups. SYMSAM **1971**, 23–28.

[52] Sims, C. C.: The Role of Algorithms in the Teaching of Algebra. In [48], 95–107.

[53] Sims, C. C.: Some Group-Theoretic Algorithms. In [48], 108–104.

[54] Sims, C. C.: Group-Theoretic Algorithms, a Survey. Proc. Internat. Congress of Mathematicians **2** (Helsinki 1978). (Lehto, O., ed.), pp. 979–985. Helsinki: Acad. Sci. Fennica 1980.

[55] Sims, C. C.: Computational Group Theory. Manuscript, distributed at the Amer. Math. Soc. Short Course on Comput. Algebra – Symb. Math. Comput., Ann Arbor, Michigan, 15 p., 1980.

[56] Stanley, R. P.: Invariants of Finite Groups and Their Applications to Combinatorics. Bull. AMS **1**, 475–511 (1979).

[57] Thackray, J. G.: Reduction of Modules in Non-Zero Characteristics. Lect. Notes, distributed at the Amer. Math. Soc. Summer Inst. on Finite Group Theory, 4 p., Santa Cruz: Univ. of Calif. 1979.

Added in second printing:

Many of the latest developments (including reports on CAYLEY, CAS and SOGOS) will be represented in the proceedings of the 1982 LMS Durham Symposium on computational group theory, to be published in 1983 by Academic Press, edited by M. Atkinson.
For further developments the reader is referred to [21].

Prof. Dr. J. Neubüser
Lehrstuhl D für Mathematik
Rheinisch-Westfälische Technische Hochschule
Templergraben 64
D-5100 Aachen
Federal Republic of Germany

Integration in Finite Terms

A. C. Norman, Cambridge

Abstract

A survey on algorithms for integration in finite terms is given. The emphasis is on indefinite integration. Systematic methods for rational, algebraic and elementary transcendental integrands are reviewed. Heuristic techniques for indefinite integration, and techniques for definite integration and ordinary differential equations are touched on only briefly.

1. Introduction

Integration and the closely related problem of summation are areas where the mathematical basis for computer algebra is particularly rich. This chapter is mainly concerned with indefinite integration but the related problems of definite integration and the solution of ordinary differential equations will be touched on briefly. In each case the precise form of integrands (or summands) and to a lesser extent the acceptable form of results has a large influence on the choice of methods available. When deciding on the class in which an integrand falls it is necessary to view it solely as a function of the variable of integration, and to disregard the form of its dependence of any other variables. Thus when it is being integrated with respect to x, $x/\log(1 - \mathrm{sqrt}(y))$ is viewed as a polynomial. If integrands do contain variables, such as the y in the above example, it may be that a closed form integral will exist only for some restricted set of values for them. The normal approach taken to this problem is to regard all parameters in integrands as representing independent values, and to ask questions about the existence of integrals on the understanding that special case values of parameters are not to be considered. In general the problem of taking an arbitrary expression dependent on a set of parameters and determining what values of the parameters make it integrable in closed form is undecidable. As an indication of how even simple integrands lead to difficult problems in this respect, consider the integral of $x^k/(1 + x)$ $+ f(k)\exp(-x^2)$ where k is an undetermined constant. The first part has an elementary integral only when k is a rational number, the second only in the trivial case when $f(k)$ vanishes. Thus after it has been shown that there is no interaction between the two parts of the integral we have an expression whose integrability depends on k being a rational solution to the equation $f(k) = 0$.

For the purposes of this chapter, then, the term rational indicates a univariate rational form in the variable of integration, and it will be supposed that the field K from which the coefficients of this form are taken does not give rise to computational problems. An expression will be called algebraic if it can be expressed as a rational function in the independent variable x and some set of quantities y_i, where each y_i is constrained to satisfy a polynomial equation. This definition encompasses, of course, the particular case of algebraic quantities that can be written in terms of radicals: $x/\mathrm{sqrt}(1 + x^2)$ will be viewed as x/y subject to $y^2 = 1 + x^2$. An elementary transcendental expression is one built up using the rational operations, trig and inverse trig functions, exponentials and logarithms. Provided we are prepared to work over a complex ground field all such expressions can be rewritten with the trig functions expanded in terms of exponentials and logarithms. When the solution to a problem is sought in elementary terms this indicates that only rational and algebraic operations together with the elementary transcendental functions are allowed. Thus at least for most of this chapter an integral that can only be expressed using an error function (say) will not be treated as elementary.

2. Indefinite Integration

Even the integration of rational functions can result in the need for the manipulation of expressions involving logarithms, and the multiple valued nature of these functions makes it necessary to tread cautiously when talking about equality of functions and equivalence of formulae. The section on computing in transcendental extension fields explains how these problems can be sidestepped — from now on it will be assumed that all integrands involving non-algebraic functions have been expressed in a way that satisfies an appropriate structure theorem.

2.1 Rational Functions

There is a complete theory of integration available for input expressions that can be expressed in the form $u(x)/v(x)$ with u and v polynomials in the integration variable x. This theory was originally mapped out by Liouville and Hermite, and its main proposition is that the integral of u/v will have the form

$$p(x)/q(x) + c_1 \log(v_1) + c_2 \log(v_2) + \cdots,$$

where p and q are again polynomials in x, and the v_i are factors of the denominator of the integrand.

If the denominator v of the integrand has a square-free decomposition $v = v_1 v_2^2 v_3^3 \cdots$ then we can take q to be $v_2 v_3^2 \cdots$ where the exponent of each square-free factor has been decreased by one. This, together with a bound on the degree of the regular part of the integrand, leads to a complete integration algorithm:

Rational Function Integration

Input: a variable x, and a rational function $u(x)/v(x)$ where u and v are polynomials over a field within which v splits maximally.

Output: the integral of u/v with respect to x.

(1) If $\deg(u) \geqslant \deg(v)$ rewrite the integrand as $w + u'/v$ where w is a polynomial as $\deg(u') < \deg(v)$. The polynomial w can be integrated by elementary means, so consider now just the regular fraction u'/v.

(2) Perform a square-free decomposition to express v as the product of factors v_i^i.

(3) Let q be the product of $v_i^{(i-1)}$. q is the denominator of the regular part of the integral. Set up a polynomial p with $\deg(p) = \deg(q) - 1$ and coefficients p_i undetermined. When values have been found for these coefficients, p/q will be the regular part of the integral.

(4) Let q' be the product of the v_i. Set up a polynomial w with $\deg(w) = \deg(q') - 1$ and undetermined coefficients w_i. The integral of w/q' will be purely logarithmic.

(5) The required integral I is now expressed as $p/q + \int w/q' \, dx$. This is linear in the unknown quantities p_i and w_i. Form the expression $u - vw/q' - v\,d(p/q)/dx$. The vanishing of this expression expresses the fact that I is indeed the integral of u/v. By construction of q and q' it is a polynomial in x. Extract the coefficients of all powers of x to obtain a set of linear equations relating the p_i and w_i.

(6) Solve the linear equations. The solution directly defines the regular part of I.

(7) Factorize q' over its splitting field, and express w/q' in partial fractions as a sum of terms $c_i/(x - r_i)$. (r_i are the roots of q'). Each term in the partial fraction decomposition then integrates directly to $c_i \log(x - r_i)$, completing the evaluation of I.

See references [4] and [5] for further discussion and analysis. In step 7 of the above algorithm the general problem of rational integration has been reduced to that of forms p/q where $\deg(p) < \deg(q)$ and q is square-free. In general the integrals of such forms can only be expressed when q has been split into linear factors, and for high degree q the computing costs of the entire algorithm are determined by this factorization and the consequent need to perform arithmetic in a field containing the roots of q. This leads to a desire to perform calculations in the smallest field that contains the required integral. The following algorithm exhibits a polynomial whose roots define the minimum field extension needed. In appropriate circumstances this will make it much more efficient than the algorithm given above.

Integrate Normal Rational Function

Input: a rational function $p(x)/q(x)$ with $\deg(p) < \deg(q)$, q square-free and monic.

Output: the integral of p/q, which will be a linear combination of logarithms.

(1) Let c be a new indeterminate, and form $r = \text{resultant}(p - c\,dq/dx, q)$. This is a polynomial in c with the same degree as q. The resultant does not vanish because p/q was initially expressed in lowest terms and q was square-free, so $\gcd(q, dq/dx) = 1$.

(2) Find all the distinct roots c_i of r. (Note: in the worst case this is just as bad as finding the distinct roots of q as required by the previous integration algorithm. However, here r may have multiple roots, and so be effectively of lower degree than q. Also even in the worst case the work done here with the roots of r is probably less than that done with the roots of q in the original algorithm). These c_i are the coefficients of the logarithms that will appear in the result.

(3) For each root c_i of r compute $v_i = \gcd(p - c_i\,dq/dx, q)$. The fact that c_i was a root of r guarantees that v_i is nontrivial. The integral is now just the sum of $c_i \log(v_i)$.

It is shown in [11] that the splitting field of the resultant r above is the smallest field in which the integral of p/q can be expressed.

There are also alternatives available to the setting up and solving of explicit sets of simultaneous equations when finding the regular part of a rational integral. These relate the determination of parts of the answer to solutions of polynomial equations of the form $aB + bA = C$, where A, B and C are known polynomials and the required solution will have $\deg(a) < \deg(A)$ and $\deg(b) < \deg(B)$. As before a rational input p/q is considered, with $\deg(p) < \deg(q)$. Based on some partial factorization of q a decomposition of the integral as $u/q' + \int v/q'' \, dx$ is proposed for some suitable q' and q'', and u and v determined through use of an extended gcd algorithm. If, for instance, q can be decomposed into rs^k for some $k > 1$, with $\gcd(r, s) = \gcd(s, ds/dx) = 1$ (note that r is not required to be square-free but s is), then the integral of p/q can be written as $u/s^{k-1} + \int v/(r \, s^{k-1}) \, dx$. On differentiating and putting all terms over a common denominator, it is found that

$$p = s(v + r \, du/dx) - (k - 1)r \, ds/dx \, u$$

and fortunately $\gcd(s, r \, ds/dx) = 1$, so this equation can be solved uniquely for u and $v + r \, du/dx$, and hence v. The effect has been to decrease the exponent k in the denominator of the integrand, and repeated application of this transformation must eventually lead to an integral with a square-free denominator. That will of course integrate to logarithms. Alternatively, p/q can be expanded using a full square-free decomposition to give a sum of terms p_i/q_i^j where j runs over the range $1 \leqslant j \leqslant i$. Reduction formulae can then be applied to each of these terms. Further details can be found in [16].

2.2 Algebraic Functions

The outline of an algorithm for integrating algebraic functions was given in [10]. If an algebraic form can be expressed as a rational function in x and some quantity y, where y^2 is linear or quadratic in x, a rational change of variables will reduce the problem of integrating it to the rational case. In general an integrand involving any more complicated algebraic quantity y, or several different algebraics y_i, will not have an integral within the class of functions we admit as being elementary. Thus the integral of $r(x, y)$ where y^2 is cubic or quartic in x will normally involve elliptic functions. The problem addressed in this section is the detection of cases where, despite the initial complex form of an integrand, an integral can be expressed using at worst logarithms. This requires a systematic way of solving those problems that could have been reduced if a suitable change of variables were found. It also calls for a way of detecting when integrals that would most naturally be expressed in terms of elliptic functions (or higher analogues) can be cast into elementary form. Note that as well as the case of integrands expressed in terms of radicals [15] this discussion covers the general case of functions defined implicitly by sets of algebraic equations.

As in the case of rational functions, the integral of any algebraic form consists of a regular part (which involves no irrational elements not present in the integrand) and a logarithmic part. The major problems in determining such integrals involve discovering the required form for the logarithmic part. In the rational case a logarithm can be associated with each place that the integrand has a non-zero residue (i.e. where it has poles), and its weight associated with the logarithm will be proportional to the value of the residue. In the algebraic case the process of locating poles is complicated by the fact that it is necessary to distinguish between the different branches of functions – for instance $(1 - q)/(1 + q)$ where $q =$ sqrt$(1 + x^2)$ has either a pole or a zero at $x = 0$, depending on the branch of the square root being considered. When such branches are taken into account, the specification of a collection of places, with multiplicity information associated with each place, is called a divisor. When related residues are grouped together to generate the smallest number of logarithms needed to express an integral, discovering the argument required for the logarithm amounts to finding a function with poles and zeros as specified by some divisor. If sufficiently complicated algebraic quantities are present it is possible to have divisors for which there is no corresponding function: this leads to a form of non-integrability not seen in either the rational or transcendental cases. In this case, matters are complicated further because it may be that some power of the divisor may correspond to a function. This corresponds to seeking a function with poles with order m_i at places p_i, and not being able to find one, but nevertheless for some multiplier n being able to produce one with order nm_i at the same places p_i. In terms of integration this leads to a logarithmic term in the result of $(c/n) \log(f)$ where c was a residue of the integrand, and f the function that was eventually constructed.

This leads to the following broad sketch of an algorithm for the integration of algebraic functions:

Algebraic Case Integration

Input: an expression r that is a rational function of the independent variable x, and a set of algebraic quantities y_i. Each y_i is defined by a polynomial equation of the form $p(x, y_1, \ldots, y_{i-1}; y_i) = 0$.

Output: the logarithmic part of the integral of r with respect to x.

(1) Find all the poles of the differential $r\,dx$ on the Riemann surface defined by the y_i. Note that since this is a closed surface it is necessary to consider the possibility of poles at infinity.

(2) Evaluate the residue of the integrand at each of these poles. Discard any poles that gave rise to zero residues. The residues will give coefficients of the logarithms that are to be generated.

(3) Form a basis for the $\mathbb{Z}$-module of residues. This collects the poles and residues into sets, each of which will correspond to a single logarithm in the result.

(4) If one of the sets of poles exhibited in (3) has multiplicities n_i at places z_i on the Riemann surface, then the required integral will normally contain a term $\log(f)$ where f has order n_i at each z_i and no poles elsewhere. In some cases such an f does not exist, and the integral will be expressible as $\log(g)/k$ (for some integer k) where g has order kn_i at each z_i.

(5) Establish a bound K for the constant k in (4). See the theses of Davenport [2] and Trager [17] for a discussion of ways in which techniques from algebraic geometry make it possible to perform this step, at least for sufficiently straightforward algebraic quantities y_i and sensible underlying constant fields.

(6) Use what is essentially an interpolation technique to try fitting a function g through the poles of order kn_i at z_i for $k = 1, \ldots, K$. If at any stage g is found, the integral includes $\log(g)/k$. If all the interpolations fail then the integral is not elementary. Again see the above-mentioned theses for technical details of the interpolation procedure.

In the special case where the algebraic quantities in an integrand can be written as (unnested) radicals, it is possible to exhibit reduction formulae that generate fragments of the algebraic part of the integral term by term, eventually leaving over a part which, if integrable at all, must just generate logarithms. This determination of the regular part of an integral can be applied even when there are transcendental components present in addition to radicals. For further discussion of the evaluation of the regular parts of algebraic integrals see [2] and [17].

2.3 Elementary Transcendental Functions

As with the case of rational and algebraic functions, if the integral of an expression involving elementary functions is itself elementary the only new irrational terms that can arise in it are logarithms. Finding what these logarithms are does not raise the issues from algebraic geometry that plagued the algebraic case — the difficulties that arise are more concerned with finding the regular part of the result. This is because the operation of differentiation can act in a complicated way on the various terms in an expression, and the major tool needed to disentangle these complications is the construction of a tower of differential fields which allow all the independent transcendentals in the integrand to be treated as separate indeterminates. The chapter on computing in transcendental extensions gives a brief survey of the techniques for setting up such representations in a well-formed way — for the present it will be convenient to assume that in all integrands to be considered all transcendental forms have been replaced by variables z_i, that each z_i is algebraically independent of x and all other z_j, and that the differential operator D acts on each z_i such that either:

$$Dz_i = (Df_i)/f_i \qquad \text{or} \qquad Dz_i = z_i Df_i,$$

where f_i in each case is rational in the independent variable x and the z_j with $j < i$. These two definitions are of course intended to capture the idea of $z_i = \log(f_i)$ and $z_i = \exp(f_i)$ respectively. It is also necessary to assume that if a rational form in the independent variable x and the z_j has a zero derivative then this rational form is really independent of x and the z_j. This hypothesis avoids the simultaneous introduction of (e.g.) both $\log(x)$ and $\log(2x)$ as supposedly independent quantities. It will be convenient to use the notation z_0 for x, and of course $Dz_0 = 1$.

The first complete integration algorithm for use in this case was developed by Risch [9]. The variant on it described here is due to Rothstein [11] and Mack. It starts by viewing an integrand as a rational function p/q where p and q are polynomials in the variable z_i with largest index i. The coefficients in these polynomials will be allowed to be rational functions of all the z_j with $j < i$, and so q can be taken as monic. p/q is gradually decomposed. Firstly, division with remainder puts $p/q = u + v/q$ with $\deg(v) < \deg(q)$. Then if q has some power of z_i as a factor and z_i is exponential, this factor is removed as follows:

Dispose of Exponentials in Denominator

Input: $v/(qz^k)$ where z is an exponential and v, q are polynomials in z not divisible by z.

Output: a decomposition of the input in the form $s/q + w/z^k$, where all s, q and w are polynomials in z but not divisible by z.

(1) Set up the equation $v = sz^k + wq$, and solve it using the extended Euclidean algorithm. This is possible because $\gcd(q, z^k) = 1$. In the solution $\deg(s) < \deg(q)$.

The term w/z_i^k produced by the application of this algorithm can be treated using the same algorithm as is applied to the polynomial part of the integrand u.

The polynomial part of the integrand can now be handled, and the treatment to be applied to it depends on whether the variable z_i was logarithmic or exponential. The logarithmic case is substantially easier to cope with. Suppose then that $z = z_i$ was logarithmic, and the polynomial part of the original integrand was $p(z)$ with degree m. Then the integral, if it exists at all in elementary form, is a polynomial in z of degree at most $m + 1$ together with a linear combination of new logarithmic terms. As before all coefficients in polynomials are rational functions in all the quantities z_j up as far as z_{i-1}.

Integration of a Polynomial in z_i, with z_i Logarithmic

Input: $p(z) = p_m z^m + \cdots + p_0$, a polynomial in $z = \log(f)$, with coefficients that are rational functions in the various z_j that are subordinate to z.

Output: the integral of p, if it exists, in the form $u_{m+1} z^{m+1} + \cdots + u_0 + c_1 \log(v_1) + \cdots$, with the c_i constants and the u and v_i in the same domain as the p_i were.

(1) Represent the terms in the result by new indeterminates, and differentiate the resulting formal expression. Equate this to the integrand.

(2) Compare coefficients of z^{m+1}, obtaining the equation

$$Du_{m+1} = 0.$$

Write the general solution to this as $u_{m+1} = k_{m+1}$ where k_{m+1} represents a constant whose value is still to be determined.

(3) Compare coefficients of powers of z from z^m down to z^1. If this is done sequentially each case leads to an equation of the form

$$p_i - (i + 1)u'_{i+1} \, Df/f = Du_i + (i + 1)k_{i+1} \, Df/f, \qquad (*)$$

where the final value that will get assigned to u_{i+1} has been written as $u'_{i+1} + k_{i+1}$ to reflect the fact that initially each term only gets determined to within an additive constant. This equation must be seen as determining u_i (more properly it is used to define some u_i', and a new undetermined constant k_i will be introduced as a constant of integration). It is required that u_i be found in the differential field that does not involve z.

(4) Use the full integration algorithm recursively on the left-hand side of (∗). This recursion is valid because the new integrand does not involve the logarithms z, and so is strictly simpler than the one originally considered. The resulting integral, if it exists in closed form, must now be matched with the integral of the right-hand side of (∗), i.e. with $u_i + (i + 1)k_{i+1} \log(f)$. Performing this match may involve the simplification of the logarithmic terms handed back by the recursive call to the integrator. If the match can be made it determines u_i to within an additive constant, and furthermore establishes a value for the constant k_{i+1} that was introduced in the previous step.

(5) Comparing the constant terms (with respect to z) in the equation of (1) produces the condition that

$$\int (p_0 - u_1' \, Df/f) \, dx = \text{constant} + \log \text{term in result}$$

and so this final integral is performed by recursion. The constant of integration that it introduces is just that for the entire integral. Note that k_1 does not get assigned a value. Its effect on the coefficient of z in the result is offset by compensating logarithmic terms (equivalent to z) that get generated by the final recursive integration.

If at any stage in the above algorithms one of the recursively evaluated integrals proves to be non-elementary, or if the form of one of these integrals is not as required, the original integral can be declared non-elementary.

The case where z is an exponential (e.g. $Dz = z \, Df$) is similar in style to the above, but messier in detail. The integral of a term $p_n z^n$ will be of the form $u_n z^n$ for some u_n independent of z. Differentiating this leads to a differential equation for u_n. Dropping subscripts this can be written:

$$Du + n \, Df u = p$$

and the integer multiplier n can be absorbed into f to give $Du + u \, Dg = p$, where $\exp(g)$ forms a regular extension of the domain in which the solution for u is sought. Note that u_n/z^n can be dealt with here be considering it as $u_n z^{-n}$, justifying a claim made earlier that such terms behave in a way similar to purely polynomial ones.

This equation can have at most one solution that is rational in the z_j, because if u' is such a solution, the general solution to the equation is $u' + k \exp(g)$. Since $\exp(g)$ is outside the desired domain, the constant k must vanish. Thus the original integration problem reduces to finding the unique solution for u, or determining that it cannot exist. This can be done by a recursive process that successively reduces the number of transcendental elements z_i in the differential field where solutions are required. Each reduction step will exhibit the form that a solution must take, and will compare coefficients of powers of some z_i to generate a family of sub-problems that can be solved to fill in the details.

The main recursion is through the solution of versions of the equation

$$Au' + Bu = C, \qquad (*)$$

where u' will now be written for Du. A, B and C may initially be rational functions in the highest monomial that they contain. Straightforward calculations explained in the references exhibit a denominator for any solution u to the equation, and give a bound to the degree of the numerator. Clearing denominators leads directly to an equation of the same form as (*) but with A, B and C now polynomials, and with the solution for u constrained to be a polynomial subject to a known degree bound. If $\deg(A) = 0$ and the equation has a solution at all then $\deg(B)$ must also be zero. In this case it is easy to compare coefficients of the highest monomial present to obtain a set of subproblems that can be solved through recursive application of the same algorithm. If A and B have a common factor G, then for a solution u to exist this factor must divide C exactly. The equation can therefore be replaced by one divided through by G. If this still leaves $\deg(A) > 0$ the following transformation can be applied to produce a new equation for a quantity v, such that finding a polynomial u of degree d satisfying the original equation can be achieved only if a v can be found satisfying the new equation and the tightened degree bound $\deg(v) < d - \deg(A)$.

Degree Bound Reduction for the Equation $Au' + Bu = C$

Input: a, b and c, polynomials in a quantity z_j with coefficients that are rational functions in variables z_i with $i < j$. $\deg(a) > 0$ and $\gcd(a, b) = 1$. A degree bound d.

Output: a polynomial u with $\deg(u)$ at most d, and $au' + bu = c$, or an indication that there is no such polynomial.

(1) Divide u by a to obtain a quotient v and a remainder r. Then naturally $u = av + r$ and $\deg(r) < \deg(a)$. Differentiating this expresses u' as $a'v + av' + r'$.

(2) Substituting the above expressions for u and u' into the equation $au' + bu = c$ and rearranging gives the equation

$$a(a'v + av' + r' + bv) + br = c$$

and where a, b and c are known. This can be solved uniquely with $\deg(r) < \deg(a)$ using the extended Euclidean algorithm.

(3) The solution gives a value for r and for $a'v + av' + r' + bv$. Rearranging the latter gives an equation for v of the same form as the original one for u:

$$av' + (a' + b)v = -r'$$

and since $\deg(v) = \deg(u) - \deg(a)$ this new equation is subject to a degree bound tighter than that relevant to the original one. It can therefore be assumed solvable by a recursive entry to the main solve process.

All that is now left of the original integration problem is the treatment of regular rational functions p/q where the denominator satisfies $\gcd(q, dq/dx) = 1$. Once again p and q can be viewed as univariate polynomials in the extension variable z_n of highest index, and without loss of generality q can be made monic. Now the integral, if elementary, will be the integral of some rational function r not involving z_n, plus the sum of a collection of logarithms $c_i \log(v_i)$. As is the case of integrating pure rational functions, the c_i are roots of resultant $(p - c\,dq/dx, q)$, and for each c_i the corresponding $v_i = \gcd(p - c_i\,dq/dx, q)$. (If the resultant does not factor to give a set of constant roots c_i the integral being considered is not elementary). Subtracting the derivative of these logarithms from p/q gives the residual function s, which since it involves one less transcendental function than p/q can be handled by a recursion. This use of resultants makes it possible to keep the algebraic number arithmetic involved in calculating the constants c_i as simple as possible.

The integration algorithm presented above works by using a recursion through the structure of the differential field in which the integrand lies. In most respects this leads to a direct and straightforward implementation: however the natural pattern of the recursion does not seem to fit in as smoothly as would be hoped with the exponential case. Partly in an attempt to remedy this, a parallel version of the integration algorithm was proposed: an initial informal report of it was made at the SYMSAC-76 meeting and further details appeared in [8]. For an integrand p/q satisfying the same constraints as in the original algorithm, the integral will be a rational u/q' plus some logarithms. Subject to a slight tightening of the constraints on what transcendentals are permitted in the integrand, q' can be obtained by forming a square-free decomposition of q, viewed as a (multivariate) polynomial in all the quantities z_i. If this decomposition involves factors q_i^i then q' will have a factor q_i^{i-1} except in the special case where q_i has as a factor one of the exponentials among the z_i, in which case that factor does not have its exponent decreased. The arguments for logarithms in the integral are just the complete set of factors of q, where the factorization treats q as a multivariate polynomial over an algebraically closed field. If a degree bound for the unknown numerator u could be found it would be easy to set up a set of linear equations defining its coefficients and the weights associated with the logarithmic terms. There is a natural formula that can be used to obtain such a bound, and when the new method was proposed it was believed that this bound, which is almost always tight, and which leads to an efficient way of generating and solving the linear equations, was correct. In degenerate cases, however, the proposed bound does not hold, and the problem of providing a simply computed degree bound for u remains open.

2.4 Composite and Heuristic Techniques

It is natural to ask how easy it would be to extend the existing integration algorithms to make them work with some of the higher transcendental functions, such as the error function, exponential integral and dilogarithm. In general such functions, being defined as integrals themselves, can be fitted into the framework of the structure theorems that are used to determine that differential fields involved in the integration process are well formed. This means that the transcendental integration algorithm given here needs only minor adjustment for it to be able to handle integrands containing these new functions. The algorithm is, however, based on the supposition that given the form of an integrand it is possible to predict the form of the corresponding integral. In all cases the integral will contain a regular part, built up out of functions present in the integrand. It will also involve some new logarithms, and the arguments for these are derived from the locations of simple poles in the integrand. Reference [14] shows that at least for some of the higher functions extra instances of them can only occur linearly in an integral. To generate these new instances of higher transcendental functions some more elaborate analysis of singularities of the integrand would be needed – for instance an error function is a reflection of an essential singularity (at infinity) whereas dilogarithms come from certain sorts of logarithmic branch points. At present there is no systematic way of recognizing the necessary sorts of behaviour.

This lack of a complete theory is offset in practice by the use of pattern matching, special purpose programs and heuristic methods. The level of success that these can achieve is clearly strongly dependent on the match between their designer's aims and the class of examples that are attempted. One strategy that has been used to good effect is to try to combine the various algorithmic integration techniques with a selection of procedures for detecting common special cases and a collection of rules for changing variables and using tabulated integrals. The first generally successful integration package [7] was of this form, and more recently [3] has had success with a variant on the same idea.

3. Definite Integration and Ordinary Differential Equations

If it is accepted that indefinite integration is under control, both in theory and in practice, it is clear that there are two problem areas that lead on from it. The solution of ordinary differential equations in general can be viewed as the obvious step beyond the consideration of the rather simple equation $dy/dx = r(x)$. The related theory can now depend not only on the functions present in coefficient functions, but also on the form and order of the differential equation. So far there has been progress for low degree equations with polynomial coefficients. Definite integration is of most interest when the limits and path of the integral are such as to lead to a closed form result despite the indefinite version of the integral not having an elementary representation. A particularly important use is in the computation of Laplace and inverse Laplace transforms – and hence in the solution of certain classes of differential equations.

3.1 Ordinary Differential Equations

There have been two lines of attack on the solution of ordinary differential equations. One seeks to apply a set of classical methods – changing variables and the like – to solve a wide class of equations. Programs based on this line, e.g. that described in [13], use heuristics to select a transformation to be applied to an equation, and consider it solved when the solution is expressed in terms of integrals. The other body of research has sought algorithmic ways of solving restricted classes of equations. Some of the special purpose methods, for instance that in [19] can also be used heuristically on problems that lie outside the domain over which they act as decision procedures.

One particular case for which a decision procedure is available is the second order linear homogeneous equation

$$ay'' + by' + c = 0, \qquad (*)$$

where a, b and c are rational functions in the independent variable x over the complex numbers. The flavour of the algorithm is rather similar to those used for integration. A solution for y is sought in Liouvillian terms, i.e. built up using rational operations, the algebraics defined by polynomial equations, exponentials and integrals. A simple change of variables reduces the original problem to one of the form

$$u'' = ru,$$

where again r is a rational function. In [6] it is shown that this equation has a Liouvillian solution if and only if the Ricatti equation

$$w' + w^2 = r \qquad (**)$$

has a solution algebraic over $C(x)$ with degree 1, 2, 4, 6 or 12. An analysis of the singularities of r, described in detail in [12] as well as [6] leads to the generation of a finite set of candidate polynomials $p(w)$ such that if $(*)$ has a Liouvillian solution at all one of the equations $p(w) = 0$ will define an algebraic solution to $(**)$.

3.2 Definite Integration

Currently there is no well-developed computational theory of definite integration. The techniques used in practical packages are combinations of heuristics. The two lines of attack that have been used have been the evaluation of an indefinite integral followed by an evaluation or limit taking process, and contour integration methods. Even when the first of these methods is available (many interesting definite integrals correspond to non-elementary indefinite ones) its correct implementation is seriously complicated by multi-valued functions. Most serious work has therefore concentrated on contour integration. This requires 3 sub-algorithms: selection of a contour, localization of poles and evaluation of residues.

It will usually be convenient to transform any given integral to one over a standard range (0 to infinity or − infinity to + infinity) before trying to select a contour. At present the most successful scheme for the selection involves having a library of commonly applicable contours and choosing from this on the basis of clues found in the integrand. References [1] and [18] illustrate this line of attack, showing that for various large classes of integrands (e.g. rational functions) a standard contour suffices and the proof that unwanted parts of the integral vanish is trivial.

The location of poles of a function is completely solved for the case of rational functions. More complicated cases lead rapidly to undecidability, and so call for a treatment that detects and exploits whatever special cases can be thought of. Given the location of the poles of a function the calculation of residues can be performed either by a process of differentiation and limit taking, or by techniques relating to Taylor and Laurent series expansion.

References

[1] Belovari, G., Campbell, J. A.: A Logic Programming Framework for Generating Contours of Symbolic Integration. Dept. Computer Science, University of Exeter, England, 1980.

[2] Davenport, J.: Ph.D. thesis, University of Cambridge. Lecture Notes on Computing, Vol. 102. Berlin-Heidelberg-New York: Springer 1981.

[3] Harrington, S. J.: A New Symbolic Integration System for Reduce. Computer Journal **22** (1979).

[4] Horowitz, E.: Ph.D. thesis, University of Wisconsin, Madison, 1969.

[5] Horowitz, E.: Algorithms for Partial Fraction Decomposition and Rational Integration. SYMSAM **1971**, 441 – 457.

[6] Kovacic, J. K.: An Algorithm for Solving Second Order Linear Homogeneous Differential Equations. Preprint, Brooklyn College, City University of New York.

[7] Moses, J.: Symbolic Integration. Report MAC-TR-47, Project MAC, MIT 1967.

[8] Norman, A. C., Moore, P. M. A.: Implementing the New Risch Integration Algorithm. MAXIMIN **1977**, 99 – 110.

[9] Risch, R. H.: The Problem of Integration in Finite Terms. Trans. AMS **139**, (1969).

[10] Risch, R. H.: The Solution of the Problem of Integration in Finite Terms. Bull. AMS **76**, (1970).

[11] Rothstein, M.: Ph.D. thesis, University of Wisconsin, Madison, 1976.

[12] Saunders, B. D.: An Implementation of Kovacic's Algorithm for Solving Second Order Linear Homogeneous Differential Equations. SYMSAC **1981**, 105 – 108.

[13] Schmidt, P.: Substitution Methods for the Automatic Solution of Differential Equations of the 1st Order and 1st Degree. EUROSAM **1979**, 164 – 176.

[14] Singer, M. F., Saunders, B. D., Caviness, B. F.: An Extension of Liouville's Theorem on Integration in Finite Terms. SYMSAC **1981**, 23 – 24.

[15] Trager, B. M.: Integration of Simple Radical Extensions. EUROSAM **1979**, 408 – 414.

[16] Trager, B. M.: Algebraic Factoring and Rational Function Integration. SYMSAC **1976**, 219 – 226.

[17] Trager, B. M.: Ph.D. thesis, MIT (forthcoming).

[18] Wang, P. S.: Symbolic Evaluation of Definite Integrals by Residue Theory in MACSYMA. Proc. IFIP-74. Amsterdam: North-Holland 1974.

[19] Watanabe, S. A.: Technique for Solving Ordinary Differential Equations Using Riemann's P-Functions. SYMSAC **1981**, 36 – 43.

Dr. A. C. Norman
Computer Laboratory
University of Cambridge
Corn Exchange Street
Cambridge CB2 3QG
United Kingdom

Summation in Finite Terms

J. C. Lafon, Strasbourg

Abstract

A survey on algorithms for summation in finite terms is given. After a precise definition of the problem the cases of polynomial and rational summands are treated. The main concern of this paper is a description of Gosper's algorithm, which is applicable for a wide class of summands. Karr's theory of extension difference fields and some heuristic techniques are touched on briefly.

The problem we study here, is the following: Given a symbolic function $f(x)$ of the main variable x, compute, if it exists, a symbolic expression S with finite terms for the definite sum

$$\sum_{x=a}^{b} f(x) = S(a, b) \qquad \text{(Definition 1)}.$$

Like the integration problem, this is a classical problem and many methods are known to solve it in particular cases (see for instance the book of Jordan), but as in the case of integration, the use of these methods is often not obvious: we have to recognize the form of $f(x)$, to remember particular solutions, to make some "intelligent" manipulations, to guess the form of a solution...

Thus, it would be very helpful if such Computer Algebra Systems like REDUCE, SCRATCHPAD or MACSYMA for instance, could be used to generate $\sum_{x=a}^{b} f(x)$ for a large set of functions $f(x)$.

This raises the following questions:

To what extent can classical methods directly be implemented in such systems? What parts of these systems do they need? Is it costly?

Can we find a general method to solve the definite summation problem (like Risch's algorithm for the integration problem)?

Can we extend these possibilities to the two related (but in anyway quite different) problems of indefinite summation ($\sum_{x=a}^{\infty} f(x)$), and finite difference equations?

1. Mathematical Framework

From a mathematical point of view, the problems of definite summation and of integration are very similar. Instead of using derivative operator and differential fields as previously, for the summation problem we use difference operators and difference fields.

Let V and Δ be respectively "the lower difference operator" and "the upper difference operator":

$$Vg(x) = g(x) - g(x - 1), \qquad \Delta g(x) = g(x + 1) - g(x).$$

If g is any function such that $Vg = f$ (or $\Delta g = f$) then $\sum_{x=a}^{b} f(x)$ is equal to $g(b) - g(a - 1)$ (or to $g(b + 1) - g(a)$).

(In a similar way, if $g(x) = \int_a^x f(t)\, dt$ we have $g'(x) = f(x)$). So, to find $\sum_{x=a}^{b} f(x)$ is equivalent to solve $\Delta g = f$ or $Vg = f$.

These operators are linear, but the product rule is more complicated than for the derivative operator:

$$V(f \cdot g) = fVg + Vf \cdot g - Vf \cdot Vg,$$

$$\Delta(f \cdot g) = f\Delta g + \Delta f \cdot g + \Delta f \cdot \Delta g.$$

An upper (lower) difference field is a field F with a map $\mathscr{F} : F \to F$ satisfying the three following conditions:

1) $\mathscr{F}$ is linear,

2) $\forall f, g, \; \mathscr{F}(f \cdot g) = f \cdot \mathscr{F}g + \mathscr{F}f \cdot g \pm \mathscr{F}f \cdot \mathscr{F}g$,

3) $\mathscr{F}$ does not satisfy any one of the following equivalent conditions:

$$\mathscr{F}f = \mp f, \quad \forall f, \quad \mathscr{F}1 = \mp 1, \quad \mathscr{F}f = \mp f \text{ for some } f, \quad \mathscr{F}1 \neq 0.$$

We can also introduce the operator of displacement E:

$$Ef(x) = f(x + 1) \qquad (E^{-1}f(x) = f(x - 1)).$$

In terms of E and E^{-1}, the equations $\Delta g = f$ and $Vg = f$ become $Eg - g = f$ and $g - E^{-1}g = f$.

A difference field is a field F with an automorphism E of F. Now, we can give a definition, more precise than Definition 1, of the problem of summation in finite terms:

"A difference field F and $f \in F$ being given, compute, if it exists, an element $g \in F$ such that

$$\Delta g = f \quad (\text{or } Vg = f).\text{"} \qquad \text{(Definition 2)}$$

(see the paper of Karr).

2. Polynomial and Rational Functions

First, let us consider the case of $f \in K[x]$ (K is a field):

$$f(x) = a_0 + a_1 x + \cdots + a_n x^n.$$

We have to solve $\Delta g = f$ or $Vg = f$. We introduce the two factorial functions $x^{(n)} = x(x + 1) \cdots (x + n - 1)$ and $x_{(n)} = x(x - 1) \cdots (x - n + 1)$. We have

$\nabla x^{(n)} = nx^{(n-1)}$ and $\Delta x_{(n)} = nx_{(n-1)}$ (similar to $(x^n)' = nx^{n-1}$). So, to solve $\Delta g = f$ (or $\nabla g = f$), we need only to express the polynomial f as a linear combination of $x^{(i)}$ ($i = 0, \ldots, n$) (or of $x_{(i)}$, $i = 0, \ldots, n$): if

$$f = \sum_{i=0}^{n} f_i x_{(i)} \tag{1}$$

then the solution g of $\Delta g = f$ is $g = \sum_{i=0}^{n} (1/i + 1) f_i x_{(i+1)}$. Expression (1) is easily computed by the use of the formula

$$x^i = \sum_{k=1}^{i} S_k^i x_{(k)} \qquad (S_k^i \text{ Stirling number}).$$

Equation $\Delta g = f$ can also be solved directly by the use of the formula

$$\Delta^{-1} x^i = \frac{B_{i+1}(x)}{i+1} \qquad (B_{i+1}(x) \text{ Bernoulli polynomial}).$$

Another method is based on the fact that $\sum_{x=a}^{b} f(x)$ with $f \in K[x]$, $\deg f = n$, is a polynomial in b of degree $n + 1$: evaluation-interpolation technique can then be used.

We turn now to the more complex case of rational functions. Let $f(x) \in K(x)$. By expansion into partial fractions, $f(x)$ can always be written in the following manner:

$$f(x) = f_0 + \sum_{i=1}^{k} \sum_{j=1}^{n_i} \frac{f_{ij}}{(x + t_i)^j}.$$

All that is needed to compute $\Delta^{-1} f(x)$ is therefore an expression for $\Delta^{-1}(1/(x + t_i)^j)$.

Such an expression is known if the use of polygamma functions is allowed: if $\psi(x) = \Gamma'(x)/\Gamma(x)$ is the digamma function, we have

$$\Delta^{-1} \frac{1}{(x + t)^n} = \frac{(-1)^{n-1}}{(n-1)!} \psi^{n-1}(x + t) \qquad (\psi^{n-1}(x) \text{ is the } n \text{ gamma function}).$$

So, the problem of summation of rational functions is solved when the field F contains the polygamma functions, but can we solve it also when F is restricted to $K(x)$? In that case, what is needed is an algorithm to decide if $\sum_{x=a}^{b} f(x)$ has a rational expression and to compute (if the answer is yes) this expression. Karr, in his first paper, has given such an algorithm which uses a kind of partial fractions decomposition of $f(x)$.

In fact, Karr shows that any rational function $f(x)$ can be written in the form:

$$f = f_0 + \sum_{i=1}^{l} \sum_{j=1}^{m_i} \sum_{k=0}^{n_{ij}} \frac{f_{ijk}}{E^k h_i^j}$$

with h_i irreducible, monic, and $h_{i_1} \neq h_{i_2}$ for $i_1 \neq i_2$,

$$\deg f_{ijk} < \deg h_i \neq 0, \quad \forall i, j, k, \qquad f_{ij0} \neq 0 \neq f_{ijn_{ij}}.$$

This partial fraction expansion is unique and may be computed if F has a complete factorization algorithm.

If

$$\sum_{k=0}^{n_{ij}} \frac{E^{n_{ij}-k}f_{ijk}}{C^{n_{ij}-k}} = 0, \qquad \forall i, j,$$

then the equation $Eg - cg = f$ can be solved in F (and Karr gives the formal expression of the solution).

3. Gosper's Algorithm

This algorithm works when $S(n) = \sum_{x=1}^{n} f(x)$ is such that $S(n)/S(n-1)$ is a rational function of n. Let $S(n) = \sum_{i=1}^{n} a_i$. Then, a_n/a_{n-1} must be also a rational function of n, and we can write

$$\frac{a_n}{a_{n-1}} = \frac{p_n q_n}{p_{n-1} r_n}, \qquad \gcd(q_n, r_{n+j}) = 1, \qquad \forall j \in \mathbb{N}.$$

Gosper then shows that S_n has the following expression:

$$S_n = \frac{q_{n-1}}{p_n} a_n f(n),$$

$f(n)$ being a polynomial satisfying the relations

$$p_n = q_{n+1} f(n) - r_n f(n-1) \tag{1}$$

and

$$p_n = (q_{n+1} - r_n) \frac{f(n) + f(n-1)}{2} + (q_{n+1} + r_n) \frac{f(n) - f(n-1)}{2}. \tag{2}$$

We can compute (by identification) $f(n)$ if we know an upper bound k for its degree.

Two cases arise:

if $\deg(q_{n+1} + r_n) \leqslant \deg(q_{n+1} - r_n) = l$, then we can take $k = \deg(p_n) - l$;

if $\deg(q_{n+1} - r_n) < \deg(q_{n+1} + r_n) = l$ then we have:

$$k = \max(k_0, \deg(p_n) - l + 1) \qquad \text{if } k_0 \text{ is an integer,}$$

or

$$k = \deg p_n - l + 1 \qquad \text{if } k_0 \text{ is not an integer,}$$

where k_0 is the root of the linear polynomial $L(k)$ such that $p_n = c_k L(k) n^{k+l-1} + 0(n^{k+l-2})$.

If we find a negative k, this means that the indefinite sum $S(n)$ does not exist.

For $a_n = 1/n(n+2)$ we can take $p_n = n+1, q_n = n-1, p_{n-1} = n, r_n = n+2$. Then

$$S_n = \frac{n}{n+1} \times \frac{1}{n(n+1)} f(n).$$

Here, we are in case 2 with $L(k) = k - 2$ and therefore $k_0 = 2$ and we find

$$f(n) = \frac{3n^2 + 5}{4} \quad \text{and} \quad S_n = \frac{3n^2 + 5}{4(n^2 + 3n + 2)}.$$

Let us present a description of Gosper's algorithm.

Input: a_n such that $S(m) = \sum_{n=0}^m a_n$ and $S(m)/S(m-1) \in Q(m)$.

Output: $S(n)$, if it exists, otherwise "false".

Method: (1) $[a_n = 0.]$ If $a_n = 0$ then $\{S_m \leftarrow 0; \text{return}\}$.

(2) [Initialize.] Set $p_n \leftarrow 1$, $q_n \leftarrow a_n$, $r_n \leftarrow a_{n-1}$, with $\gcd(a_n, a_{n-1}) = 1$.

(3) $[q_n, r_n$ shift-free.] For all $j \in \mathbb{N}$ such that $\text{res}_n(q_n, r_{n+j}) = 0$, set $g_n \leftarrow \gcd(q_n, r_{n+j})$, $q_n \leftarrow q_n/g_n$, $r_n \leftarrow r_n/g_{n-j}$, $p_n \leftarrow p_n g(n) \cdot \ \cdots \ \cdot g(n - j + 1)$.

(4) [Degree bound $(\deg 0 = -1)$, negative exit?] If $l_+ = \deg_n(q_{n+1} + r_n) \leqslant \deg_n(q_{n+1} - r_n) = l_-$, set $k = \deg p_n - l_-$, otherwise set $k = \max(k_0, \deg p_n - l_+ + 1)$; if $k < 0$ then exit with "false".

(5) $[f_n = b_k n^k + \cdots + b_0.]$ By comparing coefficients solve $p_n = q_{n+1} f_n - r_n f_{n-1}$, for $b_k, \ldots, b_0$.

(6) [Result.] Set $S_n \leftarrow q_{n-1} f_n a_n/p_n$. ∎

The subalgorithms involve resultant-, gcd-computations and the solution of a linear system. It is not known for which inputs a_n the output S_n has the required property that S_n/S_{n-1} is a rational function. There are simple examples for which this is not the case like $\sum_{n=0}^m x^n y^{m-n} = (x^{m+1} - y^{m+1})/(x - y)$.

Suppose a_n and a_{n-1} are polynomials of maximal seminorm d and maximal degree l. Then the resultant $l(j)$ in step 3 has a maximal degree of l^2 and a seminorm of $O(d^{2l})$. Hence $j \in \mathbb{N}$ could be of order $O(d)$, i.e. exponential in the length (of the coefficients of the input). However, p_n will then have j factors, at least of degree 1 each, and will have a degree which is exponential in the same sense. The same holds for the order of the linear system to be solved. For many textbook summation formulas in fact j, $\deg p_n$ and $\deg f_n$ are fairly small, so Gosper's algorithm is extremely useful in practice despite its exponential worst case behavior. We have never observed that k in step 4 was set to k_0; here some improvements may be possible.

Example: Let us apply Gosper's algorithm to $\sum_{n=0}^m n x^n$. Here $a_n/a_{n-1} = nx^n/((n-1)x^{n-1}) = nx/(n-1)$, so $p_n = 1$, $q_n = nx$, $r_n = n - 1$ in step (2). The resultant in step (3)

$$\begin{vmatrix} x & 0 \\ 1 & j - 1 \end{vmatrix} = x(j - 1)$$

yields $j = 1$. Therefore, $g_n = n$, $q_n = x$, $r_n = 1$ and $p_n = n$ after step 3. The degree bound k follows from

$$\deg(q_{n+1} + r_n) = \deg_n(x + 1) = 0 \leqslant \deg(q_{n+1} - r_n) = \deg_n(x - 1) = 0$$

as $k = \deg p_n - 1 = 1$. The set up is $f = b_1 n + b_0$ and we get b_1 and b_0 from $n = x(b_1 n + b_0) - 1 \cdot (b_1(n - 1) + b_0), b_1 = 1/(x - 1)$ and $b_0 = 1/(x - 1)^2$ in step (5). The result in step (6) is

$$S_n = (x/n)((x - 1)n - 1)/(x - 1)^2)nx^n = (nx^{n+2} - (n + 1)x^{n+1})/(x - 1)^2,$$
$$S_0 = -x/(x - 1)^2,$$

so

$$\sum_{n=0}^{m} nx^n = (mx^{m+2} - (m + 1)x^{m+1} + x)/(x - 1)^2. \quad \blacksquare$$

4. Summation in Extensions

For the integration problem, starting with the rational functions over a field k, we have considered algebraic, exponential and logarithmic extensions. Can we do the same for summation?

In a recent paper, Karr considers extensions analogous to exponential and logarithmic extensions both from a mathematical and computational point of view. Here, we give only some definitions used by Karr to define the extensions he considers.

Let G, E be an extension difference field of the difference field F, E.

This extension is affine if $G = F(t)$ with $Et = \alpha t + \beta, \alpha, \beta \in F$. An extension $F(t)$, E is first order linear if it is affine, if t is transcendental over F, and if the field of constants is not extended. An extension G, E of F, E is homogeneous if there exists an element $g \in G$ such that $Eg/g \in F$.

An extension $F(t)$, E is a $\prod$ extension of F, E if the extension is first order linear and $Et = \alpha t$. An extension $F(t)$, E is a $\sum$ extension of F, E if it is inhomogeneous and if for $n \neq 0 \ \alpha^n \in H \Rightarrow \alpha \in H$ where $Et = \alpha t + \beta$ and H is the homogeneous group.

Let $F = k(x)$ and $f(x) \in F$. We may extend F by the factorial of $f(x)$: $t(x) = \prod_{0 \leq i < x} f(i)$: this gives a $\prod$ extension. If we extend F by the indefinite sum of $f: t(x) = \sum_{0 \leq i < x} f(i)$ this gives a special case $\sum$ extension.

An extension $F(t)$, E is called a $\prod\sum$ extension of F, E if it is either a $\prod$ extension or a $\sum$ extension.

Finally, an extension F, E is a $\prod\sum$ field over k if there exists a tower of fields: $k = F_0 \subset \cdots \subset F_n = F$ such that F_i, E is a $\prod\sum$ extension of F_{i-1}, E for $i = 1, \ldots, n$.

In his paper, Karr shows how to solve arbitrary first order linear difference equations in any $\prod\sum$ field and how to choose a particular $\prod\sum$ field to compute the solution of a given equation.

5. Composite and Heuristic Methods

As in the integration case, we can try to combine the various algorithmic summation techniques with procedures to detect special cases and with the use of tabulated sums. Particular methods are available for the sums of binomial coefficients, trigonometric functions, reciprocal factorials...

For instance, to compute $\sum_{x=1}^{n} 1/x(x+2)$, instead of Gosper's algorithm we can use reciprocal factorials:

$$\Delta^{-1} \frac{1}{x(x+2)} = \frac{x+1}{x(x+1)(x+2)} = \Delta^{-1} \frac{1}{(x+1)_2} + \Delta^{-1} \frac{1}{(x+2)_3}$$

$$= \frac{-1}{x} + \frac{1}{2x(x+1)}$$

(this method works when $f(x) = p(x)/[(x+a_1) \cdots (x+a_k)]$, a_i integers in increasing order).

Moreover, indefinite sums can also be obtained by summation by parts, a technique analogous to integration by parts:

$$\Delta^{-1}(u(x)v_0(x)) = u(x)v_1(x) - \Delta^{-1}(v_1(x+1)\Delta u(x)).$$

We can apply this for instance to compute $\Delta^{-1}x\sin x$ knowing $\Delta^{-1}\sin(ax+b)$.

The implementation of these methods in a Computer Algebra system relies heavily on pattern matching, heuristics, backtracking and table look-up techniques.

A similar approach can be taken for the summation of series and for the solution of finite difference equations. For instance, an interactive computer program solving linear systems of finite difference equations has been described by Cohen and Katcoff.

For these problems, much remains to be made:

to implement existing methods,
to analyze the cost of these methods,
to find new general algorithmic methods.

References

[1] Cohen, J., Katcoff, J.: Symbolic Solution of Finite Difference Equations. ACM Trans. Math. Software 3/3, 261 – 271 (1977).
[2] Gosper, R. W., jr.: Decision Procedure for Indefinite Hypergeometric Summation. Proc. Nat. Acad. Sci. USA 75/1, 40 – 42 (1978).
[3] Jordan, C.: Calculus of Finite Differences. Sopron, Hungary: Röttig and Romwalter 1939.
[4] Karr, M.: Summation in Finite Terms. Mass. Comput. Assoc. Inc. Wakefield, Mass.: Techn. Rep. CA-7602-1911, 1976.
[5] Karr, M.: Summation in Finite Terms. J. ACM 28/2, 305 – 350 (1981).
[6] Moenck, R.: On Computing Closed Forms for Summation. MACSYMA 1977, 225 – 236.

Prof. Dr. J. C. Lafon
Centre de Calcul de l'Esplanade
Université Louis Pasteur
7, rue René Descartes
F-67084 Strasbourg
France

Quantifier Elimination for Real Closed Fields: A Guide to the Literature

G. E. Collins, Madison, Wisconsin

Abstract

This article provides a brief summary of the most important publications relating to quantifier elimination algorithms for the elementary theory of real closed fields. Especially mentioned is the cylindrical algebraic decomposition method and its relation to the facilities of computer algebra facilities.

The elementary theory of real closed fields is the first order theory with constants 0, 1, function symbols $-$, $+$, $\times$, predicate symbols $\neq$, $<$, $\leq$, $>$, $\geq$, $=$ and the axioms of a commutative field, some axioms relating the order relation with the arithmetical operations and infinitely many axioms stating the existence of roots for certain equations:

$$\bigwedge x \vee y \quad (x = y^2 \vee - x = y^2),$$

$$\bigwedge x_0 \wedge x_1 \cdots \wedge x_{2n} \vee x \quad (x_0 + x_1 x + \cdots + x_{2n} x^{2n} + x^{2n+1} = 0).$$

Many important and difficult mathematical problems can be expressed in this theory, for example the solvability of systems of multivariate polynomial equations and of nonlinear optimization problems. Hence, a decision method for this theory, i.e. an algorithm that enables one to decide whether any sentence ("closed" formula, formula without free variables) of the theory is true or false, would be of central importance for constructive mathematics.

Actually, a decision method for this theory was first discovered by Tarski 1930, published 1948 [18]. His method is a "quantifier elimination" procedure: to every formula F of the theory a quantifier free formula F' is constructed such that $F \leftrightarrow F'$ is a consequence of the axioms of the theory. Since the truth of every quantifier free closed formula of the theory can be decided effectively, the quantifier elimination procedure yields a decision algorithm for the theory. Testing the efficiency of the method by computer experiments, it became apparent that Tarski's method required too much computation to be practical except for quite trivial problems. Other methods by Seidenberg [17], Cohen [5], and an improved version of Tarski's method by Böge [4] (see also Holthusen [12]) still show exponential dependency of the time complexity on the number m of polynomials occurring in the formula and/or the maximum degree n of these polynomials (when the number r of variables is fixed).

In Collins [6, 7, 8, 9] a new quantifier elimination procedure for real closed fields has been presented whose computational complexity depends only polynomially on

m and n (for r fixed). As every other method, it depends exponentially on r. Fischer and Rabin [11] have shown that every decision method for the first order theory of the additive group of the real numbers, a fortiori for the elementary theory of a real closed field, has a maximum computing time which dominates 2^{cN} where N is the length of the input formula and c is some positive constant). The principal component of this quantifier elimination method is an algorithm by which for a finite set of polynomials in r variables with integer coefficients a "cylindrical algebraic decomposition" of the r-dimensional real space into a finite number of disjoint connected sets called "cells" is obtained. In each cell each of the polynomials is invariant in sign. The sign of a polynomial in a cell, hence, can be determined by computing its sign at a sample point belonging to the cell.

The implementation and further improvement of this method is based on the efficient design of various subalgorithms from other parts of computer algebra, for example squarefree factorization, real root isolation, computations with real algebraic numbers, construction of primitive elements for field extensions etc. In fact, much research on these problems has been motivated by their occurrence as subproblems in the quantifier elimination problem. Thus, the study of the quantifier elimination problem might well serve as a motivating approach to various parts of computer algebra. Because of the limited space available we must refer the reader to the detailed exposition of the method in Collins [8]. Recent references on quantifier elimination containing various improvements of the method described in [6] are: Arnon [1, 2], Arnon/McCallum [3], Böge [4], Müller [15], McCallum [14], Collins [10]. Examples can also be found in Tarski [18], Kahan [13], Quarles [16].

References

[1] Arnon, D. S.: A Cellular Decomposition Algorithm for Semi-Algebraic Sets. EUROSAM **1979**, 301–315. (Also issued as Technical Report No. 353, Comput. Sci. Dpt., Univ. of Wisconsin–Madison, 1979.)

[2] Arnon, D. S.: Algorithms for the Geometry of Semi-Algebraic Sets. Ph.D. Diss., Technical Report No. 436, Comput. Sci. Dpt., Univ. of Wisconsin–Madison, 1981.

[3] Arnon, D. S., McCallum, S.: Cylindrical Algebraic Decomposition by Quantifier Elimination. Proc. of the EUROCAM 82 Conference, Marseille, April 1982, to appear.

[4] Böge, W.: Decision Procedures and Quantifier Elimination for Elementary Real Algebra and Parametric Polynomial Nonlinear Optimization. Preliminary Manuscript, December 1980.

[5] Cohen, P. J.: Decision Procedures for Real and p-adic Fields. Comm. Pure and Applied Math. **22**, 131–151 (1969).

[6] Collins, G. E.: Efficient Quantifier Elimination for Elementary Algebra (Abstract). Symposium on Complexity of Sequential and Parallel Numerical Algorithms, Carnegie-Mellon University, May 1973.

[7] Collins, G. E.: Quantifier Elimination for Real Closed Fields by Cylindrical Algebraic Decomposition – Preliminary Report. EUROSAM **1974**, 80–90.

[8] Collins, G. E.: Quantifier Elimination for Real Closed Fields by Cylindrical Algebraic Decomposition. Second GI Conference on Automata Theory and Formal Languages, Lect. Notes Comput. Sci. **33**, 134–183 (1975).

[9] Collins, G. E.: Quantifier Elimination for Real Closed Fields by Cylindrical Algebraic Decomposition – A Synopsis. SIGSAM Bull. **10**, 10–12 (1976).

[10] Collins, G. E.: Advances in Cylindrical Algebraic Decomposition. Proc. of the EUROCAM 82 Conference, Marseille, April 1982, to appear.

[11] Fischer, M. J., Rabin, M. O.: Super-Exponential Complexity of Presburger Arithmetic. M.I.T. MAC Tech. Memo **43**, 1974.

[12] Holthusen, C.: Vereinfachungen für Tarski's Entscheidungsverfahren der elementaren reellen Algebra. Diplomarbeit, Univ. of Heidelberg, January 1974.

[13] Kahan, W.: An Ellipse Problem. SIGSAM Bull. **9**, 11 (1975).

[14] McCallum, S.: Constructive Triangulation of Real Curves and Surfaces. M.Sc. thesis, University of Sydney, 1979.

[15] Müller, F.: Ein exakter Algorithmus zur nichtlinearen Optimierung für beliebige Polynome mit mehreren Veränderlichen. Meisenheim am Glan: Anton Hain 1978.

[16] Quarles, D. A., jr.: Algebraic Formulation of Necessary Numerical Stability Criteria for Hyperbolic Equations with Characteristic Approximation at a Boundary. IBM Research Report RC3221, 1971.

[17] Seidenberg, A.: A New Decision Method for Elementary Algebra. Ann. of Math. **60**, 365 – 374 (1954).

[18] Tarski, A.: A Decision Method for Elementary Algebra and Geometry, 2nd revised ed. Univ. of California Press 1951.

Prof. Dr. G. E. Collins
Computer Science Department
University of Wisconsin-Madison
1210 West Dayton Street
Madison, WI 53706, U.S.A.

Real Zeros of Polynomials

G. E. Collins, Madison, Wisconsin, and **R. Loos**, Karlsruhe

Abstract

Let A be a polynomial over Z, Q or $Q(\alpha)$ where α is a real algebraic number. The problem is to compute a sequence of disjoint intervals with rational endpoints, each containing exactly one real zero of A and together containing all real zeros of A. We describe an algorithm due to Kronecker based on the minimum root separation, Sturm's algorithm, an algorithm based on Rolle's theorem due to Collins and Loos and the modified Uspensky algorithm due to Collins and Aritas. For the last algorithm a recursive version with correctness proof is given which appears in print for the first time.

1. The Problem and Applications

Let ε be any positive rational number and $A(x) \in E[x]$ be any polynomial over the Euclidean domain E being a subring of $\mathbb{C}$ with effective arithmetic. E may be Z, Q, $Q(\alpha)$ or $G = Z[i]$ in our applications in computer algebra. The problem is to compute a sequence of disjoint intervals of length less than ε, each containing exactly one real zero of A and together containing all real zeros of A. The problem is embedded in the problem to isolate by a sequence of rectangles of width less than ε all complex zeros of A. The algorithms in computer algebra for solving this problem are exact, i.e. they will strictly fulfill the given specification; therefore they cannot be compared to numerical algorithms solving a similar, but different problem.

Applications are numerous, in particular in engineering. In computer algebra zero isolation is the generating mechanism for algebraic numbers which occur in many important algorithms, in particular in cylindrical algebraic decomposition algorithms [2] which in turn are the basis for solving systems of polynomial equations and inequalities over R and for quantifier elimination algorithms in real or algebraic closed fields. Given an isolating interval $(r, s]$ with rational endpoints for a real zero α of $A(x)$ which can be assumed to be squarefree, we can rapidly decrease $|s - r|$ by bisection and sign calculation of $A(x)$. Also fast high precision algebraic number calculation [3] can be applied to isolated intervals.

2. Subalgorithms

Let $A(x) \in Z[x]$ be a polynomial of positive degree n and seminorm $d_A = d$. To have integral coefficients is not a restriction since rational (for example "floating point") coefficients can be replaced by integral ones. In fact, we will assume A to be primitive. Next, if $A = A_1^{e_1} \cdots A_r^{e_r}$, $e_i \geqslant 0$, is a complete squarefree factorization, we compute the zeros of $A_1, \ldots, A_r$ separately and attach the multiplicities $e_1, \ldots, e_r$ to the output. Using the maximum computing time of the modular gcd-algorithm (see Chapter on p.r.s.) of $O(n^3 L(d)^2)$ the squarefree factorization needs at most $O(n^4 L(\bar{d})^2)$ steps, where $\bar{d}$ is a bound for the seminorm of all factors of A. Using Mignotte's bound (see Chapter on bounds) $\bar{d} \leqslant 2^n d$, hence we have $O(n^6 + n^4 L(d)^2)$ as maximum computing time of the squarefree factorization. We will therefore assume A also to be squarefree, and leave it to the reader to replace d in subsequent formulas by $2^n d$.

A *root bound*

$$b \geqslant \max_{1 \leqslant i \leqslant n} |\alpha_i|$$

is given [10] by

$$b = 2 \max_{1 \leqslant i \leqslant n} |a_{n-i}/a_n|^{1/i}. \tag{1}$$

It is convenient to compute the smallest power $\bar{b}$ of 2 greater or equal than b; clearly $b \leqslant 2d$ and $\bar{b} \leqslant 4d$.

The *minimum root separation*

$$\text{sep}(A) = \max_{1 \leqslant i < j \leqslant k} |\alpha_i - \alpha_j|$$

was given in [5] as

$$\text{sep}(A) > \tfrac{1}{2}(e^{1/2} n^{3/2} d)^{-n}. \tag{2}$$

Therefore $|\log \text{sep}(A)^{-1}| = O(nL(nd))$. Starting with an interval (rectangle) of size at most $8d$ the number of bisections needed to isolate a zero of A is bounded by

$$2\lceil \log_2(8d/\tfrac{1}{2} \text{sep}(A)) \rceil = O(nL(nd)). \tag{3}$$

It is unknown whether (2) can be so much improved that the order in (3) is decreased.

Using $(-\bar{b}, \bar{b}]$ as starting interval and proceeding by bisections one has to evaluate A at *binary rational* endpoints e/f, where f is a power of 2. With Horner's rule this costs at most

$$nL(e)\{nL(e) + nL(f) + L(d)\}, \tag{4}$$

and if e/f is the endpoint of any interval reached in the bisection process then

$$L(e), L(f) = O(nL(nd)). \tag{5}$$

Therefore, any evaluation costs at most

$$O(n^4 L(nd)^2). \tag{6}$$

3. Kronecker's Algorithm

Kronecker [11] has given a method to isolate the real roots of an integral squarefree polynomial. He uses as rootbound [Theorem 2 in the Chapter on useful bounds] $b_K = |A|_\infty/|a_n| + 1$ and determines the minimum root separation $\text{sep}(A) = 1/s$, where s is the smallest positive integer satisfying

$$(s - 1)D \geqslant \sum_{i=0}^{n-2} |u_i| b_K^i,$$

$$(s - 1)D \geqslant \sum_{i=0}^{n-1} |v_i| b_K^i,$$

$$(s - 1)D \geqslant \sum_{i,j=0}^{n-1} |b_{ij}| b_K^i,$$

$$(s - 1)D \geqslant \sum_{i,j=0}^{2n-4} |c_{ij}| b_K^i \tag{7}$$

with

$$AU + A'V = D = \mathrm{dis}(A)$$

and

$$B(x, y) = (A(x + y) - A(x))/y = \sum_{i,j=0}^{n-1} b_{ij} x^i y^j$$

$$C(x, y) = (B(x, y) - A'(x))U(x)/y = \sum_{i,j=0}^{2n-4} c_{ij} x^i y^j.$$

Kronecker's Algorithm. Input: A, a primitive squarefree integral polynomial of positive degree.

Output: A list L of isolating intervals with rational endpoints for all real zeros of A.

(1) [Bounds.] Set $b_K = |A|_\infty/|a_n| + 1$, $d = \mathrm{sep}(A)$ by (7) and $t = \lceil b_K/d \rceil$.

(2) [Test.] $s \leftarrow \mathrm{sign}\, A(-td)$; for $k = -t + 1, \ldots, t$ do
$\{s' \leftarrow \mathrm{sign}\, A(kd)$; if $s' = 0$ then output (kd, kd) else
if $ss' < 0$ then output $((k - 1)d, kd)$; $s \leftarrow s'\}$. ∎

From $|A|_\infty \leqslant d$, (2) and (6) we obtain

$$t_{\max} = O(d^{n+1} n^{3n/2} e^{n/2} n^4 L(nd)^2) \tag{8}$$

which is exponential in n and $L(d)$. However, it would be interesting to study how many intervals have to be tested in the average by this simple algorithm.

4. Sturm Sequences for Real Zeros

In the chapter on polynomial remainder sequences the concept of a Sturm sequence was introduced in Section 3.3. Let $(u_1, \ldots, u_s)$ be a sequence of real numbers and $(u'_1, \ldots, u'_t)$ the subsequence of all non-zero real numbers. Then $\mathrm{var}(u_1, \ldots, u_s)$, the number of *sign variations*, is the number of i, $1 \leqslant i < t$, such that $u'_i u'_{i+1} < 0$. Sturm's theorem — for a simple proof see [10, 2nd ed.] — asserts that the number of real zeros of the squarefree polynomial A with Sturm sequence $(A, A', A_3, \ldots, A_r)$ in the left-open, right-closed interval $(a, b]$, $a < b$, is

$$\mathrm{var}(A(a), A'(a), \ldots, A_r(a)) - \mathrm{var}(A(b), A'(b), \ldots, A_r(b)).$$

Sturm's Algorithm. Input: A, a primitive squarefree integral polynomial of positive degree.

Output: A list of isolating intervals with rational endpoints for all real zeros of A.

(1) [Bound.] $L \leftarrow ((- b, b))$, where $b \geqslant \max |\alpha_i|$ for all zeros α_i of A.

(2) [Sturm sequence.] Compute $(A, A', A_3, \ldots, A_r)$ as a (subresultant or primitive or reduced or modular) Sturm sequence of A.

(3) [Test.] for every interval $(a, b]$ in L do
    (3.1) bisect $(a, b]$ into $(a, m]$ and $(m, b]$,
    (3.2) for each subinterval compute the number r of real zeros using Sturm's theorem.
    (3.3) if $r = 0$, drop the interval,
    if $r = 1$, output the interval,
    if $r > 1$, keep the interval in L. ∎

A subresultant Sturm sequence can be computed in time $O(n^4 L(d)^2)$ and the coefficients are bounded by d^{2n}, therefore the time of step (2) dominates the time of step (1). In step (3.2) each polynomial of the Sturm sequence is evaluated at the binary rational endpoints of an interval which costs at most $O(n^4 L(nd)^2)$ by (6). The number of bisections is by (3) $O(nL(nd))$, there are at most n intervals in L at any time and the Sturm sequence contains at most $n + 1$ polynomials, hence at most $O(n^3 L(nd))$ evaluations are needed. Therefore

$$t_{\max} = O(n^7 L(nd)^3). \tag{9}$$

Empirical observations by Heindel [7], however, indicate, that usually the time for computing the Sturm sequence dominates the time for all evaluations. An exception are unusual Sturm sequences as for Legendre or Chebyshev polynomials, where the coefficients actually decrease in the sequence, which results in significantly smaller total time. This discovery was due in part by Reitwiesner for Stirling polynomials and was first observed in [6]. For random polynomials of degree 25 with 25 bit coefficients the largest coefficient in the sequence was 1595 bit long which limits the use of Sturm's algorithm considerably.

5. Real Zero Isolation by Differentiation

In [6] a method for obtaining all real zeros of a polynomial A was devised based on the derivative sequence $A, A', \ldots, A^{(n)}$ and Rolle's theorem. The theorem states that between two distinct zeros of a differentiable real function there is at least one zero of its derivative. Conversely, if A is a squarefree polynomial and $\alpha_1 < \cdots < \alpha_k$ are the real zeros of its derivative A' then each of the intervals $(- \infty, \alpha_1)$, $(\alpha_1, \alpha_2), \ldots, (\alpha_{k-1}, \alpha_k)$, (α_k, ∞) contains at most one real zero of A which can be detected by $A(\alpha_i)$ and $A(\alpha_{i+1})$, $i = 0, \ldots, k$, having opposite signs.

The algorithm is attractive since every coefficient of $pp(A^{(i)})$ is at most $\binom{n}{i}$ times as large as some coefficient of A, so in the previous example the largest coefficient grows only to 46 bits as compared to 1595 bits in the Sturm sequence. One other advantage of the algorithm is its inductive nature yielding isolating lists of real zeros for each derivative as a byproduct which is needed, for example, in Collins' quantifier elimination algorithm [2].

On the other hand, the inductive nature weakens the input assumption A being squarefree which may not hold for A' or higher derivatives. We insist instead that together with each isolating interval I_i for a zero α_i of A' the multiplicity m_i be given. Another problem is the sign computation of $A(\alpha_i)$ given only an isolating interval I_i and the multiplicity of α_i. In most cases the converse of Rolle's theorem gives a decision, but in one case a tangent construction requires A' being monotone in I_i, i.e. A'' does not have a zero in I_i other than possibly α_i. I_i, having these properties and binary rational end points, is called a *strong isolating interval* of A'. The problem is therefore to proceed from a strong isolating list of A' to a strong isolating list of A.

Algorithm Real Zeros by Differentiation. Input: A, a non-zero primitive integral polynomial.

Output: L, a strong isolating list of the real zeros of A.

(1) [Basis.] If $\deg A = 0$ then $\{L \leftarrow (\ ); \text{return}\}$.

(2) [Recursion.] Let L' be a strong isolating list of A'; if $L' = (\ )$ then $\{L \leftarrow ((-b, b), 1); \text{return}\}$, where b is a binary rational root bound for A.

(3) [Complementary intervals.] Let $L' = (I_1, m_1, \ldots, I_k, m_k)$, $k > 0$, $I_i = (a_i, b_i]$ and set $J_i = (b_i, a_{i+1}]$ with $b_0 = -b$, $a_{k+1} = b$ for $i = 0, \ldots, k$. [By Rolle's theorem, each J_i contains at most one (simple) zero and $J_i \cup I_i \cup J_{i+1}$ contains at most 2 zeros.] If $A(a_{i+1}) = 0 \lor A(b_i + \varepsilon)A(a_{i+1}) < 0$ then include J_i, 1 in L, for $i = 0, \ldots, k$ and some $\varepsilon > 0$.

(4) [$A(\alpha_i) = 0$?] If $J_i \cup J_{i+1}$ contains 2 zeros, I_i contains no zero. Otherwise set $B = \gcd(A, A'/\gcd(A', A''))$. [All zeros of B are simple, $A(\alpha_i) = 0$ iff $B(\alpha_i) = 0$ and α_i is the only zero of B in $(b_{i-1}, b_i]$.] If $B(b_{i-1} + \varepsilon)B(b_i + \varepsilon) < 0$ then include $I_i, m_i + 1$ in L, otherwise go on.

(5) [$A(\alpha_i) \neq 0$, m_i even.] If m_i even [A is monotone in I_i] $\lor$ $J_{i-1} \cup J_i$ has 1 zero of A then if $A(a_i + \varepsilon)A(b_i) \leq 0$ then include a strong isolating subinterval of I_i for A with multiplicity 1 in L else if m_i is even, I_i contains no zero.

(6) [$A(\alpha_i) \neq 0$, m_i odd.] If $A(b_i) = 0$ then $\{$if $A(a_i + \varepsilon)A'(a_i + \varepsilon) > 0$ then include a strong isolating subinterval of I_i for A with multiplicity 1 in L; otherwise bisect I_i repeatedly and include two strong isolating subintervals with multiplicity 1 in $L\}$.

(7) [$A(\alpha_i) \neq 0$, m_i odd, $A(b_i) \neq 0$.] If $A(a_i + \varepsilon)A(b_i) < 0$ then include strong isolating subinterval of I_i with multiplicity 1 in L.

(8) [Tangent construction.] If $A(a_i + \varepsilon)A'(a_i + \varepsilon) > 0$ then there are no zeros in I_i, otherwise compute a bisection point c and if $A(a_i + \varepsilon)A(c) < 0$ then A has two

simple zeros in I_i separated by c. In the remaining case the tangent intersections with the axis $a'_i = a_i + A(a_i)/A'(a_i)$ and $b'_i = b_i + A(b_i)/A'(b_i)$ are computed. If $a'_i > b'_i$ then I_i has no zeros of A since A' is monotone in I_i. If $a'_i < b'_i$ then the subinterval not containing α_i does not contain a zero of A and I_i is replaced by the other subinterval for repeated bisection and tangent construction until a decision is reached. ■

The sign of $A(b_i + \varepsilon)$, $\varepsilon > 0$, is the sign of $A(b_i)$, if this is non-zero, otherwise it is the sign of $A'(b_i + \varepsilon)$ which can be inferred by the strong isolation list L' and $lc(A)$.

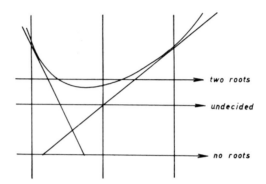

two roots

undecided

no roots

Fig. 1. Tangent construction and bisection

In order to show termination of the tangent and bisection process (Fig. 1) we have to prove the following theorems.

Theorem 1. *Let A be a non-zero squarefree integral polynomial of positive degree n and seminorm $|A|_1 = d$. Let α be any zero of A'. Then $|A(\alpha)| > (2n)^{-n}d^{-n+1}$.*

Proof. Let $\alpha_1, \ldots, \alpha_{n-1}$ be the zeros of A'. The polynomial

$$(na_n)^n \prod_{i-1}^{n-1} (x - A(\alpha_i)) = \text{res}_y(A'(y), x - A(y))$$

has degree n in x and seminorm $\leqslant (d + 1)^{n-1}(nd)^n$ which follows from Hadamard's inequality. Since A is squarefree, $|A(\alpha)| > 0$, and a minimum root bound of the resultant gives

$$|A(\alpha)| > \tfrac{1}{2}(2d)^{-n+1}(nd)^{-n} = (2n)^{-n}d^{-n+1}. \quad ■$$

Theorem 2. *Let $A(x)$ be an integral polynomial with $\deg A = n > 2$ and $|A|_1 = d$. Let*

$$T(x) = A(a) + A'(a)(x - a)$$

be the tangent to $A(x)$ at $x = a$. Then

$$|A(x) - T(x)| \leqslant 2^{n+1}\bar{a}^{n-2}d|x - a|^2 \quad \text{for} \quad |x - a| \leqslant \tfrac{1}{2},$$

where $\bar{a} = \max(|a|, 1)$.

Proof. By Taylor's theorem,

$$A(x) = \sum_{i=0}^{n} (A^{(i)}(a)/i!)(x - a)^i |A(x) - T(x)| \leqslant \sum_{i=2}^{n} (|A^{(i)}(a)|/i!)|x - a|^i.$$

But

$$|A^{(i)}(a)|/i! \leqslant |A^{(i)}|_i |a|^{n-1}/i! \leqslant \binom{n}{i} d|a|^{n-1} \leqslant 2^n d|a|^{n-1}$$

so

$$\sum_{i=2}^{n} (|A^{(i)}(a)|/i!)|x-a|^2 \leqslant 2^n d|a|^{n-2} \cdot |x-a|^2 \sum_{i=0}^{\infty} |x-a|^i$$

$$\leqslant 2^n d|a|^{n-2}|x-a|^2 \sum_{i=0}^{\infty} 2^{-i}$$

$$= 2^{n+1}|a|^{n-2}d|x-a|^2 \quad \text{if} \quad |a| \geqslant 1$$

and

$$\sum_{i=2}^{n} (|A^{(i)}(a)|/i!)|x-a|^2 \leqslant 2^n d \sum_{i=2}^{n} |x-a|^i \leqslant 2^{n+1} d|x-a|^2 \quad \text{if} \quad |a| < 1. \quad \blacksquare$$

Theorem 3. *Let A be an integral polynomial of degree $n \geqslant 2$, and seminorm $d \geqslant 2$. Let $A'(\alpha) = 0$ and $A(\alpha) \neq 0$. Let $T(x)$ be the tangent to $A(x)$ at $x = a$, let $|a| \leqslant d$ and $|\alpha - a| \leqslant (nd)^{-2n}$. Then $A(\alpha)$ and $T(\alpha)$ have the same sign.*

Proof. By Theorem 2,

$$|A(\alpha) - T(\alpha)| \leqslant 2^{n+1} d^{n-2} d(nd)^{-4} \leqslant d^{2n}(nd)^{-4n} = n^{-4n} d^{-2n}.$$

By Theorem 1,

$$|A(\alpha)| > (2n)^{-n} d^{-2n+1} > n^{-2n} d^{-2n} > |A(\alpha) - T(\alpha)|. \quad \blacksquare$$

The algorithm has [6]

$$t_{\max} = O(n^{10} + n^7 L(d)^3)$$

as maximal computing time which is the same as the Sturm sequence algorithm if A is not squarefree. However, in practice, except for unusual Sturm sequences, it is considerably faster, for random polynomials of degree 25 with 25 bits coefficients about 7 times.

6. Positive Zero Isolation by Descartes' Rule of Sign

Since the negative real zeros of $A(x)$ are the positive real zeros of $A(-x)$, and since $A(0) = 0$ iff $a_0 = 0$, we can restrict ourselves to isolate the positive zeros of A. Let $A(x) = \sum_{i=0}^{n} a_i x^i$. Descartes' rule states that the number $\text{var}(a_n, a_{n-1}, \ldots, a_0)$ exceeds the number of positive zeros of A, multiplicities counted, by an even nonnegative integer. Hence if $\text{var}(A) = 0$, A has no positive zeros, and if $\text{var}(A) = 1$, A has exactly 1 positive zero. There is a surprising theorem which Uspensky [13] attributes to Vincent in 1836 [14] which shows that after a finite number of transformations $\hat{A}(x) = A(x + 1)$ and $A^*(x) = (x + 1)^n A(1/(x + 1))$ one arrives at polynomials having sign variation 1 or 0. This suffices Uspensky to base the "most efficient method" for real zero isolation on it. His method, however, suffers from an exponential computing time since the bisection into subintervals is not balanced. This flaw was eliminated in the modified Uspensky algorithm by Collins and Akritas [4] using bisection by midpoints.

Modified Uspensky Algorithm. Input: A, a primitive squarefree integral polynomial.

Output: L, a list of isolating intervals for the positive real zeros of A.

(1) [Bound.] Let 2^k be a positive real root bound of A.

(2) [Transform.] If $k \geqslant 0$, set $B(x) = A(2^k x)$ otherwise set $B(x) = 2^{-kn} A(2^k x)$. [1 is now a root bound for B.]

(3) [Subalgorithm.] Let L' be the list of isolating intervals of the real zeros of B in $(0, 1)$.

(4) [Back.] Replace every interval (a_i, b_i) in L' by $(2^k a_i, 2^k b_i)$ and call the result L. ∎

Clearly, B has all its positive zeros in $(0, 1)$.

Subalgorithm. Input: A, a positive squarefree integral polynomial.

Output: L, a list of the zeros of A in $(0, 1)$.

(1) [Basis.] Set $A^* = (x + 1)^n A(1/(x + 1))$ [the zeros of A in $(0, 1)$ are transformed onto the zeros of A^* in $(0, \infty)$];
if $\mathrm{var}(A^*) = 0$ then $\{L \leftarrow (\ ); \text{return}\}$ else
if $\mathrm{var}(A^*) = 1$ then $\{L \leftarrow ((0, 1)); \text{return}\}$.

(2) [Midpoint.] If $A(\frac{1}{2}) = 0$ then include $(\frac{1}{2}, \frac{1}{2})$ in L and replace A by $A/(2x - 1)$.

(3) [Left half.] Set $A' = 2^n A(x/2)$ [the zeros of A in $(0, \frac{1}{2})$ are transformed onto the zeros of A' in $(0, 1)$]. Apply the subalgorithm recursively to A' with result L'; replace every interval (a_i', b_i') in L' by $(a_i'/2, b_i'/2)$ in L.

(4) [Right half.] Set $A'' = A'(x + 1)$ [the zeros of A in $(\frac{1}{2}, 1)$ are transformed onto the zeros of A'' in $(0, 1)$]. Apply the subalgorithm recursively to A'' with result L''; replace every interval (a_i'', b_i'') in L'' by $((a_i'' + 1)/2, (b_i'' + 1)/2)$ in L. ∎

Let $p(A)$ denote the number of positive zeros of the polynomial A. Partial correctness of the algorithm follows from the assertions in the algorithm and the following theorems.

Theorem 4. *Let A be a non-zero real polynomial. If $\mathrm{var}(A) = 0$ then $p(A) = 0$ and if $\mathrm{var}(A) = 1$ then $p(A) = 1$.*

Proof. If $\mathrm{var}(A) = 0$, let $a_n > 0$ then $a_{n-1} \geqslant 0, \ldots, a_0 \geqslant 0$. If $\alpha > 0$ then $a_n \alpha^n + \cdots + a_0 > 0$. Similarly, if $a_n < 0$, $A(\alpha) < 0$, so $A(\alpha) \neq 0$ for any positive α. If $\mathrm{var}(A) = 1$, let

$$a_n > 0, \quad a_{n-1} \geqslant 0, \ldots, a_{j+1} \geqslant 0, \quad a_j = -b_j < 0, \quad a_{j-1} = -b_j \leqslant 0, \ldots,$$

$$a_0 = -b_0 \leqslant 0.$$

Then

$$A(x) = x^{j+1}[a_n x^{n-j-1} + \cdots + a_{j+1} - (b_j x^{-1} + \cdots + b_0 x^{-j-1})].$$

For small positive x, $A(x) < 0$ and for large positive x, $A(x) > 0$, so by continuity, $p(A) \geqslant 1$. But $p(A) \leqslant 1$ by Descartes' rule of signs, so $p(A) = 1$. A similar argument holds, if $a_n < 0$. ∎

The converse can only be shown for $p(A) = 0$.

Theorem 5. *Let A be a non-zero real polynomial with no zeros in the right half-plane. Then* var$(A) = 0$.

Proof. Let $A = a_n \prod_{j=1}^n A_j(x)$, where each A_j is either $(x - \alpha_j)$ or $(x - \alpha_j)(x - \bar{\alpha}_j)$. In the first case $\alpha_j \leqslant 0$, so var$(A_j) = 0$. In the second case $A_j = x^2 - 2\,\text{re}(\alpha_j) + |\alpha_j|^2$, re$(\alpha_j) \leqslant 0$, so var$(A_j) = 0$. A product of polynomials without sign variations has no sign variation. ∎

The conjecture $p(A) = 1 \Rightarrow$ var$(A^*) = 1$ can be disproved. However, if A has one zero in $(0, 1)$ and no other zeros in a circle around the origin of radius $r_n = O(n^2)$ then var$(A^*) = 1$, where A^* is the test polynomial of step 1 of the subalgorithm. Then it remains to be shown that by a finite number of transformations $A'(x)$ and $A'(x + 1)$ any zero of A, except possibly one, is moved out of the circle. For, if a zero has a non-zero imaginary part, it is doubled by any of the transformations. If it has a negative zero, its absolute value is doubled, at least, without affecting the sign. The zeros in $(0, 1)$ are moved out by the transformations, except possibly one. Now suppose A has two zeros $1 - \lambda/2$ and $1 + \lambda/2$, where λ is bounded below by the minimum root separation of A. In order to keep $1 - \lambda/2$ in $(0, 1)$ a sequence of $A'(x + 1)$ transformations have to be applied recursively. So $\alpha = 1 + \lambda/2$ moves to $2(1 + \lambda/2) - 1 = 1 + \lambda$, then to $1 + 2\lambda, \ldots, 1 + 2^k\lambda$. In order to have $1 + 2^k\lambda > r_n$, one needs $k = O(nL(d))$, since $\lambda^{-1} = O(nL(d))$ and $r_n = O(n^2)$. By a similar argument it can be shown that any zero α can be made the only zero in $(0, 1)$ by a sequence of k transformations $A'(x)$ or $A'(x + 1)$. We have shown

Theorem 6. *Let A be a non-zero real polynomial of degree $n \geqslant 2$. Let $T(A)$ be $A'(x) = 2^n A(x/2)$ or $A'(x + 1)$. Then the polynomial $T^k(A)$, $k = O(nL(d))$, has at most one zero α_n in $(0, 1)$ and for all other zeros $\alpha_1, \ldots, \alpha_{n-1}$, real or complex, $|\alpha_i| > r_n = O(n^2)$.*

Theorem 7. *Let*

$$e_n = (1 + 1/n)^{1/(n-1)} - 1, \qquad n \geqslant 2.$$

Then

$$r_n = e_n^{-1} = \Theta(n^2).$$

Proof.

$$e_n = \exp\left(\frac{1}{n-1} \ln(1 + 1/n)\right) - 1.$$

Since $n \geqslant 2$ and $\ln(1 + 1/n) = 1/n - 1/2n^2 + \cdots$,

$$3/4n \leqslant 1/n - 1/2n^2 < \ln(1 + 1/n) < 1/n$$

and

$$3/4n^2 < \frac{1}{n-1} \ln(1 + 1/n) < n^{-2}(1 - 1/n)^{-1} \leqslant 2/n^2.$$

But

$$3/4n^2 < \exp(3/4n^2) - 1 < e_n < \exp(2/n^2) - 1 \leqslant (2/n^2)\exp(2/n^2)$$
$$\leqslant (2/n^2)\exp(\tfrac{1}{2}) \leqslant 4/n^2,$$

so

$$3/4n^2 < e_n < 4/n^2. \quad \blacksquare$$

Theorem 8. *Let*

$$B(x) = \prod_{i=1}^{n-1} (x + 1 + \alpha_i) = \sum_{i=0}^{n-1} b_i x^i$$

be a monic real polynomial of degree $n \geqslant 2$ with $|\alpha_i| < e_n$, $1 \leqslant i \leqslant n - 1$. Then

$$b_{n-2}/b_{n-1} > b_{n-3}/b_{n-2} > \cdots > b_0/b_1 > 0.$$

Proof. Let $0 \leqslant k \leqslant n - 1$ and

$$A(x) = \prod_{i=1}^{k} (x + \alpha_i) = \sum_{i=0}^{k} a_i x^i.$$

Then

$$\left| \prod_{i=1}^{k} (1 + \alpha_i) - 1 \right| = \left| \sum_{i=0}^{k} a_i - 1 \right| \leqslant \sum_{i=0}^{k-1} |a_i| < \sum_{j=1}^{k} \binom{k}{j} e_n^j$$
$$= (1 + e_n)^k - 1 \leqslant (1 + e_n)^{n-1} - 1 = 1/n.$$

Hence b_{n-1-k} is a sum of $\binom{n-1}{k}$ terms, each differing in absolute value from 1 by less than $1/n$. Hence

$$\binom{n-1}{k}(1 + \delta_k) = b_{n-k-1}$$

for some real δ_k with $|\delta_k| < 1/n$. Hence $b_{n-k-1} > 0$. Let $0 \leqslant k \leqslant n - 2$. Then

$$b_{n-k-2}/b_{n-k-1} = \binom{n-1}{k+1}(1 + \delta_{k+1}) \Big/ \binom{n-1}{k}(1 + \delta_k),$$

so

$$b_{n-k-1}/b_{n-k} > b_{n-k-2}/b_{n-k-1}$$

iff

$$(n - k)(1 + \delta_k)/(k(1 + \delta_{k-1})) > (n - k - 1)(1 + \delta_{k+1})/((k + 1)(1 + \delta_k))$$

iff

$$(n - k)(k + 1)/(k(n - k - 1)) > (1 + \delta_{k-1})(1 + \delta_{k+1})/(1 + \delta_k)^2.$$

But

$$(n - k)(k + 1)/(k(n - k - 1)) = 1 + n/(k(n - k - 1)) \geqslant 1 + 4n/(n - 1)^2$$

$$= (n + 1)^2/(n - 1)^2 = (1 + 1/n)^2/(1 - 1/n)^2$$

$$> (1 + \delta_{k-1})(1 + \delta_{k+1})/(1 + \delta_k)^2. \quad \blacksquare$$

Theorem 9. *Let A be a non-zero real polynomial of degree $n \geqslant 2$ with zeros $\alpha_1, \ldots, \alpha_n$ such that $0 < \alpha_n < 1$ and $|\alpha_i| > r_n$, $1 \leqslant i \leqslant n - 1$. Then $\mathrm{var}(A^*) = 1$, where $A^* = (x + 1)^n A(1/(x + 1))$.*

Proof. To each zero α_i^* of A^* corresponds the zero $\alpha = 1/(\alpha_i^* + 1)$ of A. We have to show that for $0 < \alpha_n^*$ and $|\alpha_i^*| < e_n$, $\mathrm{var}(A^*) = 1$. Set

$$A^* = \sum_{i=0}^{n} a_i x^i \quad \text{and} \quad B(x) = \sum_{i=0}^{n-1} b_i x^i = A^*/a_n(x - \alpha_n^*).$$

By Theorem 8, $b_{n-2}/b_{n-1}, \ldots, b_0/b_1 > 0$ is decreasing, hence $b_{n-2}/b_{n-1} - \alpha_n^*, \ldots, b_0/b_1 - \alpha_n^*$ is decreasing and has at most one variation.

$$\mathrm{var}(b_{n-2}/b_{n-1} - \alpha_n^*, \ldots, b_0/b_1 - \alpha_n^*) = \mathrm{var}(b_{n-2} - \alpha_n^* b_{n-1}, \ldots, b_0 - \alpha_n^* b_1) \leqslant 1,$$

since $b_i > 0$. But

$$A^*/a_n = (x - \alpha_n^*)B(x) = b_n x^n + \sum_{i=1}^{n-1} (b_{i-1} - \alpha_n^* b_i)x^i - \alpha_n^* b_0.$$

Since $b_n = 1$, $-\alpha_n^* b_0 < 0$, $\mathrm{var}(A^*) = \mathrm{var}(A^*/a_n) = 1$ or 3. But if $\mathrm{var}(A^*) = 3$ then $b_{n-2} - \alpha_n^* b_{n-1} < 0$ and, since the sequence is decreasing, $\mathrm{var}(A^*) = 1$, a contradiction. Hence $\mathrm{var}(A^*) = 1$. $\quad \blacksquare$

By Theorems 5, 6, and 9 for the test polynomial A^* of $T^k(A)$ we reach $\mathrm{var}(A^*) = 0$ or $\mathrm{var}(A^*) = 1$, i.e. the basis of the recursion in the subalgorithm, after $O(nL(d))$ recursive steps. This proves termination of the algorithm. Based on Theorem 6 it can be shown for the modified Uspensky algorithm that

$$t_{\max} = O(n^6 L(d)^2)$$

for a squarefree polynomial A. Since only simple transformations and sign calculations of coefficients are involved, the algorithm is well suited for real algebraic number coefficients in $Q(\alpha)$. Empirically the algorithm is faster than the derivative algorithm including for Chebyshev polynomials, despite the fact that they have positive zeros close to 1. The speed can be attributed to the fact that only multiplications by powers of 2 occur which can be done by shifting. The observed number of translations is proportional to n which suggests an average computing time of $O(n^4 + n^3 L(d))$ but for small n only the latter term can be empirically verified.

7. Open Problems

In [1] a survey is given also for the more general problem of complex zero isolation. Exact algorithms by Pinkert [12] based on the Routh-Hurwitz criterion and by

Collins based on the principle of argument have polynomial computing time bounds but they are not yet satisfactory from a practical point of view.

One other question is whether numerical methods for simultaneous zero calculation like Henrici's algorithm involving proximity [8] or exclusion and inclusion tests [9] can be formulated as exact algebraic isolation algorithms.

From a theoretical point of view it is an open problem whether complex zero isolation can be done in polynomial time by Weyl's [15] algorithm.

References

[1] Collins, G. E.: Infallible Calculation of Polynomial Zeros to Specified Precision, Mathematical Software III (Rice, J. R., ed.), pp. 35–68. New York: Academic Press 1977.
[2] Collins, G. E.: Quantifier Elimination for Real Closed Fields by Cylindrical Algebraic Decomposition. Lecture Notes in Computer Science, Vol. 33, pp. 134–183. Berlin-Heidelberg-New York: Springer 1975.
[3] Collins, G. E.: Real Root Subsystem, in SAC-2 System of Algebraic Algorithms. Available at request.
[4] Collins, G. E., Akritas, A. G.: Polynomial Real Root Isolation Using Descartes' Rule of Signs. SYMSAC **1976**, 272–275.
[5] Collins, G. E., Horowitz, E.: The Minimum Root Separation of a Polynomial. Math. Comp. **28**, 589–597 (1974).
[6] Collins, G. E., Loos, R.: Polynomial Real Root Isolation by Differentiation. SYMSAC **1976**, 15–25.
[7] Heindel, L. E.: Integer Arithmetic Algorithms for Polynomial Real Zero Determination. JACM **18**, 533–548 (1971).
[8] Henrici, P.: Methods of Search for Solving Polynomial Equations. JACM **17**, 273–283 (1970).
[9] Henrici, P., Gargantini, I.: Uniformly Convergent Algorithms for the Simultaneous Approximation of All Zeros of a Polynomial. In: Constructive Aspects of the Fundamental Theorem of Algebra (Dejon, B., Henrici, P., eds.), pp. 77–114. London: Wiley-Interscience 1969.
[10] Knuth, D. E.: The Art of Computer Programming, Vol. 2, 1st ed. (Seminumerical Algorithms). Reading, Mass.: Addison-Wesley 1969.
[11] Kronecker, L.: Über den Zahlbegriff. Crelle J. reine und angew. Mathematik **101**, 337–395 (1887).
[12] Pinkert, J. R.: An Exact Method for Finding the Roots of a Complex Polynomial. TOMS **2**, 351–363 (1976).
[13] Uspensky, J. V.: Theory of Equations. New York: McGraw-Hill 1948.
[14] Vincent, M.: Sur la Résolution des Equations Numériques. J. de Mathématiques Pures et Appliquées **1**, 341–372 (1836).
[15] Weyl, H.: Randbemerkungen zu Hauptproblemen der Mathematik, II. Fundamentalsatz der Algebra und Grundlagen der Mathematik. Math. Z. **20**, 131–150 (1924).

Prof. Dr. G. E. Collins
Computer Science Department
University of Wisconsin-Madison
1210 West Dayton Street
Madison, WI 53706, U.S.A.

Prof. Dr. R. Loos
Institut für Informatik I
Universität Karlsruhe
Zirkel 2
D-7500 Karlsruhe
Federal Republic of Germany

Factorization of Polynomials

E. Kaltofen, Newark, Delaware

Abstract

Algorithms for factoring polynomials in one or more variables over various coefficient domains are discussed. Special emphasis is given to finite fields, the integers, or algebraic extensions of the rationals, and to multivariate polynomials with integral coefficients. In particular, various squarefree decomposition algorithms and Hensel lifting techniques are analyzed. An attempt is made to establish a complete historic trace for today's methods. The exponential worst case complexity nature of these algorithms receives attention.

1. Introduction

The problem of factoring polynomials has a long and distinguished history. D. Knuth traces the first attempts back to Isaac Newton's Arithmetica Universalis (1707) and to the astronomer Friedrich T. v. Schubert who in 1793 presented a finite step algorithm to compute the factors of a univariate polynomial with integer coefficients (cf. [22, Sec. 4.6.2]). A notable criterion for determining irreducibility was given by F. G. Eisenstein in 1850 [12, p. 166]. L. Kronecker rediscovered Schubert's method in 1882 and also gave algorithms for factoring polynomials with two or more variables or with coefficients in algebraic extensions [23, Sec. 4, pp. 10 – 13]. Exactly one hundred years have passed since then, and though early computer programs relied on Kronecker's work [17], modern polynomial factorization algorithms and their analysis depend on major advances in mathematical research during this period of time. However, most papers which have become especially important to recent investigations do not deal with the problem per se, and we shall refer to them in the specific context.

When the long-known finite step algorithms were first put on computers they turned out to be highly inefficient. The fact that almost any uni- or multivariate polynomial of degree up to 100 and with coefficients of a moderate size (up to 100 bits) can be factored by modern algorithms in a few minutes of computer time indicates how successfully this problem has been attacked during the past fifteen years. It is the purpose of this paper to survey the methods which led to these developments. At the risk of repeating ourselves later or omitting significant contributions we shall give some main points of reference now.

In 1967 E. Berlekamp devised an ingenious algorithm which factors univariate polynomials over $\mathbb{Z}_p$, p a prime number, whose running time is of order $O(n^3 + prn^2)$, where n is the degree of the polynomial and r the number of actual factors (cf. [22, Sec. 4.6.2]). The incredible speed of this algorithm suggested factoring integer polynomials by first factoring them modulo certain small primes and then reconstructing the integer factors by some mean such as Chinese remaindering [21, Sec. 4.6.2]. H. Zassenhaus discussed in his landmark 1969 paper [60] how to apply the "Hensel lemma" to lift in k iterations a factorization modulo p to a factorization modulo p^{2^k}, provided that the integral polynomial is squarefree and remains squarefree modulo p. Readers familiar with basic field theory will know that if a polynomial over a field of characteristic 0 has repeated roots, then the greatest common divisor (GCD) of the polynomial and its derivative is nontrivial. Hence casting out multiple factors is essentially a polynomial GCD process, but we will come back to this problem in a later section. Squarefreeness is preserved modulo all but a reasonable small number of primes. Given a bound for the size of the coefficients of any possible polynomial factor, one then lifts the modular factorization to a factorization modulo p^{2^k} such that $p^{2^k}/2$ supersedes this coefficient bound. At this point factors with balanced residues modulo p^{2^k} either are already the integral factors or one needs to multiply some factors together to obtain a true factor over the integers. The slight complication arising from a leading coefficient not equal to unity will be resolved later.

D. Musser [32, 33] and, using his ideas, P. Wang in collaboration with L. Rothschild [53], generalized the Hensel lemma to obtain factorization algorithms for multivariate integral polynomials. Subsequently P. Wang has incorporated various improvements to these multivariate factorization algorithms [49, 50, 52]. In 1973 J. Moses and D. Yun found the Hensel construction suitable for multivariate GCD computations (now called the EZGCD algorithm) [31], and D. Yun has used this algorithm for the squarefree decomposition process of multivariate polynomials [58]. The classical algorithm for factoring polynomials over algebraic number fields was considered and modified by B. Trager [42] but again the Hensel approach proved fruitful [48, 56]. In 1976 M. Rabin, following an idea of E. Berlekamp [5], introduced random choices in his algorithm for factoring univariate polynomials over large finite fields whose expected complexity is at most of the order $O(n^3(\log n)^3 \log(p))$, where n is the degree and p the size of the field [41], [4]. In 1979 G. Collins published a thorough analysis of the average time behavior for the univariate Berlekamp-Hensel algorithm [10], while in the same year improved algorithms for squarefree factorization [54] and Chinese remaindering on sparse multivariate polynomials appeared [63, 64, 65].

To completely factor a univariate polynomial over the integer means, of course, to also factor the common divisor of all its coefficients. This paper does not include the topic of factorization of integers and we will not consider the previously mentioned problem as part of the polynomial factorization problem. However, some comparisons are in order. Factoring large random integers seems much harder than factoring integral polynomials. This was partially confirmed by a polynomial-time reduction from polynomial to integer factorization, which is, however, subject to an old number theoretic conjecture [3]. The problem of finding polynomially long irreducibility proofs ("succinct certificates") was first solved for prime numbers in 1975 [40] and has recently also been achieved for densely encoded integral polynomials [7]. A polynomial-time irreducibility test for prime numbers depending on the validity of the generalized Riemann hypothesis (GRH) was discovered in

1976 (cf. [22, Sec. 4.5.4]). Peter Weinberger obtained the corresponding result for densely encoded integer polynomials [55] (also see [22, p. 632, Exercise 38]). In 1971 E. Berlekamp pointed out that the modular projection and lifting algorithm may take an exponential number of trial factor combinations [5]. Except for P. Weinberger's algorithm, whose complexity analysis is subject to the GRH, the author knows of no procedure which significantly reduces this exponential behavior in contrast to the stunning advances for the integer case by L. Adleman, C. Pomerance, and R. Rumley [2]. Also, no fast probabilistic irreducibility tests for integer polynomials seem to be known, again leaving a gap for work parallel to that of M. Rabin, R. Solovay and V. Strassen (cf. [22, Sec. 4.5.4]). Little work has been done on the theoretical analysis of the multivariate versions of the Berlekamp-Hensel algorithm. Similar to the univariate case the steps involved may require an exponential number of trial factor combinations though this problem may be probabilistically controllable by virtue of the Hilbert Irreducibility Theorem. G. Viry has also shown how to replace the trial divisions of multivariate polynomials by a simple degree test which makes his algorithm the asymptotically fastest, though still exponential in the degrees, for the worst case of all known deterministic algorithms [47]. Recently the author has proven that it is only polynomially harder to factor densely encoded multivariate integer polynomials with a fixed number of variables than integer polynomials with just two variables [18].

In this paper we take the concrete approach of discussing the algorithms for coefficient domains such as Galois fields, the integers, or finite algebraic extensions of the rationals. An excellent reference for a general algebraic setting is the work of D. Musser [32, 33]. Sections 2, 3 and 4 deal with univariate polynomial factorization over the previously mentioned domains. Section 5 discusses the multivariate factorization problem over these domains with emphasis on integer coefficients. We conclude with a list of open problems in Section 6.

2. Factorization of Univariate Polynomials Over Finite Fields

The exposition of this problem given in D. Knuth's book [22, Sec. 4.6.2] is quite complete and we refer the reader who seeks an introduction to current algorithms to that work. We wish to mention here that testing a polynomial $f(x) \in \mathbb{Z}_p[x]$ of degree n for irreducibility can be achieved in $O(n^2 \log(n)^3 \log(p))$ arithmetic steps using the distinct degree factorization [41]. This bound is polynomial in $n \log(p)$ which is the order of the size needed to encode f on the tape of a Turing machine. The distinct degree factorization algorithm also produces for each occurring factor degree m the product polynomial of all factors of degree m. Though this is still an incomplete factorization, it is a further refinement of the squarefree decomposition and no corresponding algorithm is known for integer coefficients. It is the goal of any probabilistic factorization algorithm to make the expected running time polynomial in $n \log(p)$. As we have seen in the introduction, M. Rabin's algorithm, which is based on finding the roots of f in some larger Galois field of characteristic p, achieves this goal. Recently D. Cantor and H. Zassenhaus proposed a probabilistic version of Berlekamp's algorithm which takes about $O(n^3 + n^2 \log(n) \log(p)^3)$ expected steps [8]. It is not clear which one of the algorithms is in practice more efficient [6] though the root-finding algorithm has been proven to be asymptotically faster [4]. Further probabilistic improvements to the Berlekamp's algorithm are reported in [26]. There is no known deterministic complete factorization algorithm whose worst time complexity is polynomial in $n \log(p)$, except in the case that $p - 1$ is highly composite (e.g. $p = L2^k + 1$ with L of the same size as k). In that case R. Moenck devised an algorithm very similar to the root finding algorithm which factors f in $O(n^3 + n^2 \log(p) + n \log(p)^2)$ steps [29]. So far we have only addressed the problem where the coefficients lie in a prime residue field but, as one may have expected, most of the above algorithms can be modified to also work in higher dimensional Galois fields.

3. Factorization of Univariate Polynomials over the Integers

Given a polynomial $\bar{h}(x) \in \mathbb{Z}[x]$ we seek to compute its content (see chapter on "Arithmetic") and all its primitive irreducible polynomial factors $g_{ij}(x) \in \mathbb{Z}[x]$, $1 \leqslant i \leqslant r$, $1 \leqslant j \leqslant s_i$, that is

$$\bar{h}(x) = \text{cont}(\bar{h}) \prod_{i=1}^{r} \left(\prod_{j=1}^{s_i} g_{ij}(x) \right)^i$$

with all g_{ij} pairwise distinct. The complete algorithm consists of three separate steps, namely

[Factorization of $\bar{h}(x) \in \mathbb{Z}[x]$]

(C) [Content computation:] The integer GCD of all coefficients of $\bar{h}$ constitutes the $\text{cont}(\bar{h})$, $h \leftarrow \bar{h}/\text{cont}(\bar{h})$ [h is now a primitive polynomial.]

(S) [Squarefree decomposition of h:] Compute squarefree polynomials $f_i(x) \in \mathbb{Z}[x]$, $1 \leqslant i \leqslant r$, $\text{GCD}(f_j, f_k) = 1$ for $1 \leqslant j \neq k \leqslant r$ such that $h(x) = \prod_{i=1}^{r} (f_i(x))^i$.

(F) [Factor the squarefree f_i:] FOR $i = 1, \dots, r$ DO
Compute irreducible polynomials $g_{ij}(x) \in \mathbb{Z}[x]$, $1 \leqslant j \leqslant s_i$ such that $f_i(x) = \prod_{j=1}^{s_i} g_{ij}(x)$. ∎

Step (C) is a repeated integer GCD computation and shall not be discussed further.

The computational aspects of step (S) were first investigated by E. Horowitz following an idea of R. Tobey in 1969 (cf. [16]) whose algorithms were later improved by D. Musser [33], D. Yun [58] and P. Wang and B. Trager [54]. We shall briefly present D. Yun's algorithm:

[Squarefree decomposition of a primitive polynomial h:]

(S1) $g(x) \leftarrow \text{GCD}(h(x), dh(x)/dx)$, where $dh(x)/dh = h'(x)$ is the derivative of h.
$c_1(x) \leftarrow h(x)/g(x)$;
$d_1(x) \leftarrow (dh(x)/dx)/g(x) - dc_1(x)/dx$. [Assume that $h = \prod_{i=1}^{r} f_i^i$ with the f_i squarefree and pairwise relatively prime. Then

$$g = \prod_{i=2}^{r} f_i^{i-1}, \qquad c_1 = \prod_{i=1}^{r} f_i, \qquad h'/g = \sum_{i=1}^{r} \left(i f_i' \prod_{j=1, j \neq i}^{r} f_j \right)$$

which is relatively prime to g since $\text{GCD}(f_i, f_i') = 1$ (the f_i are squarefree!).

Thus

$$d_1 = \sum_{i=2}^{r} \left((i-1)f_i' \prod_{j=1, j \neq i}^{r} f_j \right).]$$

(S2) FOR $k \leftarrow 1, 2, \ldots$ UNTIL $c_k = 1$ DO
  [At this point

$$c_k = \prod_{i=k}^{r} f_i, \qquad d_k = \sum_{i=k+1}^{r} \left((i-k)f_i' \prod_{j=k, j \neq i}^{r} f_j \right).]$$

  $f_k(x) \leftarrow \text{GCD}(c_k(x), d_k(x));$
  $c_{k+1}(x) \leftarrow c_k(x)/f_k(x);$
  $d_{k+1}(x) \leftarrow d_k(x)/f_k(x) - \mathrm{d}c_{k+1}(x)/\mathrm{d}x.$ ∎

The reader should be able to derive the correctness of this algorithm from the embedded comments. It is important that the cofactor of h' in the GCD computation of step (S1) and that of d_k in step (S2) are relatively prime to the computed GCDs. This enables one to use, besides the modular GCD algorithm, the EZGCD algorithm [31] whose general version needs the above algorithm if both cofactors have a common divisor with the GCD. The relation between polynomial GCDs and squarefree decompositions is even more explicit (cf. [59]).

Step (F) is the actual heart of the algorithm. As outlined in the introduction, various substeps are needed for the Berlekamp-Hensel algorithm:

[Factorization of a primitive, squarefree polynomial f:]

(F1) [Choice of a modulus:] Find a prime number p which does not divide $\mathrm{ldcf}(f(x))$ and the resultant of $f(x)$ and $\mathrm{d}f(x)/\mathrm{d}x$. The latter is equivalent to that $f(x)$ modulo p remains squarefree and this condition is what we test. By trying various primes in connection with the distinct factorization procedure we may also attempt to minimize the number of modular factors in the next step.

(F2) [Modular factorization:] Factor $f(x)$ modulo p completely, namely compute irreducible polynomials
  $u_1(x), \ldots, u_r(x) \in \mathbb{Z}_p[x]$ such that
  $\mathrm{ldcf}(u_1) \equiv \mathrm{ldcf}(f)$ (modulo p), $u_2, \ldots, u_r$ are monic and
  $u_1(x) \cdots u_r(x) \equiv f(x)$ (modulo p).

(F3) [Factor coefficient bound:] Compute an integer $B(f)$ such that all coefficients of any possible factor of $f(x)$ in $\mathbb{Z}[x]$ are absolutely bounded by $B(f)$ (see chapter on "Useful Bounds").

(F4) [Lift modular factors:] $q \leftarrow p$;
  FOR $k \leftarrow 1, 2, \ldots$ UNTIL $q \geq 2B(f)$ DO
      $q \leftarrow q^2$; [At this point $q = p^{2^k}$.]
      Compute polynomials $u_i^{(k)}(x) \in \mathbb{Z}_q[x]$ such that
      $u_1^{(k)} \cdots u_r^{(k)} \equiv f(x)$ (modulo q),
      $\mathrm{ldcf}(u_1^{(k)}) \equiv \mathrm{ldcf}(f)$ (modulo q) and
      $u_i^{(k)} \equiv u_i$ (modulo p), where the coefficients of
      $u_i^{(k)}$ are interpreted as p-adic approximations.

(F5) [Form trial factor combinations:]

    $h(x) \leftarrow f(x);\ C \leftarrow \{2,\ldots,r\};\ s \leftarrow 0;\ j \leftarrow 1;$

    REPEAT $t \leftarrow s;$

        FOR $m \leftarrow j,\ldots,$ cardinality of C DO

            FORALL subsets $\{i_1,\ldots,i_m\}$ of C DO

                Test whether $g(x) = \mathrm{pp}(\mathrm{ldcf}(h)u_{i_1}^{(k)} \cdots u_{i_m}^{(k)}$ (modulo p^{2^k})) divides h, where k is the number of iterations in (F4) and the modulus is balanced before taking the primitive part over the integers. If so then set $s \leftarrow s + 1;\ g_s(x) \leftarrow g(x);$ $h(x) \leftarrow h(x)/g(x);\ j \leftarrow m;\ C \leftarrow C$ minus $\{i_1,\ldots,i_m\};$ and exit both FOR loops.

            END FORALL

        END FOR

    UNTIL $t = s$ [No more factors discovered in the FOR loops]

    $s \leftarrow s + 1;\ g_s(x) \leftarrow h(x)$

    [All factors are computed as $f(x) = g_1(x) \cdots g_s(x).$] ∎

We must scrutinize various steps further. By the choice of p in step (F1) $f(x)$ modulo $p = \bar{f}(x)$ is of the same degree as $f(x)$ and the inverse of $\mathrm{ldcf}(\bar{f})$ in $\mathbb{Z}_p$ exists. We factor the monic polynomial $\mathrm{ldcf}(\bar{f})^{-1}\bar{f}(x)$ first into distinct degree factors and then into irreducibles in step (F2). To satisfy the condition on the $\mathrm{ldcf}(u_1)$ we multiply u_1 by $\mathrm{ldcf}(\bar{f})$ in $\mathbb{Z}_p$. Step (F4) utilizes the "Hensel-lemma" and various lifting techniques have been investigated [60], [32], [51] (see also the chapter on "Homomorphic Images"). The following algorithm is due to P. Wang:

[Given polynomials $f(x) \in \mathbb{Z}[x]$, q relatively prime to $\mathrm{ldcf}(f)$, $u_1^*(x),\ldots,u_r^*(x) \in \mathbb{Z}_q[x]$ such that $\mathrm{ldcf}(u_1) \equiv \mathrm{ldcf}(f)$ (modulo q), $u_2^*,\ldots,u_r^*$ monic and

$$u_1^*(x) \cdots u_r^*(x) \equiv f(x) \quad (\text{modulo } q). \tag{1}$$

Furthermore given polynomials $v_1^*(x),\ldots,v_r^*(x) \in \mathbb{Z}_q[x]$ with $\deg(v_i^*) < \deg(u_i^*)$ for $1 \leqslant i \leqslant r$, and if we set $\tilde{u}_i^* = \prod_{j=1, j \neq i}^{r} u_j^*$ then

$$v_1^*(x)\tilde{u}_1^*(x) + \cdots + v_r^*(x)\tilde{u}_r^*(x) \equiv 1 \quad (\text{modulo } q).$$

The goal is to produce polynomials $u_1^{**}(x),\ldots,u_r^{**}(x), v_1^{**}(x),\ldots,v_r^{**}(x) \in \mathbb{Z}_{q^2}[x]$ which satisfy the same conditions as the single-starred polynomials if we replace the modulus q by q^2.]

(H1) Replace $\mathrm{ldcf}(u_1^*)$ by $\mathrm{ldcf}(f)$ modulo q^2;

    [Lift u_i^* by computing $\hat{u}_i^* \in \mathbb{Z}_q[x]$ such that $u_i^{**} = u_i^* + q\hat{u}_i^*$ with $\deg(\hat{u}_i^*) < \deg(u_i^*)$ for $i \geqslant 1$.]

$$t(x) \leftarrow \left(f(x) - \prod_{i=1}^{r} u_i^*(x) \right) \text{modulo } q^2;$$

    [The above replacement causes $\deg(t) < \deg(f)$. Also all coefficients of t are divisible by q because of (1).]

    $t(x) \leftarrow t(x)/q$; [Integer division, hence $t(x) \in \mathbb{Z}_q[x]$. We need to determine $\hat{u}_i^*$ with

$$\hat{u}_1^*\tilde{u}_1^* + \cdots + \hat{u}_r^*\tilde{u}_r^* = t(x). \tag{2}]$$

FOR $i \leftarrow 1, \ldots, r$ DO

$\hat{u}_i^*(x) \leftarrow$ remainder$(t(x)v_i^*(x), u_i^*(x))$ in $\mathbb{Z}_q[x]$;

$u_i^{**}(x) \leftarrow u_i^*(x) + q\hat{u}_i^*(x)$.

[Obviously the polynomials tv_i^* solve (2) but do not satisfy the degree constraint for the $\hat{u}_i^*$. Hence the $\hat{u}_i^*$ solve (2) modulo $\prod_{i=1}^{r} u_i^*$ but since all degrees are less than $\deg(f)$ there must be equality.]

(H2) [Lift v_i^* by computing $\hat{v}_i^* \in \mathbb{Z}_q[x]$ such that
$v_i^{**} = v_i^* + q\hat{v}_i^*$ and $\deg(\hat{v}_i^*) < \deg(u_i^*)$.]

$$b(x) \leftarrow \left(1 - \sum_{i=1}^{r} v_i^* \tilde{u}_i^*(x)\right) \text{ modulo } q^2/q;$$

[Again the division is integral and $b(x) \in \mathbb{Z}_q[x]$ with $\deg(b) < \deg(f)$.]

FOR $i \leftarrow 1, \ldots, r$ DO

$\hat{v}_i^*(x) \leftarrow$ remainder$(b(x)v_i^*(x), u_i^*(x))$ in $\mathbb{Z}_q[x]$;

$v_i^{**}(x) \leftarrow v_i^*(x) + q\hat{v}_i^*(x)$. ∎

In order to use the above algorithm within the loop of step (F4) we also need to initialize the $v_i(x)$ in $\mathbb{Z}_p$ with

$$1 \bigg/ \prod_{i=1}^{r} u_i(x) = \sum_{i=1}^{r} v_i(x)/u_i(x) \qquad \text{and} \qquad \deg(v_i) < \deg(u_i).$$

One can use the extended Euclidean algorithm repeatedly (see chapter on "Remainder Sequences") or use fast partial fraction decomposition algorithms [24], [1].

Step (H2) is not necessary if one only considers the first solution v_i and corrects u_i^* from modulus q to modulus pq by calculating $\hat{u}_i^*$ in $\mathbb{Z}_p[x]$. This method is referred to as "linear lifting" whereas our algorithm has quadratic p-adic convergence. We also lift all factors in parallel while earlier versions proceeded with one factor and its cofactor at a time. It is not clear which technique is preferable (cf. [57], [61]), though our parallel quadratic approach seems superior [51]. In order to prevent p^{2^k} from overshooting $B(f)$ by too much one may calculate the last correction polynomials $\hat{u}_i^*$ with a smaller modulus than q.

As we will see below, step (F5) may become the dominant step in our algorithm. Therefore one is advised to test first whether the second highest coefficient is absolutely smaller than $\deg(f)|f|_2$, the corresponding factor coefficient bound (see chapter on "Useful Bounds") [43], or whether the constant coefficient of $g(x)$ divides that of $f(x)$.

D. Musser has methodically analyzed a variation of steps (F1)−(F5), the result of which is the following [32]: Let $f = g_1 \cdots g_s$ in $\mathbb{Z}[x]$, $\deg(g_1) \leqslant \deg(g_2) \leqslant \cdots \leqslant \deg(g_s)$, and let

$$\mu = \begin{cases} \max_{i=2,\ldots,s} \{\deg(g_{i-1}), \lfloor \deg(g_i)/2 \rfloor\} & \text{if } s > 1 \\ \\ \lfloor \deg(f)/2 \rfloor & \text{if } f \text{ is irreducible.} \end{cases}$$

If f factors into r polynomials modulo p then

$$\min(2^r, r^\mu)\mu n^2(n + \log(B(f)))^2$$

dominates the complexity of the factorization problem. This bound depends intrinsically on r which is one reason why one should attempt to minimize this number in step (F2). If one does not, the algorithm is still performing quite well – on the average. An nth degree polynomial in $\mathbb{Z}_p[x]$ has an average of $\log(n)$ factors as p tends to infinity and 2^r averages $n + 1$, where r is the number of modular factors (cf. [22, Sec. 4.6.2., Exercise 5]). However, almost all integer polynomials are irreducible (cf. [22, Sec. 4.6.2, Exercise 27]), and one may not expect almost all inputs to our algorithm to behave that way since a user probably tries to factor polynomials which are expected to be composite. In this matter G. Collins has shown, subject to two minor conjectures, that if we restrict our set to those polynomials which factor over the integers into factors of degree $d_1, d_2, \ldots, d_s$ for a given additive decomposition of $n = d_1 + \cdots + d_s$, the average number of trial combinations will be below n^2. This result only holds if one processes combinations of m factors at a time as we did in step (F5) ("cardinality procedure"), because if one chooses to test combinations of a possible total degree ("degree procedure") the average behavior may be exponential in n [10].

The worst case complexity of the Berlekamp-Hensel algorithm is unfortunately exponential in n, the degree of f. This is because, as we will prove below, there exist irreducible integer polynomials of arbitrarily large degree which factor over every prime into linear or quadratic factors. This means that we must test at least $2^{n/2-1} - 1$ trial factor combinations to show that no integral factor occurs. The following theorem is attributed to H. P. F. Swinnerton-Dyer by E. Berlekamp [5, pp. 733–734].

Theorem. *Let n be an integer and $p_1, \ldots, p_n$ positive distinct prime numbers. Then the monic polynomial $f_{p_1,\ldots,p_n}(x)$ of degree 2^n whose roots are $e_1\sqrt{p_1} + \cdots + e_n\sqrt{p_n}$ with $e_i = \pm 1$ for $1 \leqslant i \leqslant n$ has integral coefficients and is irreducible in $\mathbb{Z}[x]$. Moreover, for any prime q, $f_{p_1,\ldots,p_n}(x)$ modulo q factors into irreducible polynomials in $\mathbb{Z}_q[x]$ of at most degree two.*

Proof. In the following we assume that the reader is familiar with some basic facts of Galois theory. The book by van der Waerden [44] is a standard reference whose notation we also adopt. The following abbreviations are useful:

$$f_k(x) \equiv f_{p_1,\ldots,p_k}(x) \quad \text{and} \quad K_k \equiv \mathbb{Q}(\sqrt{p_1}, \ldots, \sqrt{p_k}) \quad \text{for} \quad 1 \leqslant k \leqslant n.$$

By induction we prove that $f_n(x) \in \mathbb{Z}[x]$, $[K_n : \mathbb{Q}] = 2^n$, and that $\theta = \sqrt{p_1} + \cdots + \sqrt{p_n}$ is a primitive element of K_n. For $n = 1$ the facts are trivial. It follows from the hypothesis $f_{n-1}(x) \in \mathbb{Z}[x]$ and from $f_n(x) = f_{n-1}(x + \sqrt{p_n})f_{n-1}(x - \sqrt{p_n})$ that $f_n(x) \in \mathbb{Z}[\sqrt{p_n}, x]$ whose coefficients are symmetric in the two conjugates $\sqrt{p_n}$ and $-\sqrt{p_n}$. By the fundamental theorem on symmetric functions the coefficients must be integers. (Actually $f_n(x) = \operatorname{res}_y(f_{n-1}(x - y), x^2 - p_n)$ as is shown in the chapter on "Algebraic Domains".) From the second hypothesis, namely $[K_{n-1} : \mathbb{Q}] = 2^{n-1}$, we conclude that the set

$$B_k = \{1\} \cup \{\sqrt{p_{i_1} \cdots p_{i_j}} \mid j = 1, \ldots, k, \ 1 \leq i_1 < i_2 < \cdots < i_j \leq k\}$$

forms a basis for K_k over $\mathbb{Q}$, $1 \leq k \leq n - 1$.

We show by induction that $\sqrt{p_n}$ does not lie in the $\mathbb{Q}$-span of B_{n-1}. Assume it does, namely there exist rationals $r_0, r_{i_1, \ldots, i_j}$ such that

$$r_0 + \sum_{1 \leq i_1 < \cdots < i_j \leq n-1} r_{i_1, \ldots, i_j} \sqrt{p_{i_1} \cdots p_{i_j}} = \sqrt{p_n}. \tag{1}$$

Since p_n is a new prime, at least two coefficients on the left-hand side of (1) are non-zero. Then for the two corresponding basis elements there exists a p_k such that $\sqrt{p_k}$ occurs in one but not the other. Without loss of generality assume $p_k = p_{n-1}$. Then (1) can be rewritten as

$$s_0 + s_1 \sqrt{p_{n-1}} = \sqrt{p_n}, \qquad s_0, s_1 \in K_{n-2}. \tag{2}$$

Since s_0 and s_1 are linear combinations in B_{n-2} with a nonzero coefficient it follows that both $s_0 \neq 0$ and $s_1 \neq 0$. Squaring (2) then leads to

$$\sqrt{p_{n-1}} = (p_n - s_0^2 - s_1^2 p_{n-1})/2s_0 s_1 \in K_{n-2}$$

in contradiction to the induction hypothesis.

Therefore $[K_n : K_{n-1}] = 2$ and hence $[K_n : \mathbb{Q}] = 2^n$. We proceed to show that $K_n = \mathbb{Q}(\theta)$. Let

$$\alpha_1 = \sqrt{p_1} + \cdots + \sqrt{p_{n-1}}, \alpha_2, \ldots, \alpha_{2^{n-1}}$$

be the roots of $f_{n-1}(x)$ and consider the polynomials

$$g_1(x) = f_{n-1}(\alpha_1 + \sqrt{p_n} - x) \qquad \text{and} \qquad g_2(x) = x^2 - p_n.$$

Obviously $g_1, g_2 \in \mathbb{Q}(\theta)[x]$ and $\sqrt{p_n}$ is a common root. However, $g_1(-\sqrt{p_n}) \neq 0$ because $\alpha_1 + 2\sqrt{p_n} \neq \alpha_i$ for $1 \leq i \leq 2^{n-1}$ since $\alpha_i - \alpha_1 \in K_{n-1}$ but $\sqrt{p_n} \notin K_{n-1}$. Therefore $\mathrm{GCD}(g_1, g_2) = x - \sqrt{p_n} \in \mathbb{Q}(\theta)[x]$ and hence $\mathbb{Q}(\theta) = \mathbb{Q}(\alpha_1, \sqrt{p_n})$. By hypothesis $K_{n-1} = \mathbb{Q}(\alpha_1)$ which gives $\mathbb{Q}(\theta) = K_n$. The irreducibility of f_n now follows quickly. The minimal polynomial of θ has degree $[\mathbb{Q}(\theta) : \mathbb{Q}] = [K_n : \mathbb{Q}] = 2^n$ and therefore f_n is this irreducible polynomial.

The factorization property modulo q can be proven by the following argument. Since $\sqrt{p_i}$ modulo $q \in \mathrm{GF}(q^2)$ for $1 \leq i \leq n$ all roots of f_n modulo q lie in $\mathrm{GF}(q^2)$. If f_n modulo q had an irreducible factor of degree $m > 2$ its roots would generate the larger field $\mathrm{GF}(q^m)$ and could not be elements of $\mathrm{GF}(q^2)$. ■

The construction of $f_{p_1, \ldots, p_n}$ has been generalized using higher radicals instead of square roots [20] and it can be shown that $\log(|f_{p_1, \ldots, p_n}|) = O(2^n \log(n))$ which makes the worst case of the Berlekamp-Hensel algorithm truly exponential in its input size. Here the following remark is in place. We always assume that our algorithm operates on densely encoded polynomials. If we allow sparse encoding schemes, various primitive operations on the input polynomials such as GCD computations are NP-hard (cf. [38, 39]) and the factorization problem actually requires exponential space. In order to substantiate the last claim we consider the polynomial $x^n - 1$ whose sparse encoding requires $O(\log n)$ bits. However, R. Vaughan has shown that for infinitely many n the cyclotomic polynomials Ψ_n, which constitute irreducible factors of $x^n - 1$, have coefficients absolutely larger than $\exp(n^{\log 2/\log\log n})$ [45].

One question about our algorithm remains to be answered. That is how the choice of various primes in step (F1) can influence later steps, especially step (F5). It is clear that if a polynomial f factors modulo p_1 into all quadratic and modulo p_2 into all cubic factors, then the degrees of integral factors must be multiples of 6. Indeed if the degree sets of factorizations modulo various primes are completely incompatible we know the input polynomial to be irreducible without the need of steps (F2) – (F5). For this situation D. Musser has developed an interesting model which, given a random irreducible polynomial $f(x) \in \mathbb{Z}[x]$ of degree n, shows how to derive the average number $\mu(n)$ of factorizations modulo distinct primes $p_1, \ldots, p_{\mu(n)}$ needed to prove f irreducible [34]. His approach is based on the fact that the degrees $d_1, \ldots, d_r$ of a factorization $f \equiv g_1 \cdots g_r$ modulo p, $d_i = \deg(g_i)$ for $1 \leq i \leq r$ and p a random prime correspond to the cycle lengths of a random permutation

$$(1, \ldots, d_1)(d_1 + 1, \ldots, d_1 + d_2) \cdots (d_1 + \cdots + d_{r-1} + 1, \ldots, d_1 + \cdots + d_r)$$

of n elements. The Swinnerton-Dyer polynomials $f_{p_1,\ldots,p_n}$ of the previous theorem obviously do not satisfy this property but it remains valid for any specific polynomial f provided that the Galois group of f is the full symmetric group. Our statement is somewhat stronger than what D. Musser proves because the latter follows from the fact that almost all polynomials have the symmetric group as Galois group [13]. Our claim is a consequence of the Chebotarev Density Theorem [35, Chap. 8.3]. This theorem also applies to the Swinnerton-Dyer polynomials, and an effective version has been used to construct succinct certificates for normal polynomials, i.e., $N = \{f \mid f \in \mathbb{Z}[x], f \text{ irreducible and normal}\} \in NP$ [19]. D. Cantor has recently shown the same by more elementary means for generally irreducible polynomials, i.e., $I = \{f \mid f \in \mathbb{Z}[x], f \text{ irreducible}\} \in NP \cap \text{co-NP}$ [7]. However, in P. Weinberger's algebraic number theoretic proof showing that the Generalized Riemann Hypothesis implies $I \in P$, the Chebotarev Density Theorem again plays an important role [55].

4. Factorization of Univariate Polynomials Over Algebraic Extensions of $\mathbb{Q}$

The decidability of factoring a polynomial $f(x) \in \mathbb{Q}(\theta)[x]$, θ an algebraic number, goes back to [23, Sec. 4, pp. 12 – 13]. The same algorithm can also be found in [44, pp. 136 – 137] which has been adopted and improved for computer usage by B. Trager [42]. However, again the Hensel lemma provides a more efficient method. We shall briefly outline the ideas involved and refer the reader to P. Weinberger's and L. Rothschild's paper [56] for the details. Without loss of generality, we may assume that θ is an algebraic integer with minimal polynomial $h(\theta) \in \mathbb{Z}[\theta]$ of degree m (see chapter on "Algebraic Domains" for terminology). We seek to factor $f(x) \in \mathbb{Q}(\theta)[x]$ of degree n which we can assume to be monic. Let

$$f(x) = x^n + 1/d \sum_{i=0}^{n-1} \left(\sum_{j=0}^{m-1} b_{ij}\theta^j \right) x^i$$

with $d, b_{ij} \in \mathbb{Z}$. It can be shown [56, Sec. 8] that if $g(x) \in \mathbb{Q}(\theta)[x]$, monic, and $g(x)$ divides $f(x)$, then

$$g(x) = x^k + 1/(dD) \sum_{i=0}^{k-1} \left(\sum_{j=0}^{m-1} c_{ij}\theta^j \right) x^i$$

with $k < n$, $c_{ij} \in \mathbb{Z}$ and D the discriminant of h, $D = \mathrm{res}(h, dh/d\theta)$. Furthermore, there is an effective bound B of length polynomial in $n \log(|f|)$ such that $|c_{ij}| \leqslant B$ for $0 \leqslant i \leqslant k-1, 0 \leqslant j \leqslant m-1$ [56, Sec. 8]. Let p be a prime number such that p does not divide d and D and that $h(\theta)$ modulo $p = \bar{h}(\theta)$ is irreducible. The last condition may not be satisfiable but we shall defer that case for later. Then the coefficients of $f(x)$ modulo $p = \bar{f}(x)$ can be viewed as elements of $\mathrm{GF}(p^m)$ generated by a root of $\bar{h}(\theta)$. We factor $\bar{f}(x) = \bar{g}_1(x) \cdots \bar{g}_r(x)$ over this finite field, i.e.,

$$\bar{g}_s(x) = x^{k_s} + \sum_{i=0}^{k_s-1} \left(\sum_{j=0}^{m-1} \bar{c}_{ij}^{(s)}\theta^j \right) x^i, \quad \bar{c}_{ij}^{(s)} \in \mathbb{Z}_p \quad (1 \leqslant s \leqslant r), \quad k_1 + \cdots + k_r = n.$$

We can now lift the factors into a larger residue domain $q = p^k \geqslant 2B$ adjoined by a root of $h(\theta)$ modulo $q = \tilde{h}(\theta)$. By multiplication of dD modulo q we obtain $f(x) \equiv \tilde{g}_1(x) \cdots \tilde{g}_r(x)$ modulo $(q, \tilde{h}(\theta))$ with

$$\tilde{g}_s(x) = x^{k_s} + 1/(dD) \sum_{i=0}^{k_s-1} \left(\sum_{j=0}^{m-1} \tilde{c}_{ij}^{(s)}\theta^j \right) x^i$$

with $\tilde{c}_{ij}^{(s)}$ balanced residues in $\mathbb{Z}_q$. It remains to test whether any trial combination of factors $\tilde{g}_s(x)$ constitutes an actual factor.

If $\bar{h}(\theta)$ factors for all primes, we then can perform the lifting modulo any factor $\bar{h}^*(\theta)$ of $\bar{h}(\theta)$. To construct factors $\tilde{g}_s(x)$ modulo $(q, \tilde{h}(\theta))$ we may either use the Chinese remainder theorem [56, Sec. 10] or the lattice algorithm by A. Lenstra [27, 28]. Both algorithms are, however, of exponential complexity in the number of modular factors.

5. Factorization of Multivariate Polynomials

We shall begin this chapter with Kronecker's algorithm which, for certain coefficient domains (such as $\mathbb{C}$), is still the only one known.

[Factorization of $f(x_1, \ldots, x_v) \in D[x_1, \ldots, x_v]$ with D being a unique factorization domain.]

(K1) [Compute degree bound:] Obtain an integer d larger than all individual variable degrees of f.

(K2) [Reduction:] $\bar{f}(y) \leftarrow S_d(f) = f(y, y^d, \ldots, y^{d^{v-1}})$.

(K3) [Factorization:] Factor $\bar{f}(y)$ into irreducibles, i.e.,
$\bar{f}(y) = \bar{g}_1(y) \cdots \bar{g}_s(y)$, $\bar{g}_i(y) \in D[y]$ for $1 \leqslant i \leqslant s$.

(K4) [Inverse reduction and trial division:] For all products $\bar{g}_{i_1}(y) \cdots \bar{g}_{i_m}(y)$ (similar to step (F5) in Section 3) perform the following test:

$$g_{i_1,\ldots,i_m}(x_1, \ldots, x_v) \leftarrow S_d^{-1}(\bar{g}_{i_1} \cdots \bar{g}_{i_m})$$

where S_d^{-1} is the inverse of S_d which is additive and

$$S_d^{-1}(\lambda y^{b_1 + db_2 + \cdots + d^{v-1}b_v}) = \lambda x_1^{b_1} \cdots x_v^{b_v}$$

with $0 \leqslant b_i < d$ for $1 \leqslant i \leqslant v$, $\lambda \in \mathbb{Z}$.

Test whether $g_{i_1,\ldots,i_m}$ divides f and if so remove irreducible factor from f and proceed with co-factor. ∎

The correctness of this algorithm follows easily from the fact that no variable in any factor of f can occur with degree d or higher. The running time of the algorithm depends on of how fast the univariate polynomial $\bar{f}(y)$ can be factored, the degree of which can be substantially large.

It should be clear that step (K4) can take time exponential in the degree of f, e.g., if $D = \mathbb{C}$ and f is irreducible. Unfortunately this exponential worst case complexity remains true for $D = \mathbb{Z}$ [19]. In this case, the Hensel lemma has produced a much more efficient approach. In the following we will take a closer look at this algorithm.

The overall structure of the multivariate factorization algorithm is remarkably close to that of the univariate algorithm of Section 3. First we choose a main variable x, i.e., the input polynomial $\bar{h} \in \mathbb{Z}[y_1, \ldots, y_v, x]$. The content computation of step (C) now becomes a GCD computation in $\mathbb{Z}[y_1, \ldots, y_v]$. The squarefree decomposition performed in step (S) can also be achieved by D. Yun's algorithm if we replace the derivatives d/dx by partial derivatives $\partial/\partial x$ and the GCDs by multivariate polynomial GCDs. However, in this case P. Wang's and B. Trager's algorithm becomes more efficient [54].

The idea of their algorithm is to find an evaluation point $(b_1, \ldots, b_v)$ such that if

$$h(y_1, \ldots, y_v, x) = \prod_{i=1}^{r} f_i(y_1, \ldots, y_v, x)^i$$

is the squarefree decomposition of h, and

$$h(b_1, \ldots, b_v, x) = \hat{h}(x) = \prod_{i=1}^{\hat{r}} \hat{f}_i(x)^i$$

is that of $\hat{h}$, then $r = \hat{r}$ and $f_i(b_1, \ldots, b_v, x) = \hat{f}_i(x)$, $1 \leqslant i \leqslant r$. Under these conditions f_r divides $g = 1/(r-1)!(\partial/\partial x)^{r-1}(h)$, $\hat{f}_r$ divides $\hat{g} = 1/(r-1)!(d/dx)^{r-1}\hat{h}(x)$ and we can lift the equation

$$g(y_1, \ldots, y_v, x) \equiv \hat{f}_r(x)(\hat{g}(x)/\hat{f}_r(x)) \operatorname{modulo}(y_1 - b_1, \ldots, y_v - b_v)$$

to determine f_r from the univariate square decomposition of $\hat{h}$, provided $\hat{g}/\hat{f}_r \neq 1$. Evaluation points for which the above conditions do not hold are, as in the modular multivariate GCD algorithm, very rare.

Step (F), the complete factorization of a squarefree polynomial $f(y_1, \ldots, y_v, x)$, is again a major challenge. As in the above squarefree decomposition algorithm we evaluate the minor variables y_i at integers b_i, $1 \leqslant i \leqslant v$ then factor the resulting univariate polynomial $f(b_1, \ldots, b_v, x)$ and finally rebuild multivariate factor candidates by a Hensel lifting algorithm with respect to the prime ideal p generated by $\{(y_1 - b_1), \ldots, (y_v - b_v)\}$. Instead of presenting a complete algorithm we shall work out a simple example and refer the reader to the papers by P. Wang [53, 49, 50, 52] and D. Musser [33] for the details.

Example. Factor

$$f(y, z, x) = x^3 + ((y + 2)z + 2y + 1)x^2$$
$$+ ((y + 2)z^2 + (y^2 + 2y + 1)z + 2y^2 + y)x$$
$$+ (y + 1)z^3 + (y + 1)z^2 + (y^3 + y^2)z + y^3 + y^2.$$

The polynomial is monic and squarefree.

Step F1. Choose an evaluation point which preserves degree and squarefreeness but contains as many zero components as possible.

$$y = 0, \quad z = 0: \quad f(0, 0, x) = x^3 + x^2 \quad \text{is not squarefree,}$$
$$y = 1, \quad z = 0: \quad f(1, 0, x) = x^3 + 3x^2 + 3x + 2 \quad \text{is squarefree.}$$

Translate variables for nonzero components

$$f(w + 1, z, x) = x^3 + 3x^2 + 3x + 2 + (2x^2 + 5x + 5)w$$
$$+ (2x + 4)w^2 + w^3 + ((3x^2 + 4x + 2) + (x^2 + 4x + 5)w$$
$$+ (x + 4)w^2 + w^3)z + ((3x + 2) + (x + 1)w)z^2 + (2 + w)z^3.$$

By $f_{ij}(x)$ we denote the coefficient of $w^j z^i$.

Step F2. Factor $f_{00}(x) = g_{00}(x)h_{00}(x)$ in $\mathbb{Z}[x]$. We get

$$x^3 + 3x^2 + 3x + 2 = (x + 2)(x^2 + x + 1).$$

Step F3. Compute highest degrees of w and z in factors of

$$f(w + 1, z, x) = g(w, z, x)h(w, z, x): \quad \deg_w(g, h) \leqslant 3, \quad \deg_z(g, h) \leqslant 2.$$

Step F4. Lift g_{00} and h_{00} to highest degrees in w and z. We set

$$g(w, z, x) = g_{00}(x) + g_{01}(x)w + g_{02}(x)w^2$$
$$+ \cdots + (g_{10}(x) + g_{11}(x)w + \cdots)z + \cdots$$

and

$$h(w, z, x) = h_{00}(x) + h_{01}(x)w + h_{02}(x)w^2$$
$$+ \cdots + (h_{10}(x) + h_{11}(x)w + \cdots)z$$
$$+ (h_{20}(x) + h_{21}(x)w + \cdots)z^2 + \cdots$$

and compute $g_{01}, h_{01}, g_{02}, h_{02}, \ldots, g_{10}, h_{10}, g_{11}, h_{11}, \ldots, g_{20}, h_{20}, \ldots$ in that sequence. Since f is monic $\deg(g_{ij}) \leqslant 1$ and $\deg(h_{ij}) \leqslant 2$ for $i + j \geqslant 1$. Multiplying g times h with undetermined g_{ij}, h_{ij} we get $g_{00}h_{01} + h_{00}g_{01} = f_{01}$ whose unique solution is

$$(x + 2)(x + 2) + (x^2 + x + 1) \cdot 1 = 2x^2 + 5x + 5$$

by the extended Euclidean algorithm. In the next step we get

$$g_{00}h_{02} + h_{00}g_{02} = f_{02} - g_{01}h_{01}$$

which is solved by

$$(x + 2) \cdot 1 + (x^2 + x + 1) \cdot 0 = 2x + 4 - 1 \cdot (x + 2).$$

Finally

$$g_{00}h_{03} + h_{00}g_{03} = f_{03} - g_{01}h_{02} - g_{02}h_{01},$$

or

$$(x + 2) \cdot 0 + (x^2 + x + 1) \cdot 0 = 1 - 1 \cdot 1 - 0 \cdot (x + 2).$$

This gives factor candidates for

$$f(w + 1, 0, x) = ((x + 2) + 1 \cdot w + 0 \cdot w^2)((x^2 + x + 1) + (x + 2)w + w^2)$$

and a trial division shows them to be true factors.

We now lift z:

$$g_{00}h_{10} + h_{00}g_{10} = f_{10},$$

or

$$(x + 2)x + (x^2 + x + 1) \cdot 2 = 3x^2 + 4x + 2;$$

$$g_{00}h_{11} + h_{00}g_{11} = f_{11} - g_{01}h_{10} - g_{10}h_{01},$$

or

$$(x + 2) \cdot 0 + (x^2 + x + 1) \cdot 1 = x^2 + 4x + 5 - 1 \cdot x - 2(x + 2);$$

$$g_{00}h_{20} + h_{00}g_{20} = f_{20} - g_{10}h_{10},$$

or

$$(x + 2) \cdot 1 + (x^2 + x + 1) \cdot 0 = 3x + 2 - 2x.$$

All other equations have 0 as their right-hand sides.

The factor candidates are

$$f(w + 1, z, x) = ((x + 2) + w + (2 + w)z)((x^2 + x + 1)$$
$$+ (x + 2)w + w^2 + xz + z^2)$$

which are the actual factors. Setting $w = y - 1$ we obtain

$$f(y, z, x) = (x + yz + y + z + 1)(x^2 + (y + z)x + y^2 + z^2).$$

In factoring the above sample polynomial we followed the algorithm by P. Wang [50]. Our construction is actually a linear lifting technique. There is also the possibility of quadratic lifting [33], but in the multivariate case, the linear algorithm seems to be more efficient [57]. If more than two univariate factors are present, one can again lift each one iteratively or lift them in parallel as we demonstrated for the univariate case.

Various complications have been identified with the multivariate Hensel algorithm.

a) *The leading coefficient problem*: In our example we dealt with a monic polynomial in which case the leading coefficients of all factors are known. If a polynomial leading coefficient is present, one can impose it on one factor as in the univariate case, but this leads most likely to dense factor candidates. P. Wang describes an algorithm to predetermine the actual leading coefficients of the factors, which avoids this intermediate expression growth [50, Sec. 3].

b) *The "bad zero" problem*: In our example, y had to be evaluated at 1 in order to preserve squarefreeness. The change of variables $y_i = w_i + b_i$ for $b_i \neq 0$ can make the polynomial $f(w_1 + b_1, \ldots, w_v + b_v, x)$ dense. P. Wang suggests to compute the coefficients $f_{i_1 \cdots i_v}(x)$ of $w_1^{i_1} \cdots w_v^{i_v}$ by Taylor's formula without performing the substitution

$$f_{i_1 \cdots i_v}(x) = \frac{1}{i_1! \cdots i_v!} \left(\frac{\partial}{\partial y_1} \right)^{i_1} \cdots \left(\frac{\partial}{\partial y_v} \right)^{i_v} f(y_1, \ldots, y_v, x) \bigg|_{y_i = -b_i}.$$

See also R. Zippel's work on preserving sparseness [64, 65].

c) *The "extraneous factors" problem*: This problem is the same as in the univariate case, namely that $f(b_1, \ldots, b_v, x)$ has more factors than $f(y_1, \ldots, y_v, x)$ (in which case we call $b_1, \ldots, b_v$ "unlucky"). One immediate consequence may be that the correction coefficients $g_{i_1 \cdots i_v}(x)$, $h_{i_1 \cdots i_v}(x)$ have non-integral coefficients. In order to avoid working with denominators one can choose to work with coefficients modulo a prime which preserves the squarefreeness of $f(b_1, \ldots, b_v, x)$, and as a first step lift the coefficients. A good factor coefficient bound can be found in [14, pp. 135–139]. G. Viry has employed the initial transformation

$$\tilde{f}(y_1, \ldots, y_v, x) = f(y_1 + b_1 x, \ldots, y_v + b_v x, x)$$

with $\tilde{f}(0, \ldots, 0, x)$ squarefree. Then $\tilde{f}(y_1, \ldots, y_v, x)$ is "normalized", meaning that if

$$\tilde{f}(y_1, \ldots, y_v, x) = x^n + a_1(y_1, \ldots, y_v)x^{n-1} + \cdots + a_n(y_1, \ldots, y_v)$$

with $a_i \in \mathbb{Q}[y_1, \ldots, y_n]$ then the total degrees of the coefficients satisfy $\deg_{y_1, \ldots, y_v}(a_i) \leq i$ for $1 \leq i \leq n$. Using a polyhedron representation of polynomials introduced by A. Ostrowski [37], G. Viry then shows that a factor candidate with integral coefficients derived from the lifted factorization of $\tilde{f}(0, \ldots, 0, x)$ divides $\tilde{f}(y_1, \ldots, y_v, x)$ if and only if it is normalized. Thus the trial division can be replaced by a check for being normalized [46, 47].

Various implementation issues can be found in [30]. A good set of polynomials for benchmarking an actual implementation of the factorization algorithm can be found in [9].

Little is known about the average computing time of the multivariate Hensel algorithm. The worst case complexity can be exponential in the degree of the main variable depending on what evaluation points one chooses. However, unlike in the univariate case, it cannot happen that an irreducible polynomial factors for all possible evaluations. Actually, quite the opposite is true due to the Hilbert Irreducibility Theorem.

Theorem. *Let $f(y_1, \ldots, y_v, x_1, \ldots, x_t)$ be irreducible in $\mathbb{Z}[y_1, \ldots, y_v, x_1, \ldots, x_t]$. By $U(N)$ we denote the number of v-tuples $(b_1, \ldots, b_v) \in \mathbb{Z}^v$ such that $|b_i| \leqslant N$ for $1 \leqslant i \leqslant v$ and $f(b_1, \ldots, b_v, x_1, \ldots, x_t)$ is reducible in $\mathbb{Z}[x_1, \ldots, x_t]$. Then there exist constants α and C (depending on f) such that $U(N) \leqslant C(2N + 1)^{v - \alpha}$ and $0 < \alpha < 1$ (cf. [25, Chap. 8]).*

Unfortunately, no polynomial upper bounds on the length of C seem to be known which would make the theorem useful for "realistic" evaluations. In practice lucky evaluations seem quite frequent.

However, the situation for the theoretical study of the multivariate factorization does not appear completely hopeless. It can be shown, for instance, that if

$$f(x_1, y_2, \ldots, y_v, x_2) \in \mathbb{Z}[x_1, y_2, \ldots, y_v, x_2],$$

v arbitrary but fixed, is irreducible then one can compute integers $b_1, \ldots, b_v, c_2, \ldots, c_v$ in time polynomial in $\deg(f)\log(|f|)$ such that

$$f(x_1 - b_1, c_2(x_1 - b_2), \ldots, c_v(x_1 - b_v), x_2) \in \mathbb{Z}[x_1, x_2]$$

is also irreducible. This theorem can be extended to show that factoring multivariate polynomials with a fixed number of variables is only polynomially harder than factoring bivariate polynomials [18, 19].

We will not discuss special algorithms for coefficient domains other than the integers here. Factoring polynomials in $GF(q)[y, x]$ is very similar to factoring univariate polynomials over the integers. More general algorithms can be found in [11]. A multivariate Hensel algorithm for factoring polynomials in $\mathbb{Q}(\theta)[x_1, \ldots, x_v]$ can be found in [48]. A special problem is to test a polynomial $f(x_1, \ldots, x_v) \in \mathbb{Z}[x_1, \ldots, x_v]$ for absolute irreducibility, that is, to test $f(x_1, \ldots, x_v)$ for irreducibility in $\mathbb{C}[x_1, \ldots, x_v]$. The first criterion probably goes back to E. Noether [36] which also implies that if $f(x_1, \ldots, x_r)$ is absolutely irreducible, then $f(x_1, \ldots, x_r)$ remains irreducible modulo almost all prime numbers. Unfortunately, the first such prime number may be very large. A more efficient test for absolute irreducibility can be found in [15].

6. Conclusion

We have tried to capture the current situation of the problem of factoring polynomials. We believe that various algorithms presented here will be significantly improved in the future but we also believe that some of these ideas will persist through new developments. Following is a list of open problems which the author believes should receive attention.

1. Probabilistic univariate irreducibility test: Does there exist a probabilistic algorithm which tests $f(x) \in \mathbb{Z}[x]$ for irreducibility in expected time polynomial in $\deg(f)\log(|f|)$? Can the algorithm also find factors?

2. Deterministic univariate factorization: Does there exist an algorithm which factors $f(x) \in \mathbb{Z}[x]$ in time polynomial in $\deg(f)\log(|f|)$? (Cf. [62].)

3. Polynomial reduction from bivariate to univariate factorization: Assuming that problem 2 has a positive answer, does there exist an algorithm which factors $f(y, x) \in \mathbb{Z}[y, x]$ in time polynomial in $\deg(f) \log(|f|)$? (Cf. [19].)

4. Bivariate factorization over finite fields: Can one factor $f(y, x) \in \mathbb{Z}_p[y, x]$ in time polynomial in $p \deg(f)$? Can one test $f(y, x)$ for irreducibility in time polynomial in $\log(p) \deg(f)$?

5. Factorization of normal univariate polynomials: Given $f(x) \in \mathbb{Z}[x]$ irreducible and normal. Can one factor f over its own splitting field in time polynomial in $\deg(f) \log(|f|)$? Notice that a solution of problem 2 provides one for this problem.

6. Analysis of multivariate Hensel algorithm: Provide an effective version of the Hilbert irreducibility theorem. What is the average number of evaluation points needed to prove irreducibility or achieve a "lucky" evaluation?

Note added in proof: A. Lenstra, H. Lenstra and L. Lovász have recently solved the open problem 2. Their algorithm takes $O(\deg(f)^{12} + \deg(f)^9 \log(|f|_2)^3)$ steps. The author has recently solved the open problem 3 using ideas from [62].

Acknowledgement

The author wishes to acknowledge the support he received from Prof. B. Caviness and Prof. P. Wang while writing this paper.

References

[1] Abdali, S. K., Caviness, B. F., Pridor, A.: Modular Polynomial Arithmetic in Partial Fraction Decomposition. MACSYMA **1977**, 253 – 261.
[2] Adleman, L. M.: On Distinguishing Prime Numbers from Composite Numbers. Proc. 21st Symp. Foundations Comp. Sci. IEEE **1980**, 387 – 406.
[3] Adleman, L. M., Odlyzko, A. M.: Irreducibility Testing and Factorization of Polynomials. Proc. 22nd Symp. Foundations Comp. Sci. IEEE **1981**, 409 – 418.
[4] Ben-Or, M.: Probabilistic Algorithms in Finite Fields. Proc. 22nd Symp. Foundations Comp. Sci. IEEE **1981**, 394 – 398.
[5] Berlekamp, E. R.: Factoring Polynomials over Large Finite Fields. Math. Comp. **24**, 713 – 735 (1970).
[6] Calmet, J., Loos, R.: Deterministic Versus Probabilistic Factorization of Integral Polynomials. EUROCAM **1982** (to appear).
[7] Cantor, D. G.: Irreducible Polynomials with Integral Coefficients Have Succinct Certificates. J. of Algorithms **2**, 385 – 392 (1981).
[8] Cantor, D. G., Zassenhaus, H.: On Algorithms for Factoring Polynomials over Finite Fields. Math. Comp. **36**, 587 – 592 (1981).
[9] Claybrook, B. G.: Factorization of Multivariate Polynomials over the Integers. ACM SIGSAM Bulletin **10**, 13 (Feb. 1976).
[10] Collins, G. E.: Factoring Univariate Polynomials in Polynomial Average Time. EUROSAM **1979**, 317 – 329.
[11] Davenport, J. H., Trager, B. M.: Factorization over Finitely Generated Fields. SYMSAC **1981**, 200 – 205.
[12] Eisenstein, F. G.: Über die Irreductibilität und einige andere Eigenschaften der Gleichung, von welcher die Theilung der ganzen Lemniscate abhängt. J. f. d. reine u. angew. Math. **39**, 160 – 179 (1850).
[13] Gallagher, P. X.: Probabilistic Galois Theory. AMS Proc. Symp. in Pure Math., Analytic Number Theory **1972**, 91 – 102.
[14] Gelfond, A. O.: Transcendental and Algebraic Numbers. New York: Dover 1960.

[15] Heintz, J., Sieveking, M.: Absolute Primality of Polynomials is Decidable in Random Polynomial Time in the Number of Variables. Springer Lecture Notes Comp. Sci., Vol. 115, pp. 16 – 28. Berlin-Heidelberg-New York: Springer 1981.

[16] Horowitz, E.: Algorithms for Partial Fraction Decomposition and Rational Function Integration. SYMSAM **1971**, 441 – 457.

[17] Johnson, S. C.: Tricks for Improving Kronecker's Method. Bell Laboratories Report 1966.

[18] Kaltofen, E.: A Polynomial Reduction from Multivariate to Bivariate Integer Polynomial Factorization. ACM Proc. Symp. on Theory Comp. **1982**, 261 – 266.

[19] Kaltofen, E.: On the Complexity of Factoring Polynomials with Integer Coefficients. Ph.D. Thesis, RPI (1982), in preparation.

[20] Kaltofen, E., Musser, D. R., Saunders, B. D.: A Generalized Class of Polynomials that Are Hard to Factor. SYMSAC **1981**.

[21] Knuth, D. E.: The Art of Computer Programming, Vol. 2. Seminumerical Algorithms. Reading, Mass.: Addison-Wesley 1969.

[22] Knuth, D. E.: The Art of Computer Programming, Vol. 2. Seminumerical Algorithms. 2nd ed. Reading, Mass.: Addison-Wesley 1981.

[23] Kronecker, L.: Grundzüge einer arithmetischen Theorie der algebraischen Größen. J. f. d. reine u. angew. Math. **92**, 1 – 122 (1882).

[24] Kung, H. T., Tong, D. M.: Fast Algorithms for Partial Fraction Decomposition. SIAM J. Comp. **6**, 582 – 593 (1977).

[25] Lang, S.: Diophantine Geometry. New York: Interscience 1962.

[26] Lazard, D.: On Polynomial Factorization. EUROCAM **1982**, to appear.

[27] Lenstra, A. K.: Lattices and Factorization of Polynomials. ACM SIGSAM Bulletin **15**, 15 – 16 (August 1981).

[28] Lenstra, A. K.: Lattices and Factorization of Polynomials. EUROCAM **1982**, to appear.

[29] Moenck, R. T.: On the Efficiency of Algorithms for Polynomial Factoring. Math. Comp. **31**, 235 – 250 (1977).

[30] Moore, P. M. A., Norman, A. C.: Implementing a Polynomial Factorization Problem. SYMSAC **1981**, 109 – 116.

[31] Moses, J., Yun, D. Y. Y.: The EZGCD Algorithm. Proc. 1973 ACM National Conf., 159 – 166.

[32] Musser, D. R.: Algorithms for Polynomial Factorization. Ph.D. Thesis and TR # 134, Univ. of Wisconsin 1971.

[33] Musser, D. R.: Multivariate Polynomial Factorization. J. ACM **22**, 291 – 308 (1976).

[34] Musser, D. R.: On the Efficiency of a Polynomial Irreducibility Test. J. ACM **25**, 271 – 282 (1978).

[35] Narkiewicz, W.: Elementary and Analytic Theory of Algebraic Numbers. Warsaw: Polish Science Publ. 1974.

[36] Noether, E.: Ein algebraisches Kriterium für absolute Irreduzibilität. Math. Ann. **85**, 26 – 33 (1922).

[37] Ostrowski, A. M.: On Multiplication and Factorization of Polynomials I. Aequationes Math. **13**, 201 – 228 (1975).

[38] Plaisted, D. A.: Some Polynomial and Integer Divisibility Problems are NP-Hard. SIAM J. Comp. **7**, 458 – 464 (1978).

[39] Plaisted, D. A.: The Application of Multivariate Polynomials to Inference Rules and Partial Tests for Unsatisfiability. SIAM J. Comp. **9**, 698 – 705 (1980).

[40] Pratt, V. R.: Every Prime Has a Succinct Certificate. SIAM J. Comp. **4**, 214 – 220 (1975).

[41] Rabin, M. O.: Probabilistic Algorithms in Finite Fields. SIAM J. Comp. **9**, 273 – 280 (1980).

[42] Trager, B. M.: Algebraic Factoring and Rational Function Integration. SYMSAC **1976**, 219 – 226.

[43] Trotter, H. F.: Algebraic Numbers and Polynomial Factorization. AMS Short Course Series, Ann Arbor 1980.

[44] van der Waerden, B. L.: Modern Algebra, Vol. 1. Engl. transl. by F. Blum. New York: Frederick Ungar Publ. Co. 1953.

[45] Vaughan, R. C.: Bounds for the Coefficients of Cyclotomic Polynomials. Michigan Math. J. **21**, 289 – 295 (1975).

[46] Viry, G.: Factorisation des Polynômes à Plusieurs Variables à Coefficient Entiers. RAIRO Informatique Théorique **12**, 305 – 318 (1978).

[47] Viry, G.: Factorisation des Polynômes à Plusieurs Variables. RAIRO Informatique Théorique **14**, 209−223 (1980).
[48] Wang, P. S.: Factoring Multivariate Polynomials over Algebraic Number Fields. Math. Comp. **30**, 324−336 (1976).
[49] Wang, P. S.: Preserving Sparseness in Multivariate Polynomial Factorization. MACSYMA **1977**, 55−61.
[50] Wang, P. S.: An Improved Multivariate Polynomial Factoring Algorithm. Math. Comp. **32**, 1215−1231 (1978).
[51] Wang, P. S.: Parallel *p*-adic Constructions in the Univariate Polynomial Factoring Algorithm. MACSYMA **1979**, 310−318.
[52] Wang, P. S.: Analysis of the *p*-adic Construction of Multivariate Correction Coefficients in Polynomial Factorization: Iteration vs. Recursion. EUROSAM **1979**, 291−300.
[53] Wang, P. S., Rothschild, L. P.: Factoring Multivariate Polynomials over the Integers. Math. Comp. **29**, 935−950 (1975).
[54] Wang, P. S., Trager, B. M.: New Algorithms for Polynomial Square-Free Decomposition over the Integers. SIAM J. Comp. **8**, 300−305 (1979).
[55] Weinberger, P. J.: Finding the Number of Factors of a Polynomial. 1981, submitted.
[56] Weinberger, P. J., Rothschild, L. P.: Factoring Polynomials over Algebraic Number Fields. ACM Trans. Math. Software **2**, 335−350 (1976).
[57] Yun, D. Y. Y.: Hensel Meets Newton − Algebraic Construction in an Analytic Setting. In: Analytic Computational Complexity (Traub, J., ed.). New York: Academic Press 1976.
[58] Yun, D. Y. Y.: On Squarefree Decomposition Algorithms. SYMSAC **1976**, 26−35.
[59] Yun, D. Y. Y.: On the Equivalence of Polynomial GCD and Squarefree Factorization Problems. MACSYMA **1977**, 65−70.
[60] Zassenhaus, H.: On Hensel Factorization I. J. Number Theory **1**, 291−311 (1969).
[61] Zassenhaus, H.: A Remark on the Hensel Factorization Method. Math. Comp. **32**, 287−292 (1978).
[62] Zassenhaus, H.: Polynomial Time Factoring of Integral Polynomials. ACM SIGSAM Bulletin **15**, 6−7 (1981).
[63] Zippel, R. E.: Probabilistic Algorithms for Sparse Polynomials. EUROSAM **1979**, 216−226.
[64] Zippel, R. E.: Probabilistic Algorithms for Sparse Polynomials. Ph.D. Thesis, M.I.T., 1979.
[65] Zippel, R. E.: Newton's Iteration and the Sparse Hensel Algorithm. SYMSAC **1981**, 68−72.

E. Kaltofen, M. S.
Department of Computer
and Information Sciences
University of Delaware
Newark, DE 19711, U.S.A.

Generalized Polynomial Remainder Sequences

R. Loos, Karlsruhe

Abstract

Given two polynomials over an integral domain, the problem is to compute their polynomial remainder sequence (p.r.s.) over the same domain. Following Habicht, we show how certain powers of leading coefficients enter systematically all following remainders. The key tool is the subresultant chain of two polynomials. We study the primitive, the reduced and the improved subresultant p.r.s. algorithm of Brown and Collins as basis for computing polynomial greatest common divisors, resultants or Sturm sequences. Habicht's subresultant theorem allows new and simple proofs of many results and algorithms found in different ways in computer algebra.

1. The Problem

1.1 Polynomial Division and Determinants

The problem we want to study in this chapter is the efficient computation of polynomial sequences based on polynomial division. Let F be a field and $A, B \in F[x]$ be polynomials of degree m and n respectively with $m \geqslant n$. There exist unique polynomials Q and R such that $A = BQ + R$ with $R = 0$ or $\deg R < \deg B$. Q and R can be computed by the following algorithm

$$QR(A, B; Q, R).$$

(1) Set $A^{(0)} = A$.

(2) For $k = 1, \ldots, m - n + 1$, set
   (2.1) $q^{(k)} = \operatorname{lc}(A^{(k-1)})/\operatorname{lc}(B)$,
   (2.2) $A^{(k)} = A^{(k-1)} - x^{m-n+1-k}q^{(k)}B.$ ■

On termination $Q = \sum_{i=0}^{m-n} q^{(i+1)}x^i$ and $R = A^{(m-n+1)}$ are *quotient* $\operatorname{quot}(A, B)$ and *remainder* $\operatorname{rem}(A, B)$ of A and B.

If we consider $a_m, a_{m-1}, \ldots, a_0$ as the first row and the coefficients of $x^{m-n}B$, $x^{m-n-1}B, \ldots, B$ as the second, third, $\ldots$, and $(m - n + 2)$th row of an $m - n + 2 \times m + 1$ matrix

$$M = \left. \begin{pmatrix} a_m & a_{m-1} & \cdots & & & a_0 & \\ b_n & b_{n-1} & \cdots & b_0 & & & \\ & b_n & \cdots & & b_0 & & \\ & & \ddots & & & \ddots & \\ & & & b_n & \cdots & & b_0 \end{pmatrix} \right\} \begin{array}{l} \\ \\ m - n + 1 \text{ rows} \\ \\ \end{array}$$

we observe that the first row is iteratively transformed by elementary row operations in step (2.2) into the coefficients of the remainder. If we interchange the first row with all others we arrive at an upper triangular matrix M^-, which shows that the division and remainder algorithm is a special instance of the Gaussian elimination algorithm:

$$M^- = \left.\begin{pmatrix} b_n & \cdots & & b_0 & & \\ & \ddots & & & \ddots & \\ & & b_n & & \cdots & b_0 \\ & & a_{n-1}^{(m-n+1)} & \cdots & a_0^{(m-n+1)} \end{pmatrix}\right\} m-n+1 \text{ rows.}$$

This observation motivates the following two definitions.

Definition. Let $A_i = \sum_{j=0}^{n_i} a_{ij} x^j$, $1 \leqslant i \leqslant k$, be a sequence of polynomials over an integral domain I. Then the $k \times l$ *matrix associated* with $A_1, \ldots, A_k$ is $\text{mat}(A_1, \ldots, A_k) = (a_{i,l-j})$, where $l = 1 + \max_{1 \leqslant i \leqslant k}(n_i)$. We have

$$M = \text{mat}(A, x^{m-n}B, \ldots, B) \quad \text{and} \quad M^- = \text{mat}(x^{m-n}B, \ldots, B, \text{rem}(A, B)).$$

The next definition connects a polynomial with any matrix.

Definition. Let M be a $k \times l$ matrix over I, $k \leqslant l$. The *determinant polynomial* $A = \text{detpol}(M)$ is

$$|M^{(k)}| x^{l-k} + \cdots + |M^{(l)}|,$$

where $M^{(j)}$ denotes the submatrix of M consisting of the first $k-1$ columns and the jth column of M, $k \leqslant j \leqslant l$. Note that $\deg A \leqslant l-k$ or $A = 0$. With this notation we have $\text{detpol}(M) = (-1)^{m-n+1} \text{detpol}(M^-)$ and since M^- is upper triangular we get

$$\text{detpol}(M) = (-b_n)^{m-n+1} \text{rem}(A, B). \tag{1}$$

The transformation by the division process on M can be described as follows

$$M^{(0)} = M,$$

$$M_{i,j}^{(k)} = \begin{vmatrix} 1 & M_{k,j}^{(k-1)}/M_{k,k}^{(k-1)} \\ M_{i,k}^{(k-1)} & M_{i,j}^{(k-1)} \end{vmatrix}, \quad 1 \leqslant k \leqslant m-n+1,$$

where $M_{k,k}^{(k-1)} = b_n$ for all k. Since n is the degree of B, no pivoting is required. The problem with both division and Gaussian elimination is that there are important cases as the ring of integers and of polynomials over the integers where division in the coefficient domain cannot be performed. For Gaussian elimination Bareiss [0] discovered a method to have always exact divisions. He sets

$$M_{11}^{(-1)} = 1,$$

$$M^{(0)} = M,$$

$$M_{i,j}^{(k)} = \begin{vmatrix} M_{k,k}^{(k-1)} & M_{k,j}^{(k-1)} \\ M_{i,k}^{(k-1)} & M_{i,j}^{(k-1)} \end{vmatrix} \Big/ M_{k-1,k-1}^{(k-2)}. \tag{2}$$

If M is a $p \times q$ matrix the transformations take place for $k < i < j \leqslant q$, in the lower triangle we get 0 and the rows with $i \leqslant k$ are not changed. Also one needs pivoting to ensure $M_{k,k}^{(k-1)} \neq 0$ for all k. Then the division in (2) is always exact as can be proved by induction on k using (2) a second time and commutativity. Taking $p = q$ we have:

$$|M| = |M^{(1)}|/M_{11}^{p-1} = |M^{(2)}|M_{11}^{p-2}/M_{22}^{(1)p-2}M_{11}^{p-1}$$

$$= |M^{(2)}|/M_{11}M_{22}^{(1)p-2}$$

$$\vdots$$

$$= |M^{(p-1)}|/M_{11}M_{22}^{(1)} \cdots M_{p-1,p-1}^{(p-2)}$$

$$= M_{p,p}^{(p-1)},$$

since $M^{(p-1)}$ is upper triangular. In this way determinants and solutions to linear systems can be computed by the ring operations and exact division only. It is obvious from (2) that removing the exact division would result in an exponential growth of the $M_{i,j}^{(k)}$ considered as integers or polynomials.

Let us apply Bareiss' method to the division problem. In abuse of notation we write detpol$(A_1, \ldots, A_k)$ instead of detpol(M), if $M = \text{mat}(A_1, \ldots, A_k)$. Then we have to compute

$$\text{detpol}(x^{m-n}B, \ldots, B, A).$$

In the kth step the kth row is multiplied by b_n^{k-1}, $1 < k \leqslant m - n + 1$ and the last row becomes

$$\bar{A}^{(k)} = b_n \bar{A}^{(k-1)} - x^{m-n+1-k} \text{lc}(\bar{A}^{(k-1)})B \tag{3}$$

after having taken out the exact divisor b_n^{k-1}. This means because of the near triangular form we can save the transformations of the B coefficients totally and replace (2.1) and (2.2) in algorithm QR by (3). We call the resulting algorithm PQR and $\bar{A}^{(m-n+1)} = \text{prem}(A, B)$ the *pseudo-remainder* and $Q = \sum_{i=0}^{m-n} \text{lc}(\bar{A}^{(i)})x^i = \text{pquot}(A, B)$ the *pseudo-quotient* of A and B. The pseudo-remainder algorithm is therefore a specialization of Bareiss' algorithm with no division at all! We have

$$\text{detpol}(x^{m-n}B, \ldots, B, A) = \text{prem}(A, B) = \text{rem}(b_n^{m-n+1}A, B) \tag{4}$$

but this should not be taken as a computation rule. Since in the division process successive $\text{lc}(A^{(i)})$ are annihilated by Gaussian elimination there is a further saving of computation possible if $\text{lc}(A^{(i)})$ is already 0. This will always happen if for any i both coefficients at the ith position (starting to count with the leading coefficient) of A and B are both 0, but the determinant in equation (2) may also vanish if this is not the case. We change (3) to

$$\text{if } \text{lc}(A^{k-1}) = 0 \quad \text{then set} \quad A^{(k)} = A^{(k-1)},$$

$$\text{else perform (3).}$$

We call the resulting algorithm SPQR and $A^{(m-n+1)} = \text{sprem}(A, B)$ the *sparse pseudo-remainder* of A and B. We have

$$b_n^e \text{sprem}(A, B) = \text{prem}(A, B), \quad \text{for some} \quad e \geqslant 0,$$

where e counts the number of times the sparsity condition is true.

Let F be the quotient field of the integral domain I. Then for all pairs of remainders R_1, R_2 we have defined so far there are non-zero elements r_1, $r_2 \in I$ such that $r_1 R_1 = r_2 R_2$. In general let A, $B \in I[x]$ (or A, $B \in F[x]$). A and B are called *similar*, $A \sim B$, iff there exist $a, b \in I$, $ab \neq 0$, such that $aA = bB$. $C \sim 0$ iff $C = 0$ and $\sim$ is an equivalence relation. As a special case, A and B are *associates*, iff they are similar and the coefficients of similarity are units (divisors of the identity).

We now turn to the problem central to this chapter. We wish to iterate the polynomial remainder process.

Definition. A sequence of polynomials $A_1, \ldots, A_r$ is a *polynomial remainder sequence* (p.r.s.) of $A_1 \neq 0$ and $A_2 \neq 0$ over I in case $r \geq 2$, $A_i \in I[x]$ and

$$A_{i+2} \sim \text{prem}(A_i, A_{i+1}) \neq 0, \qquad 1 \leq i \leq r - 2, \tag{5}$$

and

$$\text{prem}(A_{r-1}, A_r) = 0. \tag{6}$$

The key tool in studying p.r.s.'s are subresultant chains in which p.r.s.'s can be embedded up to similarity. The matrix M is suited to perform the division process once; for repeated divisions it is natural to treat the coefficients of A and B more equally. This motivates the following definition.

Definition. Let A, $B \in I[x]$ with $\deg A = m > 0$ and $\deg B = n > 0$. For k, $0 \leq k < \min(m, n)$, set

$$M_k = \text{mat}(x^{n-k-1}A(x), \ldots, A(x), x^{m-k-1}B(x), \ldots, B(x)).$$

Then $S_k = \text{sres}_k(A, B) = \text{detpol}(M_k)$ is the kth *subresultant* of A and B.

Since M_k has $m + n - 2k$ rows and $m + n - k$ columns,

$$\deg S_k \leq (m + n - k) - (m + n - 2k) = k.$$

S_0 is a polynomial of degree 0 which is the resultant of A and B (see the chapter on Algebraic Extensions, Section 3).

In the next section we will give some applications of p.r.s. In Section 3 we will investigate the structure of the subresultant chain A, B, $S_{n-1}, \ldots, S_0$. Since $A_{i+2} \sim \text{prem}(A_i, A_{i+1})$ there exist $e_i, f_i \in I$, both non-zero, and $Q_i(x) \in I[x]$ such that

$$e_i A_i = Q_i A_{i+1} + f_i A_{i+2}. \tag{7}$$

(7) in turn implies $A_{i+2} \sim \text{prem}(A_i, A_{i+1})$. We will establish for any p.r.s. with $n_i = \deg A_i$

$$A_i \sim S_{n_{i-1}-1} \sim S_{n_i}, \qquad 3 \leq i \leq r \tag{8}$$

and

$$S_k = 0, \qquad n_i < k < n_{i-1} - 1, \qquad 3 \leq i \leq r. \tag{9}$$

Moreover, in the fundamental theorem of p.r.s.'s the coefficients of similarity in (8) will be given explicitly in terms of the e_i and f_i of (7). In Section 4 several algorithms based on these results will be given, including several important polynomial gcd algorithms. In Section 5 some computing times are presented and we close in Section 6 with a short history.

2. Applications of Polynomial Remainder Sequences

The most important applications are polynomial gcd-algorithms which are used in rational function arithmetic to keep the "quolynomials" simplified. If only a test for relative primeness is required the modular algorithm usually needs only a few primes to find this out. The non-modular gcd-algorithms produce for relative prime inputs sequences of maximal length. GCD applications occur in squarefree factorization [24, 35] and in shiftfree factorizations [21]. The half extended GCD-algorithm is used to compute inverses in $Q(\alpha)$ where the field polynomial is irreducible. The extended polynomial GCD algorithm is used to find polynomial solutions $\bar{u}$, $\bar{v}$ to polynomial equations $\bar{u}A + \bar{v}B = H$, which occur in the Hensel construction for example (see the Chapter on Homomorphic Images). An important polynomial GCD over $Q(\alpha)$ occurs in the primitive element algorithm (see the Chapter on Algebraic Extensions) and in van der Waerden's factorization algorithm over $Q(\alpha)[x]$, [32, 31].

Fast polynomial GCD algorithms were obtained by Moenck [22] carrying over a method from integer gcd's due to Lehmer [22], Knuth [15] and Schönhage [26]. Applications are made in Strassen [29] to continued fraction expansions and are possible to Padé approximations [17].

Negative polynomial p.r.s.'s are generalized Sturm sequences which can be used for $B = dA/dx$ to count the number of real roots of A in any real interval. They occur also in Tarski's quantifier elimination algorithm for real closed fields [30]. Habicht's work on p.r.s. was motivated by elimination theory [12].

The subresultant chain is the key to analyze coefficients and computing times for any p.r.s. algorithm, it is hard to see how this can be otherwise done than by bounds on the subdeterminants of Sylvester's matrix. Even the number of unlucky primes, and therefore correctness and termination, in the modular algorithms depend on the subresultant properties.

Polynomial p.r.s. algorithms are also a key method to calculate resultants, which have – for example as discriminants – too many applications in algebraic number theory to enumerate them all (see the Chapter on Algebraic Extensions). In fact any calculation on symmetric polynomials can be done by resultant calculations [19].

It is not surprising that every advanced computer algebra system offers a variety of efficient p.r.s. algorithms. The computational needs have resulted in algebraic discoveries and rediscoveries on a century old subject.

3. The Fundamental Theorem of p.r.s.'s

We have already seen when we introduced the concept of pseudoremainder in Section 1 that the sparsity of the input polynomials could introduce systematically factors from the coefficient domain I into the output. In this Section we study first the effect of sparse subresultants on the structure of the subresultant chain. Secondly we trace the effect of similarity coefficients e_i and f_i in the generalized remainder (1.7) on generalized remainder sequences. Both results are stated in the fundamental theorem on polynomial remainder sequences. The approach we take for proving the theorem is new in the computer science literature, but it is partly contained in the algebraic literature (see Section 6).

We use Kronecker's method of indeterminate coefficients, in our case for the input polynomials $A = \sum_{i=0}^{n+1} a_i x^i$ and $B = \sum_{i=0}^{n} b_i x^i$ as elements in

$$Z[x; a_{n+1}, \ldots, a_0, b_n, \ldots, b_0] = Z[x; a, b].$$

As Habicht points out this ansatz is only seemingly a special case since we can specialize leading coefficients of A or of B, but not of both, to be zero if one specializes $Z[x; a, b]$ to $I[x]$; as a result we can study subresultant chains for a given n once and for ever. A second advantage of the method of indeterminates is, that it separates sparsity effects and similarity relations which hold already in the indeterminate case from the effects due to special coefficients in the domain I. This is important in relating properties of p.r.s.'s to the underlying axioms, but it is also important in understanding the structure of p.r.s.'s despite horrifying products of powers of similarity coefficients which have hidden simple divisibility properties for a long time. For example, the improved subresultant gcd-algorithm follows in a few steps from the indeterminate case and was known to Habicht 35 years ago. Finally note that A and B are always primitive in the indeterminate case.

3.1 Subresultant Chains

Let $A = \sum_{i=0}^{n+1} a_i x^i$ and $B = \sum_{i=0}^{n} b_i x^i \in Z[x; a, b]$ and let e and f be indeterminates. Then

$$\mathrm{prem}(eA, fB) = ef^2 \, \mathrm{prem}(A, B), \tag{1}$$

since both sides equal $\mathrm{detpol}(xfB, fB, eA)$ by (1.4). Let us repeat the elementary matrix operations which lead us to the pseudo-remainder definition, this time in the matrix M_k of the subresultant definition:

$$S_k = \mathrm{detpol}(x^{n-1-k}A, \ldots, A, x^{n-k}B, \ldots, B)$$

$$= \mathrm{detpol}(x^{n-1-k}b_n^2 A, \ldots, b_n^2 A, x^{n-k}B, \ldots, B)b_n^{-2(n-k)}$$

$$= \mathrm{detpol}(x^{n-1-k}\,\mathrm{prem}(A, B), \ldots, \mathrm{prem}(A, B), x^{n-k}B, \ldots, B)b_n^{-2(n-k)}$$

$$= \mathrm{detpol}(x^{n-k}B, \ldots, B, x^{n-1-k}\,\mathrm{prem}(A, B), \ldots, \mathrm{prem}(A, B))(-b_n)^{-2(n-k)}$$

$$= \mathrm{detpol}(x^{n-2-k}B, \ldots, B, x^{n-1-k}\,\mathrm{prem}(A, B), \ldots, \mathrm{prem}(A, B))b_n^{-2(n-k)+2}.$$

Therefore

$$b_n^{2(n-k-1)}S_k = \mathrm{sres}_k(B, \mathrm{prem}(A, B)), \qquad 0 \leqslant k < n - 1. \tag{2}$$

According to the derivation we get for

$$k = n - 1, \qquad S_{n-1} = \mathrm{detpol}(xB, B, \mathrm{prem}(A, B))b_n^{-2},$$

where the matrix is upper triangular. Hence

$$S_{n-1} = \text{prem}(A, B). \tag{3}$$

It is convenient to extend the definition of $S_{n+1} = A$ and $S_n = B$ provided we define the *principal subresultant coefficients* $\text{psc}_k(A, B)$ as the leading coefficient $\text{lc}(S_k)$, $0 \leqslant k \leqslant n$, except for $k = n + 1$, where we set $\text{psc}_{n+1}(A, B) = 1$. We use the abbreviation

$$R_k = \text{psc}_k(A, B), \qquad 0 \leqslant k \leqslant n + 1. \tag{4}$$

The convention $S_n = B$ is consistent with (2) and (3) since we get for $k = n - 1$ in (2) $S_{n-1} = \text{sres}_{n-1}(S_n, S_{n-1})$ by (3).

The next theorem relates any subresultant to its two predecessors in the chain and exhibits the kth subresultant up to similarity as $(n - k)$th iterated remainder function of the starting polynomials.

Habicht's Theorem. *Let A and B be polynomials of degree $n + 1$ and n with indeterminate coefficients. With $S_{n+1} = A$, $S_n = B$, let*

$$S_{n+1}, S_n, \ldots, S_0$$

be the subresultant chain of A and B. Then for all j, $0 < j \leqslant n$,

$$R_{j+1}^{2(j-r)} S_r = \text{sres}_r(S_{j+1}, S_j), \qquad 0 \leqslant r < j, \tag{5}$$

$$R_{j+1}^2 S_{j-1} = \text{prem}(S_{j+1}, S_j). \tag{6}$$

Proof. For $j = n$ (5) is the definition of S_r since $R_{n+1} = 1$.

Now suppose (5) has been shown for any fixed j. Then we get from (5) for $r = j - 1$ and with (3)

$$R_{j+1}^2 S_{j-1} = \text{sres}_{j-1}(S_{j+1}, S_j) = \text{prem}(S_{j+1}, S_j) \tag{7}$$

which is (6). Replacing in (2) A by S_{j+1} and B by S_j yields in (5) with (7)

$$
\begin{aligned}
S_r &= R_{j+1}^{-2(j-r)} \text{sres}_r(S_{j+1}, S_j) \\
&= R_{j+1}^{-2(j-r)} R_j^{-2(j-1-r)} \text{sres}_r(S_j, \text{prem}(S_{j+1}, S_j)) \\
&= R_{j+1}^{-2(j-r)} R_j^{-2(j-1-r)} \text{sres}_r(S_j, R_{j+1}^2 S_{j-1}) \\
&= R_j^{-2(j-1-r)} \text{sres}_r(S_j, S_{j-1})
\end{aligned}
$$

which shows that (5) holds with j replaced by $j - 1$. Therefore the argument can be repeated until $j = 1$ is reached. ∎

Now we are ready to specialize $Z[x; a, b]$ to $I[x]$ for some integral domain I. Let A be of degree n_1 and B be of degree n_2. If $n_1 > n_2$ then we set $n_1 = n + 1$ and specialize $b_n = \cdots b_{n_2+1} = 0$. If $n_1 \leqslant n_2$ then we set $n = n_2$ and specialize $a_{n+1} = a_n = \cdots = a_{n_1+1} = 0$. Therefore,

$$n = \begin{cases} n_1 - 1, & \text{if } n_1 > n_2, \\ n_2, & \text{if } n_1 \leqslant n_2. \end{cases} \tag{8}$$

Over I we have $\deg S_j \leqslant j$ since R_j and subsequent leading terms may vanish, where we also have $R_j \neq \mathrm{lc}(S_j)$. If $\deg S_j = r < j$ we call S_j *defective* of degree r and otherwise *regular* following the terminology which was in use for more than one hundred years. A subresultant chain or a p.r.s. is called *regular* if all its elements are regular and otherwise *defective*. The subresultant theorem connects the sparsity in the leading terms of S_j with the gap structure of the chain.

Subresultant Theorem. *Let* $S_{n+1}, S_n, \ldots, S_0$ *be a subresultant chain in* $I[x]$ *of* S_{n+1} *and* S_n. *Let* S_{j+1} *be regular and* S_j *be defective of degree* $r < j$ *(with* $\deg(0) = -1$*).* *Then*

$$S_{j-1} = S_{j-2} = \cdots = S_{r+1} = 0, \qquad -1 \leqslant r < j < n, \qquad (9)$$

$$R_{j+1}^{j-r} S_r = \mathrm{lc}(S_j)^{j-r} S_j, \qquad 0 \leqslant r \leqslant j < n, \qquad (10)$$

$$(-1)^{j-r} R_{j+1}^{j-r+2} S_{r-1} = \mathrm{prem}(S_{j+1}, S_j), \qquad 0 < r \leqslant j < n. \qquad (11)$$

Proof. If the number of vanishing leading terms in S_j exceeds the number of rows of S_{j+1} in the associated matrix $\mathrm{sres}_k(S_{j+1}, S_j)$ in (5) we get 0's in the main diagonal:

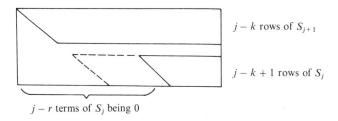

$j - k$ rows of S_{j+1}

$j - k + 1$ rows of S_j

$j - r$ terms of S_j being 0

Therefore, $j - r > j - k$ implies $k > r$ and since $j > k$ we have from (5) $S_k = 0$, $r < k < n$ which proves (9).

If the number $j - r$ equals $j - k$ we have in (5)

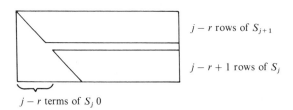

$j - r$ rows of S_{j+1}

$j - r + 1$ rows of S_j

$j - r$ terms of S_j 0

Hence by (5)

$$R_{j+1}^{2(j-r)} S_r = \mathrm{sres}_r(S_{j+1}, S_j) = R_{j+1}^{(j-r)} \mathrm{lc}(S_j)^{(j-r)} S_j$$

by the definition of the operators sres_r and detpol. Since I is an integral domain, (10) follows.

Finally, if the number of vanishing leading terms in S_j $j - r$ is 1 less than the number $j - k$ of rows of S_{j+1}, we have by (5)

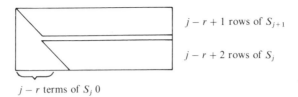

$j - r + 1$ rows of S_{j+1}

$j - r + 2$ rows of S_j

$j - r$ terms of S_j 0

Here we expand the determinant along the first $j - r$ main diagonal entries and get from (5)

$$R_{j+1}^{2(j-r+1)}S_{r-1} = \operatorname{sres}_{r-1}(S_{j+1}, S_j)$$

$$= R_{j+1}^{j-r} \operatorname{detpol}(S_{j+1}, x^{j-r+1}S_j, \ldots, S_j)$$

$$= R_{j+1}^{j-r}(-1)^{j-r+2} \operatorname{detpol}(x^{j+1-r}S_j, \ldots, S_j, S_{j+1})$$

$$= R_{j+1}^{j-r}(-1)^{j-r} \operatorname{prem}(S_{j+1}, S_j)$$

from which (11) follows. ∎

Note that the assumption $\deg A = n_1 \geq n_2 = \deg B$ would be strictly less general. For the three cases $n_1 > n_2$, $n_1 = n_2$ and $n_1 < n_2$ Habicht displays the gap structure of a subresultant chain by diagrams in which a line of length $k + 1$ denotes a polynomial of degree k:

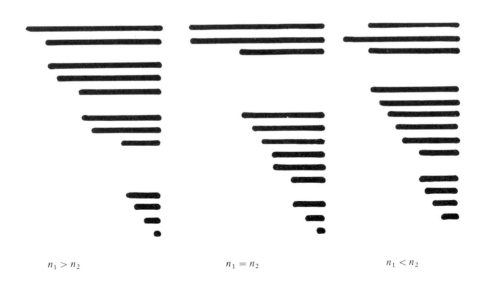

$n_1 > n_2$ $n_1 = n_2$ $n_1 < n_2$

3.2 The Fundamental Theorem

Our next goal is the relation of any p.r.s. defined by (1.7) to the subresultant chain starting with the same polynomials. With the notation of the Subresultant Theorem we have

gap structure	subresultants related by: p.r.s.

$S_{n_{i-1}}$

A_i

$\updownarrow$ (10)

S_{n_i}

S_{n_j-1} $\leftarrow$ (14) $\rightarrow$ A_{i+1}

missing link $\updownarrow$ (10) $\nearrow$ (15) $\nearrow$

$S_{n_{i+1}}$

A_{i+2}

Only regular subresultant chains are p.r.s., which can never contain zero polynomials as defective chains do. The general relation is given by the

Fundamental Theorem of p.r.s. *Let $A_1, \ldots, A_r$ be a p.r.s. over I satisfying for $e_i, f_i \in I$, both non-zero*

$$e_i A_i = Q_i A_{i+1} + f_i A_{i+2} \qquad (1 \leqslant i \leqslant r - 2). \tag{12}$$

Let $n_1, n_2, \ldots, n_r$ be the degree sequence and let $c_1, c_2, \ldots, c_r$ be the leading coefficient $\mathrm{lc}(A_i)$ sequence of (A_1, A_2). For any j, $2 \leqslant j \leqslant r - 1$,

$$S_k = 0, \qquad 0 \leqslant k < n_r, \qquad n_{j+1} < k < n_j - 1, \tag{13}$$

$$\left\{ \prod_{i=1}^{j-1} e_i^{n_{i+1}-n_j+1} \right\} S_{n_j-1} = \left\{ \prod_{i=1}^{j-1} (-1)^{(n_i-n_j+1)(n_{i+1}-n_j+1)} f_i^{n_{i+1}-n_j+1} c_{i+1}^{n_i-n_{i+2}} \right\}$$

$$\times c_j^{-n_j+n_{j+1}+1} A_{j+1} \qquad (1 < j < r), \tag{14}$$

$$\left\{ \prod_{i=1}^{j-1} e_i^{n_{i+1}-n_j+1} \right\} S_{n_{j+1}} = \left\{ \prod_{i=1}^{j-1} (-1)^{(n_i-n_j+1)(n_{i+1}-n_j+1)} f_i^{n_{i+1}-n_j+1} c_{i+1}^{n_i-n_{i+2}} \right\}$$

$$\times e_{j+1}^{n_j-n_{j+1}-1} A_{j+1} \qquad (1 < j < r). \tag{15}$$

Proof. (13) is (9). Let us repeat the steps leading to (2) with $A = A_i$, $B = A_{i+1}$ for $0 \leqslant k < n_{i+2}$:

$\mathrm{sres}_k(A_i, A_{i+1})$

$$= \mathrm{detpol}(x^{n_{i+1}-k-1} A_i, \ldots, A_i, x^{n_i-k-1} A_{i+1}, \ldots, A_{i+1})$$

$$= (e_i^{-1} f_i)^{n_{i+1}-k} \mathrm{detpol}(x^{n_{i+1}-k-1} A_{i+2}, \ldots, A_{i+2}, x^{n_i-k-1} A_{i+1}, \ldots, A_{i+1})$$

$$= (e_i^{-1} f_i)^{n_{i+1}-k} c_i^{n_i-n_{i+2}} (-1)^{(n_i-k)(n_{i+1}-k)} \mathrm{sres}_k(A_{i+1}, A_{i+2}).$$

Iterating this for $i = 1, \ldots, j - 2$ gives

$$\left\{ \prod_{i=1}^{j-2} e_i^{n_{i+1}-k} \right\} S_k = \left\{ \prod_{i=1}^{j-2} (-1)^{(n_i-k)(n_{i+1}-k)} f_i^{n_{i+1}-k} c_{i+1}^{n_i-n_{i+2}} \right\}$$

$$\times \mathrm{sres}_k(A_{j-1}, A_j) \qquad (0 \leqslant k < n_j). \tag{16}$$

For $k = n_j - 1$ we have simply by (12):

$$\mathrm{sres}_{n_j-1}(A_{j-1}, A_j)$$
$$= \mathrm{detpol}(A_{j-1}, x^{n_j-1-n_j}A_j, \ldots, A_j)$$
$$= (e_{j-1}^{-1}f_{j-1})(-1)^{n_j-1-n_j+1}\mathrm{detpol}(x^{n_j-1-n_j}A_j, \ldots, A_j, A_{j+1})$$
$$= (e_{j-1}^{-1}f_{j-1})(-c_j)^{n_j-1-n_j+1}A_{j+1},$$

since the matrix is upper triangular, which proves (14).

Since $n_{j+1} < n_j$ we can set $k = n_{j+1}$ in (16) and get

$$\mathrm{sres}_k(A_{j-1}, A_j) = \mathrm{detpol}(x^{n_j-n_{j+1}-1}A_{j-1}, \ldots, A_{j-1}, x^{n_j-1-n_{j+1}-1}A_j, \ldots, A_j)$$
$$= (e_{j-1}^{-1}f_{j-1})^{n_j-n_{j+1}+1}(-1)^{(n_j-1-n_{j+1}+1)(n_j-n_{j+1})}$$
$$\times c_j^{n_j-1-n_{j+1}+1}c_{j+1}^{n_j-n_{j+1}+1-1}A_{j+1}$$

what transforms (16) into (15). ∎

3.3 Polynomial Remainder Sequences

If $A, B \in E[x]$, where E is some Euclidean domain in which division is effective, algorithm QR in Section 1 can be used to produce the p.r.s. of A and B. This means we have in (12)

$$e_i = f_i = 1, \qquad 1 \le i \le r, \tag{17}$$

and call this the *Euclidean p.r.s.*

If we work over an integral domain I, we may produce the p.r.s. of A and B over I by working over Q, the quotient field of I, being a Euclidean domain. In fact, one gains a little bit by setting

$$A_1 = A, \qquad A_2 = B/\mathrm{lc}(B),$$
$$e_i = 1, \qquad f_i = \mathrm{lc}(A_{i+2}), \qquad 1 \le i \le r - 2. \tag{18}$$

We call this the *monic Euclidean p.r.s.*

Avoiding rational arithmetic in Q and staying over I leads to

$$e_i = (\mathrm{lc}\, A_{i+1})^{n_i-n_{i+1}+1}, \qquad f_i = 1, \qquad 1 \le i \le r - 1. \tag{19}$$

We call this the *pseudo p.r.s.* of A and B. This sequence suffers from an exponential coefficient growth (see Section 5), which occurs not only in exceptional cases.

In an attempt to avoid coefficient growth one arrives at

$$e_i = (\mathrm{lc}\, A_{i+1})^{n_i-n_{i+1}+1}, \qquad f_i = \mathrm{cont}(A_{i+2}), \qquad 1 \le i \le r - 2. \tag{20}$$

This is the *primitive p.r.s.*, since the remainders are kept primitive. Similar to the monic Euclidean p.r.s., gcd-operations in the coefficient domain may become the dominant factor.

A surprisingly simple sequence was discovered by Collins [7]

$$e_i = (\mathrm{lc}\, A_{i+1})^{n_i-n_{i+1}+1}, \qquad f_1 = 1, \qquad f_{i+1} = e_i, \qquad 1 \le i \le r - 1. \tag{21}$$

This is the *reduced p.r.s.* In fact one has to prove

The Reduced p.r.s. Theorem. *If $A_1 = A$ and $A_2 = B \in I[x]$ then $A_i \in I[x]$, where A_i is defined by (12) and (21), $1 \leqslant i \leqslant r$.*

Proof. If we substitute (21) in the Fundamental Theorem, we get with the notations used there up to a sign by (14) with

$$d_{j-2} = \left\{ \prod_{i=1}^{j-2} c_{i+1}^{(n_i - n_{i+1})(n_{i+1} - n_{i+2} - 1)} \right\},$$

$$A_{j+1} = \pm d_{j-2} S_{n_j - 1} \tag{22}$$

and by (15) with j replaced by $j - 1$

$$c_j^{\delta_{j-1} - 1} A_j = \pm d_{j-2} S_{n_j}, \qquad \text{with} \qquad \delta_{j-1} = n_{j-1} - n_j \tag{23}$$

which implies

$$c_j^{\delta_{j-1}} = \pm d_{j-2} R_{n_j}. \tag{24}$$

From (22), (23), (24) and the Subresultant Theorem (11) with $j + 1$ replaced by n_j and r replaced by n_{j+1} we obtain

$$c_j^{\delta_{j-1} - 1} \operatorname{prem}(A_j, A_{j+1}) = \pm d_{j-2}^{\delta_j} \operatorname{prem}(S_{n_j}, S_{n_j - 1})$$

$$= \pm d_{j-2}^{\delta_j + 2} R_{n_j}^{\delta_{j+1}} S_{n_{j+1} - 1}$$

$$= \pm d_{j-2} c_j^{\delta_{j-1}(\delta_j + 1)} S_{n_{j+1} - 1}.$$

Once more, we apply (22) with $d_{j-1} = c_j^{\delta_{j-1}(\delta_j - 1)} d_{j-2}$

$$\operatorname{prem}(A_j, A_{j+1}) = \pm c_j^{\delta_{j-1} + 1} A_{j+2}, \qquad 1 < j \leqslant r - 2. \tag{25}$$

By (22) $A_{j+2} \in I[x]$ and f_j is a divisor of $\operatorname{prem}(A_j, A_{j+1})$. ∎

In 1967 Collins [7] introduced the *subresultant p.r.s.* defined by (12) (first method)

$$e_i = c_{i+1}^{\delta_i + 1}, \qquad f_i = (-1)^{\delta_i + 1} c_i^{\delta_i - 1 + 1} \prod_{j=2}^{i} c_j^{h_{ij}},$$

with

$$h_{i+k,i} = \begin{cases} \delta_{i-1}(\delta_i - 1) & \text{for} \quad k = 0, \\ (-1)^k \delta_{i-1} \delta_{i+k} \prod_{j=1}^{i+k-1} (\delta_j - 1), & k > 0. \end{cases}$$

The proof that $\operatorname{prem}(A_i, A_{i+1}) = f_i A_{i+2}$ can be found in Collins [7]. More important, we have with $n_0 = n_1 + 1$

$$A_j = S_{n_{j-1} - 1}, \qquad 1 \leqslant j \leqslant r. \tag{26}$$

However, there is a more direct way to compute the *subresultant p.r.s.* (improved method) by

$$e_i = c_{i+1}^{\delta_i + 1}, \qquad 1 \leqslant i < r,$$

$$f_1 = 1, \qquad f_i = -c_i(-R_{n_i})^{\delta_i}, \qquad 2 \leqslant i \leqslant r - 2, \tag{27}$$

with the notation $R_{n_i} = \operatorname{lc}(S_{n_i})$, $i > 1$.

Theorem. *Let*

$$e_i A_i = Q_i A_{i+1} + f_i A_{i+2}, \qquad 1 \leqslant i \leqslant r - 1,$$

with e_i, f_i from (27). Then

$$A_i = S_{n_{i-1}-1}, \qquad 1 \leqslant j \leqslant r \quad with \quad n_0 = n_1 + 1.$$

Proof. By definition $A_1 = S_{n_1}$, $A_2 = S_{n_1-1}$. Assume, the Theorem has been shown for i and $i + 1$. Then (10) becomes with $j + 1$ replaced by n_{i-1} and r replaced by n_i

$$R_{n_{i-1}}^{\delta_{i-1}-1} S_{n_i} = c_i^{\delta_{i-1}-1} A_i, \tag{10'}$$

and

$$R_{n_{i-1}}^{\delta_{i-1}-1} R_{n_i} = c_i^{\delta_{i-1}}. \tag{28}$$

(11) becomes with $j + 1$ replaced by n_i and r replaced by n_{i+1}

$$(- R_{n_i})^{\delta_i+1} S_{n_{i+1}-1} = \text{prem}(S_{n_i}, A_{i+1}). \tag{11'}$$

Hence by (10') and (11')

$$f_i A_{i+2} = \text{prem}(A_i, A_{i+1}) = (R_{n_{i-1}}/c_i)^{\delta_{i-1}-1} \text{prem}(S_{n_i}, A_{i+1})$$

$$= (R_{n_{i-1}}/c_i)^{\delta_{i-1}-1}(- R_{n_i})^{\delta_i+1} S_{n_{i+1}-1}$$

$$= f_i S_{n_{i+1}-1}$$

by (28), therefore $A_{i+2} = S_{n_{i+1}-1}$. ∎

Any p.r.s. with $e_i, f_i \in I$, I being ordered,

$$e_i A_i = Q_i A_{i+1} + f_i A_{i+2}, \qquad e_i f_i < 0, \qquad 1 \leqslant i \leqslant r - 1,$$

computes a *negative p.r.s.* or *generalized Sturm sequence* of A_1 and A_2. An instance is *Tarski's remainder sequence* [30] with

$$f_i = - 1,$$

$$e_i = c_{i+1}^{2(\delta_i + 1)}. \tag{29}$$

3.4 Extended Polynomial Remainder Sequences

The defining equation (12) for a p.r.s. can be written

$$\begin{pmatrix} A_i \\ A_{i+1} \end{pmatrix} = \begin{pmatrix} Q_i/e_i & f_i/e_i \\ 1 & 0 \end{pmatrix} \begin{pmatrix} A_{i+1} \\ A_{i+2} \end{pmatrix}$$

or by induction

$$\begin{pmatrix} A_1 \\ A_2 \end{pmatrix} = \begin{pmatrix} Q_1/e_1 & f_1/e_1 \\ 1 & 0 \end{pmatrix} \cdots \begin{pmatrix} Q_i/e_i & f_i/e_i \\ 1 & 0 \end{pmatrix} \begin{pmatrix} A_{i+1} \\ A_{i+2} \end{pmatrix}$$

or

$$\begin{pmatrix} 0 & 1 \\ e_i/f_i & - Q_i/f_i \end{pmatrix} \cdots \begin{pmatrix} 0 & 1 \\ e_1/f_1 & - Q_1/f_1 \end{pmatrix} \begin{pmatrix} A_1 \\ A_2 \end{pmatrix} = \begin{pmatrix} A_{i+1} \\ A_{i+2} \end{pmatrix} \tag{30}$$

by inverting the matrix product. Hence with $U_1 = 1$, $V_1 = 0$, $U_2 = 0$, $V_2 = 1$

$$U_i A_1 + V_i A_2 = A_i, \qquad 1 \leqslant i \leqslant r. \tag{31}$$

$(U_1, \ldots, U_r)$ is the *first cosequence* and $(V_1, \ldots, V_r)$ is the *second cosequence* of the p.r.s. $(A_1, \ldots, A_r)$.

Let us consider the subresultant chain. From the definition of S_j in Section 1 we have

$$S_j(x) = \begin{vmatrix} a_m & \cdots & a_0 & & & & x^{n-j-1}A \\ & \ddots & & & \ddots & & \vdots \\ & & a_m & & & a_0 & \vdots \\ & & & & & \vdots & \\ & & a_m & \cdots & a_{j+1} & & A \\ b_n & \cdots & b_{j+1} & \cdots & b_0 & & x^{m-j-1}B \\ & \ddots & & & & \ddots & \vdots \\ & & & & & b_0 & \vdots \\ b_n & & \cdots & & b_{j+1} & & B \end{vmatrix}. \tag{32}$$

Therefore

$$S_j = U_j A + V_j B, \tag{33}$$

where U_j is S_j except for the last column, which is top down

$$x^{n-j-1} \quad \cdots \quad 1 \quad 0 \quad \cdots \quad 0$$

and V_j is S_j except for the last column, which is top down

$$0 \quad \cdots \quad 0 \quad x^{m-j-1} \quad \cdots \quad 1,$$

hence $\deg U_j \leqslant n - j - 1$, $\deg V_j \leqslant m - j - 1$. From this construction $U_j, V_j \in I[x]$ is immediate; since (33) may contain zero elements we call $\{U_j\}$ and $\{V_j\}$ *first* and *second cochains*.

4. Algorithms

One can test whether two non-zero polynomials $A, B \in I[x]$ are *similar* by checking

$$\text{lc}(B)A = \text{lc}(A)B \tag{1}$$

which follows from the definition of similarity. We furthermore assume having available an algorithm for $R = \text{prem}(A, B)$ and exact division $Q = \text{equot}(A, B)$ given by (1, 3) and QR in Section 1.

The algorithm given by (1.1) is *Bareiss' algorithm* which uses equot over I. It may be used for determinant calculation, solving linear systems and matrix inversion over integral domains. It would also provide for a *detpol* algorithm and based on this for a *subresultant* algorithm based directly on the definitions. However, we can compute cheaper the entire subresultant chain, than a single subresultant by Bareiss' method.

4.1 The Subresultant Chain Algorithm

Based on (3.8) and the Subresultant Theorem (9) – (11) we compute the chain of all subresultants.

Subresultant Chain Algorithm. Input: $A, B \in I[x]$, both non-zero.

Output: $A, B, S_{n-1}, \ldots, S_0$, the subresultant chain of A and B.

(1) [Initialize (3.8)] $m \leftarrow \deg(A)$; $n \leftarrow \deg(B)$; if $m > n$ then
$j \leftarrow m - 1$ else $j \leftarrow n$; $S_{j+1} \leftarrow A$; $S_j \leftarrow B$; $R_{j+1} \leftarrow 1$.

(2) [Degree] if $S_j = 0$ then $r \leftarrow -1$ else $r \leftarrow \deg(S_j)$.

(3) [Gap] for $k = j - 1, j - 2, \ldots, r + 1$ do $S_k \leftarrow 0$.

(4) [Similar polynomial (3.10)] If $j > r \geqslant 0$ then $S_r \leftarrow (\mathrm{lc}(S_j)^{j-r} S_j)/R_{j+1}^{j-r}$.

(5) [Finished] If $r \leqslant 0$ then return.

(6) [Remainder (3.11)] $S_{r-1} \leftarrow \mathrm{prem}(S_{j+1}, S_j)/(- R_{j+1})^{j-r+2}$;
$j \leftarrow r - 1$; $R_{j+1} \leftarrow \mathrm{lc}(S_{j+1})$; go to 2. ∎

Note that both divisions in (4) and (6) are exact, also S_0 is the *resultant* of A and B computed by a non-modular algorithm. For computing the resultant, step (3) can be dropped.

4.2 The Generalized p.r.s. Algorithm

Given a remainder definition by (3.12), the following algorithm generalizes Euclid's algorithm.

General p.r.s. Algorithm

Input: $A, B \in I[x]$, both non-zero, $\deg A \geqslant \deg B$.
functions to compute the similarity coefficients of (3.16).

Output: The generalized p.r.s. of A and B.

(1) [Initialize] $A_1 \leftarrow A$; $A_2 \leftarrow B$; $i \leftarrow 1$.

(2) [Finished?] If $A_{i+1} = 0$ then return.

(3) [Quotient and remainder] Compute Q_i and A_{i+2} such that
$e_i A_i = Q_i A_{i+1} + f_i A_{i+2}$.

(4) [Swap] $i \leftarrow i + 1$; go to 2. ∎

We get the *extended p.r.s. algorithm* by initializing in step (1)

$$\begin{pmatrix} u'_0 & v'_0 \\ u_0 & v_0 \end{pmatrix} \leftarrow \begin{pmatrix} 1 & 0 \\ 0 & 1 \end{pmatrix}$$

and adding after step (3)

$$\begin{pmatrix} u'_i & v'_i \\ u_i & v_i \end{pmatrix} \leftarrow \begin{pmatrix} u_{i-1} & v_{i-1} \\ (u'_{i-1}e_i - u_{i-1}Q_i)/f_i & (v'_{i-1}e_i - v_{i-1}Q_i)/f_i \end{pmatrix}.$$

Then we have on exit $U_i A_1 + V_i A_2 = A_i$ with

$$U_i = u'_{i-1} \quad \text{and} \quad V_i = v'_{i-1}$$

since in step (3) the matrix on the right-hand side is

$$\begin{pmatrix} 0 & 1 \\ e_i/f_i & -Q_i/f_i \end{pmatrix} \begin{pmatrix} u'_{i-1} & v'_{i-1} \\ u_{i-1} & v_{i-1} \end{pmatrix},$$

where, however, the division is not exact.

In the *half-extended p.r.s. algorithm* the computation of v'_i, v_i and V_i is omitted.

4.3 The Primitive p.r.s. Algorithm

In the general p.r.s. algorithm we have (with e_i implicit in the prem-definition)

(3) $R \leftarrow \mathrm{prem}(A_i, A_{i+1}); \quad f_i \leftarrow \mathrm{cont}(R); \quad A_{i+2} \leftarrow \mathrm{pp}(R).$

If the quotient is required it is better to have an algorithm $\mathrm{PQR}(A, B; Q, R)$ instead of forming Q after R. Here prem could be improved to sprem according to (1.4).

4.4 The Reduced p.r.s. Algorithm

Here we have in addition $e_0 \leftarrow 1$ in step (1) and

(3) $e_i \leftarrow (\mathrm{lc}\, A_{i+1})^{n_i - n_{i+1} + 1}; \quad R \leftarrow \mathrm{prem}(A_i, A_{i+1});$
$A_{i+2} \leftarrow R/e_{i-1}$ with exact division.

4.5 The (Improved) Subresultant p.r.s. Algorithm

Input: $A, B \in I[x]$, both non-zero, $\deg A \geqslant \deg B$.

Output: The subresultant p.r.s. of A and B.

(1) [Initialize] $A_i \leftarrow A; \; A_2 \leftarrow B; \; i \leftarrow 1; \; R_{n_1} \leftarrow 1.$

(2) [Exit] If $A_{i+1} = 0$ then return.

(3) [Remainder] $A_{i+2} \leftarrow \mathrm{prem}(A_i, A_{i+1})/(-c_i(-R_{n_i})^{\delta_i});$
$i \leftarrow i + 1; \; R_{n_i} \leftarrow (R_{n_{i-1}} c_i^{\delta_i - 1})/R_{n_{i-1}}^{\delta_i - 1}; \text{ go to 2.} \quad \blacksquare$

The algorithm follows from (3.27) and (3.28), which shows that all divisions in step (3) are exact. This algorithm seems to be the best non-modular p.r.s. algorithm and is in particular suited for sparse multivariate polynomials.

4.6 Non-Modular Polynomial GCD and Resultant Algorithms

For every p.r.s. $A, B, A_3, \ldots, A_r$ with the same starting polynomials we have

$$\gcd(A, B) \sim A_r \sim S_{n_{r-1} - 1}. \tag{2}$$

It is therefore possible to use any p.r.s. algorithm for gcd computation, if we make the gcd of two polynomials canonical (see for a definition of gcd and the unit-function I the section on Canonical simplification of rational expressions in the chapter on Algebraic Simplification). Obviously, for any non-zero polynomials

$$A \sim B \qquad \text{iff} \qquad A \sim \text{pp}(B), \tag{3}$$

where pp(B) denotes the *primitive part* of B, defined as $B/\text{cont}(B)$, where the *content* of B is the gcd of its coefficients. The definitions imply

$$A = \text{cont}(A)\text{pp}(A) \tag{4}$$

and $\text{gcd}(\text{cont}(A), \text{pp}(A)) = 1$ for any non-zero polynomial. Using the multiplication and commutative properties of any gcd function over a gcd-domain U with a *canonical simplifier* $\text{can}(A) = l(A)A$ and a multiplicative set of canonical elements (see Collins [8]):

$$\text{gcd}(A, B) = \text{gcd}(\text{cont}(A), \text{cont}(B)) \, \text{gcd}(\text{pp}(A), \text{pp}(B)). \tag{5}$$

This relation suggests to define a canonical polynomial GCD-function in the following way:

Definition. Let $A, B \in U[x]$. Set $\text{GCD}(0, 0) = 0$, $\text{GCD}(A, 0) = \text{can}(A)$ if $A \neq 0$ and $\text{GCD}(0, B) = \text{can}(B)$ for $B \neq 0$. For $A, B \neq 0$ let $c = \text{gcd}(\text{cont}(A), \text{cont}(B)) \in U$ and let $A, B, A_3, \ldots, A_r$ be any p.r.s. of A and B. Then

$$\text{GCD}(A, B) = \text{can}(c \, \text{pp}(A_r)). \tag{6}$$

Generalized Polynomial GCD-Algorithm. Input: $A, B \in U[x]$.

Output: $C = \text{GCD}(A, B)$.

(1) [Initialize] $c \leftarrow \text{gcd}(\text{cont}(A), \text{cont}(B))$; $i \leftarrow 1$; $A_1 \leftarrow \text{pp}(A)$; $A_2 \leftarrow \text{pp}(B)$.

(2) [Exit?] If $A_{i+1} = 0$ then $\{C \leftarrow c \, \text{pp}(A_i); C \leftarrow \text{can}(C); \text{return}\}$; if $\deg A_{i+1} = 0$ then $\{C \leftarrow \text{can}(c); \text{return}\}$.

(3) [Remainder] Compute Q_i, A_{i+2} such that $e_i A_i = Q_i A_{i+1} + f_i A_{i+2}$.

(4) [Swap] $i \leftarrow i + 1$; go to 2. ∎

It is assumed that $\text{cont}(0) = \text{pp}(0) = 0$ and $l(0) = 1$. If step (3) is modified as in the primitive, reduced- and subresultant p.r.s. algorithm we get the corresponding GCD-algorithms. In the primitive algorithm the computation of $\text{pp}(A_i)$ in step (2) can be saved. The modifications to the extended and half-extended GCD-algorithms are obvious.

For any two polynomials $A, B \in I[x]$ with $\text{gcd}(A, B) \sim 1$ every p.r.s. algorithm may be used to compute $S_0 = \text{res}(A, B)$, the resultant of A and B. In the general case the similarity coefficients for $A_r \sim S_0$ have to be computed from the Fundamental Theorem explicitly, say $a A_r = b S_0$ and $\text{res}(A, B) = (a A_r)/b$. This final exact division is not necessary if one uses the subresultant-p.r.s. or subresultant-chain algorithm directly.

4.7 Modular GCD and Resultant Algorithms

A comprehensive survey is given by Collins in [8], see also the chapter on Computing with Homomorphic Images. We show here only, how the correctness of the method depends on the Subresultant Theorem.

Let U and U^* be canonical gcd-domains and
$h: U \to U^*$ be a homomorphism which we extend to
$h: U[x] \to U^*[x]$.

Theorem. *Let* $A, B \in U[x]$, *both non-zero, with* $\deg(\gcd(A, B)) = k$ *and let* $\mathrm{lc}(A) = a$, $\mathrm{lc}(B) = b$. *If* $h(a) \neq 0$ *or* $h(b) \neq 0$ *then* $\deg(\gcd(h(A), h(B))) > k$ *iff* $h(R_k) = 0$, *where* $R_k = \mathrm{lc}(S_k)$ *and* $S_k = \mathrm{sres}_k(A, B)$.

Proof. Assume $h(a) \neq 0$. Then $h(S_k) = h(a)^r \mathrm{sres}_k(h(A), h(B))$ by (3.32) where $r = \deg B - \deg h(B)$. If $\deg \gcd(h(A), h(B)) > k$ then by (3.9) of the Subresultant Theorem $\mathrm{sres}_k(h(A), h(B)) = 0$, hence $h(S_k) = 0$ and $h(R_k) = 0$. Conversely, if $\deg \gcd(h(A), h(B)) = k$, then by the Subresultant Theorem $\mathrm{sres}_k(h(A), h(B)) \neq 0$ and $\deg \mathrm{sres}_k(h(A), h(B)) = k$. If $h(R_k) = 0$ then $\deg h(S_k) < k$, so $h(S_k) \not\sim \mathrm{sres}_k(h(A), h(B))$, which is a contradiction. Hence $\deg \gcd(h(A), h(B)) \neq k$. Set $G = \gcd(A, B)$ and $A = \bar{A}G$, hence $h(A) = h(\bar{A})h(G) \neq 0$ since $h(a) \neq 0$. Therefore $\deg h(G) \leqslant \deg \gcd(h(A), h(B))$. But also $h(a) = h(\mathrm{lc}(\bar{A}))h(\mathrm{lc}\, G)) \neq 0$ implies $h(\mathrm{lc}(G)) \neq 0$ and $\deg h(G) = \deg G = k$, hence $\deg \gcd(h(A), h(B)) > k$. ∎

In using the modular method primes or irreducible polynomials m are therefore unlucky moduli if $m \mid \mathrm{lc}(A)$, $m \mid \mathrm{lc}(B)$ or $m \mid R_k$. Since there are only finitely many prime divisors of $\mathrm{lc}(A)$, $\mathrm{lc}(B)$ and R_k the last condition will be detected if the degree of an image gcd is *non-minimal* with respect to the other moduli according to the Theorem.

Important improvements of the standard modular technique are due to W. S. Brown [2], in particular for multivariate polynomials and also for the computation of the *cofactors* $A/\gcd(A, B)$ and $B/\gcd(A, B)$.

The essence for computing gcd's and resultants by modular methods consists in choosing Euclidean domains as image domains where a *natural* Euclidean p.r.s. algorithm is available. Variants of the modular gcd-algorithms include either the Chinese remainder algorithm [2, 10] or Hensel's lemma, resulting in the Yun-Moses EZGCD (for *e*xtended *Z*assenhaus) algorithm [36, 23]. A proof of correctness is given in Knuth [17].

5. Computing Times

For simplicity and comparison with empirical times, we assume *classical* arithmetic and denote by $L(d)$ the maximum length of the semi-norm, by n the maximum degree in any variable in all input polynomials. We are usually not only interested in the computing time of the algorithms but also in the complexity parameters of the outputs. We also will assume that arithmetic in $GF(p)$ is done using single precision primes, i.e. $L(p) \sim 1$. Sometimes we use also the assumptions that $n < \beta$ and $d < \beta^\beta$ without mentioning.

5.1 Bareiss' Algorithm

Over $GF(p)$ we have for a $k \times k$ matrix $t(k) = O(k^3)$. Over Z in the ith step $L(a^{(i)}) = O(iL(a))$, $L(\det(A)) = O(kL(a))$, where $a^{(i)} = \max |a_{i_j}^{(i)}|$ and $t(k, a) = O(k^5 L(a)^2)$, for $Z[x]$ $L(a)$ has to be replaced by $knL(d)$ and for $Z[x_1, \ldots, x_r]$ by $(kn)^r L(d)$.

5.2 Division and Remainder Algorithm

Let d be the maximum of d_A, d_B, and let $m = \deg A, n = \deg B$. The time to compute $\mathrm{lc}(B)^{m-n+1} A = QB + R$ over $Z[x]$ is dominated by $O(n(m-n+1)^2 L(d)^2)$. Also $L(d_Q), L(d_R) = O((m-n+1)L(d))$ and $\deg Q = m-n$ and $\deg R < n$ for $R \neq 0$.

5.3 Subresultants

By Hadamard's inequality (see chapter on bounds) for determinants we can bound the size of any subresultant coefficient semi-norm by

$$L(d_{S_k}) = O(nL(nd)), \qquad 0 \leqslant k \leqslant n.$$

By definition $\deg S_k \leqslant k$.

The subresultant chain algorithm computes all S_k's over $GF(p)$ in $O(n^2)$ steps, over Z in $O(n^4 L(d)^2)$ and over $Z[x_1, \ldots, x_{r-1}]$ in $O(n^{2r+2} L(d)^2)$ steps. The range bounds apply to the (improved) subresultant p.r.s. algorithm and to the non-modular resultant algorithm based on the chain computation, which seem to be the most reasonable non-modular algorithms.

5.4 The Primitive p.r.s. Algorithm

By definition $L(d_{A_i}) \leqslant L(d_{S_{n_i-1}-1})$. Heindel [14] has shown that the time to compute a regular (all $\delta_i = n_i - m_{i+1} = 1$) primitive p.r.s. is $O(n^4 L(d)^2)$ and that in the defective case it is $O(n^5 L(d)^2)$. In the multivariate case the time of gcd-operations on the polynomial coefficients is prohibitive. In [13] Hearn has given a variant of the primitive p.r.s. algorithm using sprem and trial-divisions with powers of previous leading coefficients facilitating the subsequent content-computation. Using ordinary arithmetic the algorithm is *not* exponential except in the number of variables.

5.5 The Reduced p.r.s. Algorithm

Since the reduced p.r.s. algorithm knows about the quadratic principal subresultant coefficients it takes them (partly) out by exact divisions which are faster than gcd-calculations on the coefficients. So in practice the reduced p.r.s. algorithm behaves in general better than the primitive p.r.s. algorithm. Nevertheless, in the irregular case it can have exponential coefficient growth and computing time (as the first subresultant p.r.s. algorithm). By (22) with all $\delta_i = 2$ we get

$$c_{j+1} = \pm \left\{ \prod_{i=1}^{j-2} c_{i+1}^2 \right\} \mathrm{lc}(S_{n_j-1}),$$

so with $\mathrm{lc}(S_{n_j-1}) = s_j$ we get

$$c_3 = \pm c_2^2 s_2, \qquad c_4 = \pm c_2^2 c_3^2 s_3 = \pm c_2^{2+2^2} s_2^2 s_3, \ldots.$$

For any given degree sequence $n_1, \ldots, n_r$ a p.r.s. can be constructed (with all $e_i = 1$ and f_i arbitrary) by

$$\begin{pmatrix} A_{r-1} \\ A_r \end{pmatrix} = \begin{pmatrix} Q_{r-1} & f_{r-1} \\ 1 & 0 \end{pmatrix} \begin{pmatrix} A_r \\ 0 \end{pmatrix}$$

and using the first equation of (3.30) inductively, so A_1 and A_2 are defined by its *Euclidean representation* $(Q_1, \ldots, Q_{r-1}, A_r)$ and $(f_1, \ldots, f_{r-2})$. So, applying the reduced p.r.s. to A_1 and A_2 with the given degree sequence will result in coefficient growth as described above.

5.6 The Pseudo Polynomial Remainder Algorithm

The p.p.r.s. and Tarski's p.r.s. algorithm have an exponential growth rate for the length of the coefficients or the degrees of the coefficients and therefore an exponential computing time.

5.7 GCD and Resultants Algorithm

The computing times for the non-modular gcd-algorithms are codominant with the computing times of the corresponding p.r.s.-algorithms.

Let us consider the maximum computing time of the modular gcd-algorithm for univariate polynomials. Since $L(R_k) = O(nL(d))$ there are at most $O(nL(d))$ unlucky primes and at most $O(nL(d))$ lucky primes needed. For each prime $h_p(A)$ and $h_p(B)$ costs $O(nL(d))$, $\gcd(h_p(A), h_p(B))$ costs $O(n^2)$ and the Chinese remainder algorithm $O(n^2 L(d))$ at most, so the maximum computing time is $O(n^3 L(d)^2)$. For $Z[x_1, \ldots, x_r]$ it is $O(n^{2r+1} L(d)^2)$. Collins conjectures that the average computing time is $O(n^{r+1} L(d) + n^r L(d)^2)$. If A and B are relatively prime the number of lucky primes to detect this is in the average only $O(L(d))$, so the average time in this case is $O(n^2 + nL(d))$. The computing times of the EZGCD-algorithm are not known.

For resultants the time for the non-modular algorithm is codominant with the time for the subresultant chain algorithm. The maximum computing time for the modular resultant algorithm [9] is $O(n^{2r+1} L(d) + n^{2r} L(d)^2)$.

6. On the History

Around 1570 the problem to compute a polynomial gcd was posed by P. Nonius [25], a Portuguese mathematician. The question arose in the context of solving the cubic and quartic equation.

In 1585 Steving gave an answer and computed the p.r.s. $x^3 + x^2$, $x^2 + 7x + 6$, $-6x^2 - 6x$, $6x + 6$ by the Euclidean algorithm for "algebraic multinomials". It comes close to our definition of a p.r.s. [27].

Kronecker attributes the introduction of the concept of an indeterminate (as opposed to an "unknown" or variable) to Gauss [18]. The gcd of two polynomial functions is in general different from the gcd of two polynomials. Also Kronecker used already the notion of a *regular* sequence in the context of Sturm sequences.

The polynomials of a p.r.s. over a field could be expressed as determinants of the coefficients of A and B in the last century [34], but neither the similarity coefficients nor a name for subresultants seem to appear in the literature.

The concept of a subresultant (without this name) is defined in a paper by Habicht [11] in 1948 up to a sign. He generalizes the approach van der Waerden's [32, see also 34], to resultants by asking for conditions that two polynomials have a gcd of given degree j. He calls the leading subresultant coefficients "Nebenresultanten" and discovers working with indeterminate coefficients that powers of them are introduced systematically in the subresultant chain (see his Theorem). He states the Subresultant Theorem over fields and suggests its use as a computation rule for generalized Sturm sequences. As we have done, the (improved) Brown Collins subresultant algorithm can be based directly on Habicht's results. The importance of his results for computer algebra was first recognized by V. Strassen [28]; we have simplified the proofs by avoiding the cochains U_j and V_j (3.33) and using ideas from Collins proof of the Fundamental Theorem. Coming from this theorem Brown [3] provided in 1978 the missing link (3.10) to the subresultant p.r.s. algorithm, not being aware of Habicht's work. For the history of Brown's discovery see [4]. Research in computer algebra on p.r.s. algorithms started with two papers by Collins [6, 7], the motivation seems to have originated from Tarski's Decision Machine [30]. Tarski's p.r.s. algorithm has the worst growth rate of all p.r.s. algorithms considered here. In 1964 Brown seems to have implemented a variant of the primitive p.r.s. algorithm in his ALPAK-system. The papers by Collins present for the first time similarity coefficients between p.r.s.'s in full detail, in particular for the reduced and the first subresultant p.r.s. algorithm. As an abstraction and generalization the Fundamental Theorem was found independently by Brown and Collins and is the basis to study the coefficient growth and computing time of any p.r.s. Aside from its computational importance it seems more to hide than to display Habicht's Subresultant Theorem. The reason for this is that the coefficient relations in the subresultant *chain* stay simple between neighboring non-zero subresultants. In a defective subresultant *p.r.s.* the A_i's are not neighbors anymore and the coefficient relations must be expressed by products over the whole initial p.r.s. segment. For example, relation (3.28) which follows immediately from the Theorem becomes expressed by the leading coefficients c_j of the *p.r.s.* (as opposed to the *chain*)

$$R_{n_j} = \prod_{i=1}^{j-1} c_{i+1}^{\delta_i \prod_{k=i+1}^{j-1}(1-\delta_k)}$$

which was known to Collins ([7], Theorem 2).

The idea to use modular methods for polynomial g.c.d. calculations seems to have appeared first in print in Knuth [16]. The method was fully worked out and underwent important improvements by Brown [2] and Collins [10]. The Yun-Moses EZ-algorithm appeared 1973; its sparseness conservation was studied by Wang [33].

Acknowledgement

The proof of the Fundamental Theorem, the material on species of p.r.s., the modular gcd and on computing times are based on unpublished notes of lectures given by G. E. Collins in 1974/1975 in Kaiserslautern.

References

[0] Bareiss, E. H.: Sylvester's Identity and Multistep Integer-Preserving Gaussian Elimination. Math. Comp. **22**, 565 – 578 (1968).

[1] Brown, W. S., Traub, J. F.: On Euclid's Algorithm and the Theory of Subresultants. JACM **18**, 505 – 514 (1971).

[2] Brown, W. S.: On Euclid's Algorithm and the Computation of Polynomial Greatest Common Divisors. JACM **18**, 476 – 504 (1971).

[3] Brown, W. S.: The Subresultant PRS Algorithm. ACM TOMS **4**, 237 – 249 (1978).

[4] Brown, W. S.: On the Subresultant PRS Algorithm. Abstract, SYMSAC **1976**, 271.

[5] Cantor, M.: Vorlesungen über Geschichte der Mathematik, II. New York: Johnson; Stuttgart: Teubner 1963 (reprint of the 2nd edition 1900).

[6] Collins, G. E.: Polynomial Remainder Sequences and Determinants. Am. Math. Monthly **73**, 708 – 712 (1966).

[7] Collins, G. E.: Subresultants and Reduced Polynomial Remainder Sequences. JACM **14**, 128 – 142 (1967).

[8] Collins, G. E.: Computer Algebra of Polynomials and Rational Functions. Am. Math. Mon. **80**, 725 – 755 (1973).

[9] Collins, G. E.: The Calculation of Multivariate Polynomial Resultants. JACM **19**, 515 – 532 (1971).

[10] Collins, G. E.: The SAC-1 Polynomial GCD and Resultant System, Univ. of Wisconsin Comp. Sci. Dept. Tech. Report No. 145, 1 – 93 (Feb. 1972).

[11] Habicht, W.: Eine Verallgemeinerung des Sturmschen Wurzelzählverfahrens. Comm. Math. Helvetici **21**, 99 – 116 (1948).

[12] Habicht, W.: Zur inhomogenen Eliminationstheorie. Comm. Math. Helvetici **21**, 79 – 98 (1948).

[13] Hearn, A. C.: Non-modular Computation of Polynomial GCD's Using Trial Divisions. EUROSAM **1979**, 227 – 239.

[14] Heindel, L. E.: Integer Arithmetic Algorithms for Polynomial Real Zero Determination. JACM **18**, 533 – 548 (1971).

[15] Knuth, D. E.: The Analysis of Algorithms. Proc. Internat. Congress Math. (Nice, 1970), Vol. 3, pp. 269 – 274. Gauthier-Villars 1971.

[16] Knuth, D. E.: The Art of Computer Programming, Vol. II, 1st ed., Reading, Mass.: Addison-Wesley. 1969.

[17] Knuth, D. E.: The Art of Computer Programming, Vol. II, 2nd ed., Reading, Mass.: Addison-Wesley. 1981.

[18] Kronecker, L.: Über den Zahlbegriff, Crelle J. reine und angew. Mathematik **101**, 337 – 355 (1887).

[19] Lauer, E.: Algorithms of Symmetric Polynomials. SYMSAC **1976**, 242 – 247.

[20] Lehmer, D. H.: Euclid's Algorithm for Large Numbers. Amer. Math. Monthly **45**, 227 – 233 (1937).

[21] Loos, R.: Computing Rational Zeros of Integral Polynomials by *p*-adic Expansion. SIAM J. Comp. (to appear).

[22] Moenck, R.: Fast Computation of GCD's. Proc. Fifth Sympos. on Theory of Computing, ACM, New York, pp. 142 – 151 (1973).

[23] Moses, J., Yun, D. Y. Y.: The EZ GCD Algorithm, Proc. of ACM Annual Conference, Atlanta, pp. 159 – 166 (August 1973).

[24] Musser, D. R.: Multivariate Polynomial Factoring. JACM **22**, 291 – 308 (1975).

[25] Nuñez, P.: Livro de Algebra em Arithmatica e Geometrica, problem No. LIII, in [27].

[26] Schönhage, A.: Schnelle Berechnung von Kettenbruchentwicklungen. Acta Inform. **1**, 130 – 144 (1971).

[27] Stevin, S.: Les oevres mathematiques de Simon Stevin (Girard, A., ed.). Leyden: 1634.

[28] Strassen, V.: Private communication, 1976.

[29] Strassen, V.: The Computational Complexity of Continued Fractions. SYMSAC **1981**, 51 – 67.

[30] Tarski, A.: A Decision Method for Elementary Algebra and Geometry, 2nd ed. Berkeley: Univ. of California Press 1951.

[31] Trager, B. M.: Algebraic Factoring and Rational Function Integration. SYMSAC **1976**, 219 – 226.

[32] van der Waerden, B. L.: Modern Algebra, Vol. II. New York: Frederick Ungar Publ. Co. 1953.

[33] Wang, R. S.: An Improved Multivariable Polynomial Factorising Algorithm. Math. Comp. **32**, 1215 – 1231 (1978).

[34] Weber, H.: Lehrbuch der Algebra, 2. Aufl., §54, Theorem IV. Braunschweig: 1898.

[35] Yun, D. Y. Y.: On Square-Free Decomposition Algorithms. SYMSAC **1976**, 26 – 35.
[36] Yun, D. Y. Y.: The Hensel Lemma in Algebraic Manipulation, Ph. D. Thesis, Dept. of Math., MAC-TR-138, M.I.T., November 1974.

Prof. Dr. R. Loos
Institut für Informatik I
Universität Karlsruhe
Zirkel 2
D-7500 Karlsruhe
Federal Republic of Germany

Computing by Homomorphic Images

M. **Lauer,** Karlsruhe

Abstract

After explaining the general technique of computing by homomorphic images, the chinese remainder algorithm and the Hensel lifting construction are treated extensively. Chinese remaindering is first presented in an abstract setting. Then the specialization to Euclidean domains, in particular Z, $K[y]$ and $Z[y_1, \ldots, y_r]$ is treated. The lifting construction is first also presented in an abstract form from which Hensel's Lemma derives by specialization. After introducing Zassenhaus' quadratic lifting construction, again, the case of Z and $Z[y_1, \ldots, y_r]$ is considered. For both techniques, chinese remaindering as well as the lifting algorithms, a complete computational example is presented and the most frequent applications are discussed.

0. Introduction

This chapter gives an exposition of Chinese remainder and Hensel lifting algorithms which are the most important algebraic algorithms of the so-called modular method. Finite fields belong to the few algebraic structures which can exactly and completely be represented on a digital computer. As a general rule, all problems involving a considerable amount of arithmetic on long integers or rationals should be considered for application of the modular method. Similarly, multivariate problems can be reduced to univariate or even numerical ones.

The problem type of this chapter is the transformation of tedious tasks to domains where they can be solved efficiently and to transform the solutions obtained in the image domains back to the domain of the original problem.

There are further methods, as Kronecker's famous trick to transform multivariate polynomials to univariate polynomials which are not treated in this chapter. Moenck [12, 13] has given an application of this idea. Also, the numerical representation of polynomials as long integers for applying fast integer arithmetic – an idea proposed by Schönhage [18] recently – may be considered in this context.

1. Homomorphic Images

Let R be a commutative ring with identity element and I an ideal in R. For $a \in R$ the residue class $a + I$ is denoted by $\bar{a}$. Following [1, Section 4], an ample set for R/I is a set $A \subseteq R$ which contains exactly one element from each residue class $\bar{a}$. An ample function for R/I is a function $\alpha: R \to A$ from R onto some ample set A such that $\alpha(a) = \alpha(b)$ just in case $a \equiv b \pmod{I}$. Given an ample set A, we can represent each residue class by its unique element in A. Given algorithms for addition, subtraction and multiplication in R and given an algorithm which realizes an ample function α, we obtain algorithms for addition, subtraction and multiplication in R/I. More precisely, R/I is replaced by A which is made to a ring isomorphic to R/I by defining $a +_A b = \alpha(a + b)$ and $a \cdot_A b = \alpha(a \cdot b)$. In practical cases of algebraic algorithms there are "natural" choices of ample sets and ample functions. For example, let $R = \mathbb{Z}$ (the ring of integers) and $I = (m)$, $m \in \mathbb{Z}$. Then $\{0, 1, \ldots, m - 1\}$ is an ample set for R/I, and the ample function maps each $a \in R$ onto the positive remainder of the division of a by m. Another ample set is $\{a \in \mathbb{Z} \mid -m/2 < a \leqslant m/2\}$, and the associated ample function is also easily realized. As a second example, let $R = K[y]$, where K is a field, and $I = (g)$, $g \in K[y]$. Then $\{a \in K[y] \mid \deg(a) < \deg(g)\}$ is an ample set for R/I, and the ample function is again "remaindering". Note that in both examples (for apparent reasons related to efficiency of computation) the ample sets are chosen such that they contain the "smallest" element of each residue class.

Now suppose that we want to solve some problem in R. The problem involves some elements $r_1, r_2, \ldots, r_m \in R$ ($m \geqslant 1$), and $a_1, a_2, \ldots, a_l \in R$ ($l \geqslant 1$) is the solution to this problem. For example, $a_1, a_2, \ldots, a_l$ may be the complete factorization of an integral polynomial r_1 ($m = 1$), or $l = 1$ and a_1 is the greatest common divisor of two polynomials r_1, r_2. Assume further that I is an ideal in R, and that the natural modular homomorphism from R onto R/I commutes with the solution of our problem. This means that we get the same result whether we first solve the problem in R and then apply the modular homomorphism to the solution or first apply the modular homomorphism and then solve the "image"-problem in R/I. Then I is called lucky.

Let us now consider two extreme cases of the ideal I. In the first case we know in advance that all elements a_i of the solution of our problem lie in the ample set associated with I. For example, if $R = \mathbb{Z}[y_1, y_2, \ldots, y_r]$, $r \geqslant 1$, it may be possible to give a priori bounds on the degrees and norms of the a_i: if it is known that $|a_i|_\infty < q$ and $\deg_{y_j}(a_i) < n_j$, $j = 1, \ldots, r - 1$, let $I = (2q, f_1, f_2, \ldots, f_{r-1})$ where each f_j is an element of R which contains only the variable y_j and has degree n_j. Then I has this property. It is then "sufficient" to solve our problem modulo such an ideal I if we also know that the solution mod I is congruent to the solution in R (congruent mod I).

In the second case the ideal I has the property that the image problem in R/I can be solved rather efficiently. For example, if $R = \mathbb{Z}[y_1, y_2, \ldots, y_r]$, $r \geqslant 1$, I may be chosen such that $R/I \simeq \mathrm{GF}(p)[y_r]$, or even $R/I \simeq \mathrm{GF}(p)$, where p is a prime integer. In $\mathrm{GF}(p)[y_r]$ or $\mathrm{GF}(p)$ many problems can be solved very efficiently; cf. the contributions on polynomial factorization and on polynomial remainder sequences in this volume. The reasons for this are obvious: $\mathrm{GF}(p)$ is a field, and hence $\mathrm{GF}(p)[y_r]$ is a Euclidean Domain. Also the length of the "numerical" coefficients is bounded by $p - 1$ (or $p/2$, according to the ample set). In particular, if p is chosen to be single precision, then all numerical coefficients are single precisions and hence do not contribute to the computational complexity of the algorithms. Furthermore the finiteness of $\mathrm{GF}(p)$ is used extensively by some algorithms; Berlekamp's algorithm for the complete factorization of univariate squarefree polynomials is a well-known example.

The point now is that there exist methods which – under certain conditions – bring these two cases together. One method is based on the abstract Chinese Remainder Theorem. It uses solutions modulo several ideals $I_1, I_2, \ldots, I_n$, where each I_j belongs to the second kind, and constructs a solution modulo $I = I_1 \cap I_2 \cap \cdots \cap I_n$; k and each I_j are chosen such that I belongs to the first kind.

The second method is based on Newton's method for computing the zeros of an equation. It computes ("lifts") from a solution modulo I^t a solution modulo I^{t+1} (or I^{2t} in the "quadratic" version). Starting with a solution modulo I_1 this process is iterated to yield a solution modulo I_1^k, where I_1 and k are chosen such that I_1 belongs to the second kind and I_1^k belongs to the first kind.

These two methods will subsequently be described in more detail.

2. Chinese Remainder Algorithms

An in-depth study of the mathematical and algorithmic aspects of the Chinese Remainder Theorem and its connection with interpolation is given in [8]. We partially follow this exposition.

2.1 An Abstract Chinese Remainder Theorem

A set $\{I_1, I_2, \ldots, I_n\}$ of ideals in R is called pairwise spanning, if

$$I_j + I_k = R, \qquad \forall j \neq k. \tag{2.1}$$

As an immediate consequence for any pair I_j, I_k of these ideals ($j \neq k$) there exist $n_j^{(k)} \in I_j$, $n_k^{(j)} \in I_k$ such that

$$n_j^{(k)} + n_k^{(j)} = 1. \tag{2.2}$$

Theorem (Abstract Chinese Remainder Theorem). *Let the set* $\{I_1, I_2, \ldots, I_n\}$ *of ideals in R be pair-wise spanning. Then the map*

$$\phi : u \mapsto (u + I_1, u + I_2, \ldots, u + I_n)$$

is a homomorphism from R onto $\prod_{j=1}^{n} R/I_j$ *with kernel* $I_1 \cap I_2 \cap \cdots \cap I_n$.

Proof. It is easily verified that ϕ is a homomorphism with kernel $I_1 \cap I_2 \cap \cdots \cap I_n$. So it remains to prove that ϕ is surjective. Thus for arbitrary $(u_1 + I_1, u_2 + I_2, \ldots, u_n + I_n) \in \prod_{j=1}^{n} R/I_j$ we must establish $u \in R$ such that

$$\phi(u) = (u_1 + I_1, u_2 + I_2, \ldots, u_n + I_n), \tag{2.3}$$

or equivalently

$$u \equiv u_j \pmod{I_j}, \qquad j = 1, 2, \ldots, n. \tag{2.4}$$

This system of congruences will subsequently be called the CRP (Chinese Remainder Problem).

There are two "natural" approaches to this problem. In the first one tries to find $L_k \in R$, $k = 1, 2, \ldots, n$, with

$$L_k \equiv \delta_{jk} \pmod{I_j}.$$

Then

$$u = u_1 L_1 + u_2 L_2 + \cdots + u_n L_n$$

is the desired solution of (2.4). It is called the *Lagrangian solution of the CRP*.

The second approach is recursive. Having already found a solution $u^{(k-1)}$ for

$$u^{(k-1)} \equiv u_j \pmod{I_j} \qquad \text{for} \qquad j = 1, 2, \ldots, k-1,$$

we try to use this to find $u^{(k)}$. Starting with $u^{(1)} = u_1$, $u = u^{(n)}$ is a solution of (2.4) which is called the *Newtonian solution of the CRP*.

We will expose both solutions.

(i) Lagrangian solution. We are looking for $L_k \in R$, $k = 1, 2, \ldots, n$, with

$$L_k \equiv \delta_{jk} \pmod{I_j}. \tag{2.5}$$

With $n_j^{(k)}$ as in (2.2) let

$$L_k = \prod_{\substack{j=1 \\ j \neq k}}^{n} n_j^{(k)}. \tag{2.6}$$

Since $n_j^{(k)} \in I_j$,

$$L_k \equiv 0 \pmod{I_j}, \qquad j \neq k, \tag{2.7}$$

by the defining properties of an ideal. Modulo I_k we have

$$L_k = \prod_{\substack{j=1 \\ j \neq k}}^{n} n_j^{(k)}$$

(by (2.2))
$$= \prod_{\substack{j=1 \\ j \neq k}}^{n} (1 - n_k^{(j)})$$

$$\equiv 1 \pmod{I_k}, \tag{2.8}$$

since each $n_k^{(j)} \in I_k$. (2.7) and (2.8) together establish (2.5). The complete Lagrangian solution of (2.4) reads as follows:

$$u = \sum_{k=1}^{n} u_k L_k, \tag{2.9}$$

where

$$L_k = \prod_{\substack{j=1 \\ j \neq k}}^{n} n_j^{(k)} \tag{2.10}$$

and $n_j^{(k)}$ as in (2.2).

(ii) Newtonian solution. As mentioned above we assume a solution $u^{(k-1)}$ with

$$u^{(k-1)} \equiv u_j \pmod{I_j} \qquad \text{for} \qquad j = 1, 2, \ldots, k-1. \tag{2.11}$$

Let

$$a_k = u_k - u^{(k-1)} \tag{2.12}$$

and

$$u^{(k)} = u^{(k-1)} + a_k \prod_{j=1}^{k-1} n_j^{(k)} \tag{2.13}$$

with $n_j^{(k)}$ as in (2.2). Then, since each $n_j^{(k)} \in I_j$,

$$u^{(k)} \equiv u^{(k-1)} \pmod{I_j},$$

(by (2.11)) $\equiv u_j \pmod{I_j}$ for $j = 1, 2, \ldots, k-1.$ \hfill (2.14)

Modulo I_k we have

$$u^{(k)} = u^{(k-1)} + a_k \prod_{j=1}^{k-1} n_j^{(k)}$$

(by (2.2)) $= u^{(k-1)} + a_k \cdot \prod_{j=1}^{k-1} (1 - n_k^{(j)})$

(since $n_k^{(j)} \in I_k$) $\equiv u^{(k-1)} + a_k \cdot 1 \pmod{I_k}$

(by (2.12)) $\equiv u^{(k-1)} + (u_k - u^{(k-1)})$

$$\equiv u_k \pmod{I_k}. \tag{2.15}$$

(2.14) and (2.15) together establish that

$$u^{(k)} \equiv u_j \pmod{I_j}   \text{for}   j = 1, 2, \ldots, k.$$

The complete Newtonian solution of (2.4) reads as follows:

$$u = u^{(n)},$$

where

$$u^{(k)} = u^{(k-1)} + a_k \cdot q_k, \tag{2.16}$$

$$q_k = \prod_{j=1}^{k-1} n_j^{(k)}, \tag{2.17}$$

$$a_1 = u^{(1)} = u_1 \tag{2.18}$$

and

$$a_k = u_k - u^{(k-1)}, \tag{2.19}$$

so

$$u = \sum_{k=1}^{n} a_k q_k. \tag{2.20}$$

This completes the proof of the theorem. $\blacksquare$

As stated in the theorem the kernel of ϕ is $I_1 \cap I_2 \cap \cdots \cap I_n$ which means that the solution u of (2.4) is only unique modulo the ideal $I_1 \cap I_2 \cap \cdots \cap I_n$.

Both proofs of the theorem are constructive, and in the next section we will derive algorithms out of them. Before doing so let us refer back to Section 1 and the abstract problem we wanted to solve.

Suppose that the set $\{I_1, I_2, \ldots, I_n\}$ of ideals in R is pair-wise spanning and that each I_j is lucky and of the second kind. So the solutions $a_i^{(j)}$ of the image problems in R/I_j can "easily" be obtained. Using an algorithm which is either based on the Lagrangian or the Newtonian proof of the theorem we can compute $a_i^{(*)}$ with

$$a_i^{(*)} \equiv a_i^{(j)} \pmod{I_j}, \quad j = 1, \ldots, n, \quad i = 1, \ldots, l. \tag{2.21}$$

Since each I_j is lucky,

$$a_i \equiv a_i^{(j)} \pmod{I_j},$$

so (2.21) together with the "uniqueness" statement above implies that

$$a_i \equiv a_i^{(*)} \pmod{I_1 \cap I_2 \cap \cdots \cap I_n}.$$

If also $I_1 \cap I_2 \cap \cdots \cap I_n$ belongs to the first kind we are "done".

2.2 Chinese Remainder Algorithms in Euclidean Domains

The algebraic setting relevant to symbolic computation is $Z[y_1, y_2, \ldots, y_r]$ which is embedded in $Q(y_1, \ldots, y_{r-1})[y_r]$ which in turn is a Euclidean Domain. Therefore we will specialize our attention to this case.

For many applications in symbolic computation the required number of ideals (moduli) will not be known in advance. In this case an algorithm based on the Lagrangian solution has one bad property, in that all L_k's of (2.10) have to be recomputed if the number of ideals is changed (generally it is increased by 1). (The algorithm is off-line.)

On the other hand the Newtonian solution in Euclidean Domains is very favourable in this regard: increasing the number of moduli from n to $n + 1$ simply entails in passing from $u^{(n)}$ to $u^{(n+1)}$, i.e. adding one iteration step (i.e. the algorithm is on-line). For these reasons we omit the description of a Lagrangian algorithm.

An Euclidean Domain R is a Principle Ideal Domain, and a set $\{I_1, I_2, \ldots, I_n\}$ of ideals in R with $I_j = (m_j)$ is pair-wise spanning if their generators are pair-wise relatively prime, i.e.

$$\gcd(m_j, m_k) = 1, \quad \forall j \neq k.$$

Then for each $n_j^{(k)}$ from (2.2) there exists $s_j^{(k)} \in R$ with

$$n_j^{(k)} = s_j^{(k)} m_j. \tag{2.22}$$

$s_j^{(k)}$ and $s_k^{(j)}$ are computable by the extended Euclidean Algorithm (cf. chapter on polynomial remainder sequences) as solutions of:

$$s_j^{(k)} m_j + s_k^{(j)} m_k = 1. \tag{2.23}$$

Using (2.22) the Newtonian solution (2.16) – (2.20) becomes

$$u^{(k)} = u^{(k-1)} + a_k \cdot q_k, \tag{2.24}$$

$$q_k = \prod_{j=1}^{k-1} m_j, \tag{2.25}$$

$$a_1 = u^{(1)} = u_1, \tag{2.26}$$

and

$$a_k = (u_k - u^{(k-1)}) \prod_{j=1}^{k-1} s_j^{(k)}, \tag{2.27}$$

so

$$u = \sum_{k=1}^{n} a_k q_k. \tag{2.28}$$

According to the uniqueness considerations discussed at the end of Section 2.1, a solution u is only unique modulo the ideal $I_1 \cap I_2 \cap \cdots \cap I_n$ which now is equal to the (principal) ideal $(\prod_{j=1}^{n} m_j)$. Therefore an algorithm which computes u will compute each $u^{(k)}$ (cf. (2.24)) modulo the ideal $(\prod_{j=1}^{k} m_j)$, i.e. as an element of the ample set associated with $(\prod_{j=1}^{k} m_j)$. For "natural" ample sets this means that a_k (cf. (2.27)) is computed modulo m_k. This is the reason for taking $\prod_{j=1}^{k-1} s_j^{(k)}$ into the definition of a_k in (2.27) which results in a considerable computational simplification. If the solution of a problem contains several elements $a_1, a_2, \ldots, a_l$ the Chinese Remainder Algorithm will be used several times, but always with the same ideals. Hence another computational gain is made if all entities which depend only on the ideals are precomputed and then given as additional inputs to the Chinese Remainder Algorithm. These are

$$q_k = \prod_{j=1}^{k-1} m_j, \qquad k = 2, 3, \ldots, n \tag{2.29}$$

and

$$s_k = \prod_{j=1}^{k-1} s_j^{(k)} \pmod{m_k}, \qquad k = 2, 3, \ldots, n. \tag{2.30}$$

Finally we note from (2.30) that only the $s_j^{(k)} \pmod{m_k}$ with $j < k$ are needed.

By (2.23)

$$s_j^{(k)} m_j \equiv 1 \pmod{m_k},$$

and hence

$$s_j^{(k)} = m_j^{-1} \pmod{m_k}, \tag{2.31}$$

which can be computed using a "half-extended" Euclidean Algorithm. With this in mind the algorithm now becomes

Algorithm CRA (Chinese Remainder Algorithm).

Inputs: $m_1, m_2, \ldots, m_n$ (pair-wise relatively prime),
$u_1, u_2, \ldots, u_n,$
$q_2, q_3, \ldots, q_n$ (q_k as in (2.29)),
$s_2, s_3, \ldots, s_n$ (s_k as in (2.30)).

Outputs: $u \pmod{\prod_{j=1}^{n} m_j}$ with: $u \equiv u_j \pmod{m_j}$.

(1) [Initialize]

  $u := u_1 \bmod m_1$.

(2) [Incorporate other ideals]

  for $k = 2, 3, \ldots, n$ do

  begin $v := u \bmod m_k$;

    $a := (u_k - v) \cdot s_k \bmod m_k$;

    $u := u + a \cdot q_k$

  end. ∎

(*Remark*. A similar algorithm is attributed in [4], p. 274, to H. L. Garner at 1958. It does not precompute the s_k but only the $s_j^{(k)}$, and then computes "inside" the a_k by evaluating (2.27) using Horner's scheme.)

As mentioned above the required number of moduli (ideals) will often not be known in advance. Instead after each new u_k is generated a Chinese Remainder Algorithm for two moduli will be reapplied to incorporate m_k and u_k, and a test will then be made to decide whether additional moduli are required.

In such a Chinese Remainder Algorithm for two ideals the first ideal will typically be the intersection of several original ideals and hence have a "large" generator m_1 (the product of the original moduli). The second ideal will be an original ideal. For this reason, it will not be uncommon to have $u_1 \equiv u_2 \pmod{m_2}$ which means that $u = u_1$ is the solution to this two-ideal CRP.

Regarding this we obtain

Algorithm CRA2 (Chinese Remainder Algorithm for 2 moduli).

Inputs: m_1, m_2 (relatively prime),

    u_1, u_2,

    s $(s \cdot m_1 \equiv 1 \pmod{m_2})$.

Output: $u \pmod{m_1 \cdot m_2}$ with: $u \equiv u_j \pmod{m_j}$.

(1) $[u = u_1 ?]$

    $u_1' = u_1 \bmod m_2$;

    $d := u_2 - u_1' \bmod m_2$;

    if $d = 0$ then $\{u := u_1; \text{exit}\}$.

(2) [General case.]

    $a := d \cdot s \bmod m_2$;

    $u := u_1 + a \cdot m_1$. ∎

In essentially this way the (integer) Chinese Remainder Algorithm is realized for example in SAC-2 (cf. chapter on arithmetic).

Algorithm CRA can now be replaced by a repeated application of CRA2 in an obvious manner.

2.3 Chinese Remainder Algorithms in $\mathbb{Z}$, $K[y]$, and $\mathbb{Z}[y_1, y_2, \ldots, y_r]$

The case $R = \mathbb{Z}$ is of interest for symbolic computation because it is common in algebraic calculations that very large integer coefficients are generated. So typically the moduli are chosen to be large primes which are just smaller than the word size on the computer which is used. (More precisely, the primes are smaller than the base for the representation of large integers.)

The computing time for algorithm CRA is $O(n^2)$ if all inputs are single-precision. The computing time for CRA2 is $O(\log m_1)$ if m_2, u_2 and s are single precision and u_1 is "reduced" mod m_1.

We now give an example. Let

$$m_1 = 7, \qquad m_2 = 11 \qquad m_3 = 13, \qquad m_4 = 15,$$
$$u_1 = -1, \qquad u_2 = -2, \qquad u_3 = 2, \qquad u_4 = 6.$$

We want to compute u such that

$$u \equiv u_j \pmod{m_j}, \qquad j = 1, \ldots, 4.$$

We choose $\{a \in \mathbb{Z} \mid -m/2 < a \leqslant m/2\}$ as an ample set associated with (m). So, since u is only unique modulo $m = \prod_{j=1}^{4} m_j = 15015$, u will satisfy $-m/2 < u \leqslant m/2$.

Since we know the number of moduli we may use Algorithm CRA. The (precomputed) constants used by CRA are

$$q_2 = 7, \qquad q_3 = 77, \qquad q_4 = 1001 \qquad \text{(cf. (2.29)),}$$
$$s_1^{(2)} = -3, \qquad s_1^{(3)} = 2, \qquad s_1^{(4)} = -2,$$
$$s_2^{(3)} = 6, \qquad s_2^{(4)} = -4 \qquad \text{(cf. (2.31)),}$$
$$s_3^{(4)} = 7,$$
$$s_2 = -3, \qquad s_3 = -1, \qquad s_4 = -4 \qquad \text{(cf. (2.30)).}$$

Following CRA the variables v, a, u are successively computed as

$$u = -1,$$
$$k = 2: \quad v = -1 \bmod 11 = -1,$$
$$a = (-2 - (-1)) \cdot (-3) \bmod 11 = 3,$$
$$u = -1 + 3 \cdot 7 = 20,$$
$$k = 3: \quad v = 20 \bmod 13 = 7,$$
$$a = (2 - 7) \cdot (-1) \bmod 13 = 5,$$
$$u = 20 + 5 \cdot 77 = 405,$$
$$k = 4: \quad v = 405 \bmod 15 = 0,$$
$$a = (6 - 0) \cdot (-4) \bmod 15 = 6,$$
$$u = 405 + 6 \cdot 1001 = 6411.$$

Hence $u = 6411$ is the solution of the above CRP.

Next let $R = K[y]$ where K is a field. If $m(y) \in K[y]$ we choose $\{a(y) \in K \,|\, \deg(a) < \deg(m)\}$ as an ample set associated with the ideal $(m(y))$. Now we take a special set of moduli, namely $m_j(y) = y - d_j \in K[y]$ with $d_j \in K$, $j = 1, 2, \ldots, n$. Clearly these are pair-wise relatively prime if $d_j \neq d_k$ for $j \neq d$. Furthermore, it is well-known that, for $u(y) \in K[y]$,

$$u(y) \bmod (y - d_j) = u(d_j).$$

In other words,

$$u(y) \equiv u_j \quad (\bmod (y - d_j)), \qquad u_j \in K,$$

is equivalent to

$$u(d_j) = u_j.$$

Thus we see that in $K[y]$ the CRP with the special set of moduli $m_j(y) = y - d_j$ and the special "residues" $u_j \in K$ turns out to be the usual Interpolation Problem. Moreover, the Lagrangian and the Newtonian Solution of this special CRP are the Lagrangian and the Newtonian Solution, respectively, of the Interpolation Problem (whence the naming for the CRP solutions). Also, the well-known uniqueness of the interpolating polynomial follows immediately from the uniqueness mod $\prod_{j=1}^{n} m_j(y) = \prod_{j=1}^{n} (y - d_j)$ of the solution of the CRP.

As an example we take $K = \mathbb{Q}$, the set of rational numbers. Let

$$m_1 = y - 1, \qquad m_2 = y - 4, \qquad m_3 = y - 6, \qquad m_4 = y - 11,$$

$$u_1 = 10, \qquad u_2 = 334, \qquad u_3 = 1040, \qquad u_4 = 5920.$$

We want to compute $u(y)$ such that

$$u(y) \equiv u_j \quad (\bmod m_j), \qquad j = 1, \ldots, 4,$$

i.e.

$$u(1) = 10, \qquad u(4) = 334 \quad \text{etc.}$$

Furthermore, $\deg(u) < 4$.

Again we use CRA and first precompute the needed constants:

$$q_2 = y - 1,$$

$$q_3 = y^2 - 5y + 4 \qquad\qquad \text{(cf. (2.29)),}$$

$$q_4 = y^3 - 11y^2 + 34y - 24.$$

From (2.23) we obtain

$$s_j^{(k)} \cdot (y - d_j) + s_k^{(j)} \cdot (y - d_k) = 1,$$

whence

$$s_j^{(k)} = \frac{1}{d_k - d_j}. \tag{2.32}$$

Hence

$$s_1^{(2)} = \tfrac{1}{3}, \qquad s_1^{(3)} = \tfrac{1}{5}, \qquad s_1^{(4)} = \tfrac{1}{10},$$

$$s_2^{(3)} = \tfrac{1}{2}, \qquad s_2^{(4)} = \tfrac{1}{7},$$

$$s_3^{(4)} = \tfrac{1}{5},$$

$$s_2 = \tfrac{1}{3}, \qquad s_3 = \tfrac{1}{10}, \qquad s_4 = \tfrac{1}{350}.$$

Following Algorithm CRA the variables v, a, u are successively computed as

$$u(y) = 10,$$

$k = 2$: $\qquad v = u(y) \bmod y - 4 = u(4) = 10,$

$\qquad\qquad a = (334 - 10) \cdot \tfrac{1}{3} = 108,$

$\qquad\qquad u(y) = 10 + 108 \cdot (y - 1) = 108y - 98,$

$k = 3$: $\qquad v = u(6) = 550,$

$\qquad\qquad a = (1040 - 550) \cdot \tfrac{1}{10} = 49,$

$\qquad\qquad u(y) = 108y - 98 + 49 \cdot (y^2 - 5y + 4)$

$\qquad\qquad\quad = 49y^2 - 137y + 98,$

$k = 4$: $\qquad v = u(11) = 4520,$

$\qquad\qquad a = (5920 - 4520) \cdot \tfrac{1}{350} = 4,$

$\qquad\qquad u(y) = 49y^2 - 137y + 98 + 4 \cdot (y^3 - 11y^2 + 34y - 24)$

$\qquad\qquad\quad = 4y^3 + 5y^2 - y + 2.$

Hence $u(y) = 4y^3 + 5y^2 - y + 2$ is the solution of the above CRP.

This example shows an important aspect of Algorithm CRA. Although we choose $K = \mathbb{Q}$ the variables v and a and the coefficients of $u(y)$ all take integral values! The only fractions that appear are the s_k which by (2.32) all have the numerator 1. If we set $r_k = 1/s_k$ and in CRA replace the multiplication with s_k by the division by r_k, then all these divisions were exact. The reason for this is that all evaluation points d_j were in $\mathbb{Z}$ (rather than in $\mathbb{Q}$) and that the final result was in $\mathbb{Z}[y]$. This fact is stated in the following theorem.

Theorem. *Let D be an integral domain and consider the Interpolation Problem in D[y]. (This means that the evaluation points d_j are chosen in D, and that the solution polynomial is known (from a priori considerations) to have coefficients in D.) Then during the execution of Algorithm CRA all values of the variables v and a and all coefficients of u(y) lie in D.*

A proof for this "*D*-closure property" is given in [8]. Of course this remains true if Algorithm CRA is replaced by a repeated application of CRA2.

We finally consider the case $R = \mathbb{Z}[y_1, y_2, \ldots, y_r]$ which is the domain of greatest interest for symbolic computation. We could choose I such that it is of the second kind (cf. Section 1) which means that $R/I \simeq \mathrm{GF}(p)[y]$; for example

$I = (p, y_1 - d^{(1)}, y_2 - d^{(2)}, \ldots, y_{r-1} - d^{(r-1)})$. But then we cannot embed the problem into a CRP for Euclidean domains which are the only domains for which efficient Chinese Remainder Algorithms are known.

We can use these efficient algorithms if instead we make a step-wise approach from R to $GF(p)[y]$:

$$\mathbb{Z}[y_1, y_2, \ldots, y_r] \underset{\bmod p_i}{\to} GF(p_i)[y_1, y_2, \ldots, y_r] \to$$

$$\underset{\bmod(y_r - d_j^{(r)})}{\to} GF(p_i)[y_1, y_2, \ldots, y_{r-1}] \to \cdots \to$$

$$\underset{\bmod(y_2 - d_k^{(2)})}{\to} GF(p_i)[y_1] \quad (\text{possibly} \underset{\bmod(y_1 - d_t^{(1)})}{\to} GF(p_i)).$$

For the reversal steps we write

$$GF(p)[y_1, y_2, \ldots, y_l] = GF(p)[y_1, y_2, \ldots, y_{l-1}][y_l]$$

whence the step

$$GF(p)[y_1, y_2, \ldots, y_{l-1}] \to GF(p)[y_1, y_2, \ldots, y_{l-1}][y_l]$$

is precisely an Interpolation Problem in $GF(p)[y_1, y_2, \ldots, y_{l-1}][y_l]$. Since $GF(p)[y_1, y_2, \ldots, y_{l-1}]$ is an integral domain the "D-closure"-Theorem above guarantees the existence of efficient fraction-free interpolation algorithms.

So an algorithm which solves a problem in $\mathbb{Z}[y_1, y_2, \ldots, y_r]$ and which employs Chinese Remainder Algorithms (a "modular" algorithm) typically works as follows (cf. Section 2.4.):

The problem is taken mod p_i and hence mapped onto a problem in $GF(p_i)[y_1, y_2, \ldots, y_r]$ for several primes p_i. For each image problem a (sub-) algorithm for solving the problem in $GF(p_i)[y_1, y_2, \ldots, y_r]$ is invoked. The solutions which are returned by these invocations are combined using a Chinese Remainder Algorithm to produce the final solution in $\mathbb{Z}[y_1, y_2, \ldots, y_r]$. (See Section 2.4 for more details of this step.)

Analogously, the algorithm working in $GF(p)[y_1, y_2, \ldots, y_r]$ maps the problem mod $y_r - d_j^{(r)}$ onto a problem in $GF(p)[y_1, y_2, \ldots, y_{r-1}]$ for several $d_j^{(r)} \in GF(p)$. For each image problem it invokes itself recursively. The solutions which are returned by these recursive calls are combined using a Chinese Remainder Algorithm (which now specializes to an interpolation algorithm) to produce the result in $GF(p)[y_1, y_2, \ldots, y_r]$. The recursion base is $r = 1$ or $r = 0$, according to the problem. It is chosen such that the solution of the image problem in the corresponding domain (i.e. $GF(p)[y_1]$ or $GF(p)$) is meaningful. For example, if we want to compute the determinant of a matrix the entries of which are elements of $GF(p)[y_1, y_2, \ldots, y_r]$ we will choose $r = 0$. At the other hand factorization or gcd calculation is meaningless in $GF(p)$, so for these problems we will have $r = 1$ as a recursion base.

Another problem which arises in this context is that the solution in $GF(p)[y_1]$ may be unique only up to units; the gcd problem gives an example. This is the "leading coefficient problem" because it is usually solved by imposing appropriate leading coefficients to the solution in $GF(p)[y_1]$.

Of course at each stage on the way from $GF(p)[y_1]$ to $\mathbb{Z}[y_1, y_2, \ldots, y_r]$ we need a termination condition, i.e. we have to decide whether enough moduli are processed. One way to achieve this is by giving a priori bounds (computable from the "inputs") on the degrees and norms of the solution (cf. Section 1). Such bounds can quite often be given; for example Hadamard's inequality bounds the norm of a determinant (cf. [4], p. 414).

If the bound is not sharp (and many bounds are not) then an algorithm which uses this bound for termination ordinarily uses far more moduli than actually necessary. So in place of this one may use some "direct" test to decide whether the solution obtained so far (i.e. the solution modulo the product of the used moduli) is the final solution. For gcd calculation we may use a trial division. If this test for termination is rather costly we may defer it until the incorporation of a modulus (by a Chinese Remainder Algorithm) leaves the result unchanged from the previous application.

At the end of this section let us make some remarks on the "luckiness" of ideals; cf. Section 1. The "definition" of luckiness we gave there is the best we can hope for. It may be too restrictive for some applications since there are instances of problems for which every modulus may be unlucky. The standard example is the factorization of $x^4 + 1$ which factors modulo every prime even though it is irreducible over the integers. Roughly speaking, an ideal I is called lucky if the solution mod I can be used to reconstruct the solution in R. In $\mathbb{Z}[y_1, y_2, \ldots, y_r]$ a necessary condition usually is that the mod I operation does not affect the degrees of the input polynomials or properties like squarefreeness. Although the number of unlucky ideals is almost always negligible, for every problem careful considerations must be given to their detection. It may even be the case that the unluckiness cannot be detected at once. If for example in the gcd problem in $\mathbb{Z}[y]$ a certain prime p yields a gcd mod p of smaller degree than all previous primes then all these previous primes were unlucky, which means that the result obtained so far is invalid, and that the whole computation has to start again with this prime! (Fortunately this will eventually be detected if, for example, a trial division is used as a test for termination since there exist only finitely many unlucky primes; cf. [1].)

2.4 A Complete Sample Application: Polynomial Division with Remainder

In this section we will give a sample application for the application of Chinese Remainder Algorithms. We do not want to present a highly efficient algorithm. Instead, it should be simple, clear, and illustrative.

We choose the problem of computing the quotient and the remainder of two polynomials in $\mathbb{Z}[y_1, y_2, \ldots, y_r]$, that is given A, B, we look for Q, R such that

$$A = B \cdot Q + R,$$

with $\deg_{y_r}(R) < \deg_{y_r}(B)$.

We assume that we know a priori that Q and R exist. If this is not true we may multiply A by the appropriate power of the leading coefficient of B in order to make our algorithm applicable.

We first give the specifications for the Chinese Remainder Algorithms we are going to use and which are actually implemented with similar specifications in SAC-2.

For the step $GF(p_i)[y_1, y_2, \ldots, y_r] \rightarrow \mathbb{Z}[y_1, y_2, \ldots, y_r]$ we use

$$IPCRA(M, m, m', r, A, a).$$

[Integral polynomial Chinese Remainder Algorithm. M and m are the two moduli with $\gcd(M, m) = 1$. $m' = M^{-1} \bmod m$. A and a are integral polynomials in r variables. The output $A^* \in \mathbb{Z}[y_1, y_2, \ldots, y_r]$ satisfies $A^* \equiv A \bmod M$ and $A^* \equiv a \bmod m$, and is reduced mod $M \cdot m$.]

Algorithm IPCRA essentially applies Algorithm CRA2 of Section 2.2 (working in $\mathbb{Z}$) to each pair of corresponding integer coefficients of A and a.

In $GF(p)[y_1, y_2, \ldots, y_r]$ we make a slight change in the strategy we outlined in Section 2.3 in that we take $y_1 - d_j^{(1)}$ as moduli (instead of $y_r - d_j^{(r)}$). Of course algebraically nothing is changed, but evaluation and interpolation are simplified. So we use

$$MPINT(p, B, b, b', r, A, A_1).$$

[Modular polynomial interpolation. p is a prime, B is a univariate polynomial, $b \in GF(p)$ with $B(b) \neq 0$, and $b' = B(b)^{-1}$ (in $GF(p)$). $A \in GF(p)[y_1, y_2, \ldots, y_r]$ with $\deg_{y_1}(A) < \deg(B)$. $A_1 \in GF(p)[y_2, y_3, \ldots, y_r]$. The output

$$A^* \in GF(p)[y_1, y_2, \ldots, y_r]$$

satisfies

$$A^*(y_1, y_2, \ldots, y_r) \equiv A(y_1, y_2, \ldots, y_r) \bmod B(y_1),$$

$$A^*(b, y_2, \ldots, y_r) = A_1(y_2, y_3, \ldots, y_r), \quad \text{and} \quad \deg_{y_1}(A^*) \leqslant \deg(B).]$$

Analogously to IPCRA, MPINT applies Algorithm CRA2 of Section 2.2 (working in $GF(p)[y_1]$) to each pair of corresponding "innermost" coefficients of A (which are in $GF(p)[y_1]$) and A_1 (which are in $GF(p)$).

Now we are ready to formulate

$$IPQRM(r, A, B; Q, R).$$

[Integral polynomial quotient and remainder, modular algorithm. The inputs are r and A, $B \in \mathbb{Z}[y_1, y_2, \ldots, y_r]$. The outputs Q, $R \in \mathbb{Z}[y_1, y_2, \ldots, y_r]$ satisfy $A = B \cdot Q + R$ and $\deg_{y_r}(R) < \deg_{y_r}(B)$, and must be known to exist.]

begin
$Q := 0; R := 0; P := 1;$
repeat repeat $p := $ next prime;
$\bar{A} := A \bmod p;$
$\bar{B} := B \bmod p$

until $\deg_{y_r}(\bar{A}) = \deg_{y_r}(A)$ and $\deg_{y_r}(\bar{B}) = \deg_{y_r}(B)$;
$\quad$ MPQRM$(r, p, \bar{A}, \bar{B}; \bar{Q}, \bar{R})$;
$\quad$ $p' := P^{-1} \bmod p$;
$\quad$ $Q := $ IPCRA$(P, p, p', r, Q, \bar{Q})$;
$\quad$ $R := $ IPCRA$(P, p, p', r, R, \bar{R})$;
$\quad$ $P := P \cdot p$;
until $A = B \cdot Q + R$
end. $\blacksquare$

$$\text{MPQRM}(r, p, A, B; Q, R).$$

[Modular polynomial quotient and remainder, modular algorithm. The specifications are as in IPQRM except that $A, B, Q, R \in $ GF$(p)[y_1, \ldots, y_r]$.]

begin
if $r = 1$ then compute Q, R directly
else begin
$Q := 0$; $R := 0$; $M(y_1) := 1$;
repeat repeat $m := $ next evaluation point;
$\qquad\qquad$ $\bar{A}(y_2, \ldots, y_r) := A(m, y_2, \ldots, y_r)$;
$\qquad\qquad$ $\bar{B}(y_2, \ldots, y_r) := B(m, y_2, \ldots, y_r)$
$\qquad$ until $\deg_{y_r}(\bar{A}) = \deg_{y_r}(A)$ and $\deg_{y_r}(\bar{B}) = \deg_{y_r}(B)$;
$\qquad$ MPQRM$(r - 1, p, \bar{A}, \bar{B}; \bar{Q}, \bar{R})$;
$\qquad$ $m' := M(m)^{-1} \bmod p$;
$\qquad$ $Q := $ MPINT$(p, M, m, m', r, Q, \bar{Q})$;
$\qquad$ $R := $ MPINT$(p, M, m, m', r, R, \bar{R})$;
$\qquad$ $M := M \cdot (y_1 - m)$
until $A = B \cdot Q + R$ end
end. $\blacksquare$

We finally trace a sample run of IPQRM in $\mathbb{Z}[y_1, y_2]$. To have a better visual distinction of the variables we set $x = y_1$ and $y = y_2$. The domain in which the algorithms are working at a specific time is indicated by the column in which the corresponding trace line begins. Lines beginning in the first column indicate $\mathbb{Z}[x, y]$, in the second column GF$(p)[x, y]$, and in the third column GF$(p)[y]$. So let us go.

IPQRM$(2, A, B; Q, R)$
$A = (x + 1)y^3 + (x^2 + 3x + 2)y^2 + (2x^2)y + (2x^3 + 3x^2 - 2x + 3)$
$B = (x + 1)y^2 + (2x^2 - x)$
$Q = 0$
$R = 0$
$P = 1$
$p = 3$
$\bar{A} = (x + 1)y^3 + (x^2 - 1)y^2 + (- x^2)y + (- x^3 + x)$
$\bar{B} = (x + 1)y^2 + (- x^2 - x)$

$\text{MPQRM}(2, 3, \bar{A}, \bar{B}; \bar{Q}, \bar{R})$

$A = \bar{A}$

$B = \bar{B}$

$Q = 0$

$R = 0$

$M = 1$

$m = 0$

$\bar{A} = y^3 - y^2$

$\bar{B} = y^2$

$\qquad \text{MPQRM}(1, 3, \bar{A}, \bar{B}; \bar{Q}, \bar{R})$

$\qquad A = \bar{A} = y^3 - y^2$

$\qquad B = \bar{B} = y^2$

$\qquad Q = y - 1 \quad$ directly in $\text{GF}(3)[y]$

$\qquad R = 0 \qquad\quad$ directly in $\text{GF}(3)[y]$

$\bar{Q} = y - 1$

$\bar{R} = 0$

$m' = 1$

$Q = y - 1 \quad (= \text{MPINT}(3, 1, 0, 1, 2, Q, \bar{Q}))$

$R = 0 \qquad (= \text{MPINT}(3, 1, 0, 1, 2, R, \bar{R}))$

$M = x$

"$A = B \cdot Q + R$" = false

$m = 1$

$\bar{A} = -y^3 - y$

$\bar{B} = -y^2 + 1$

$\qquad \text{MPQRM}(1, 3, \bar{A}, \bar{B}; \bar{Q}, \bar{R})$

$\qquad A = \bar{A} = -y^3 - y$

$\qquad B = \bar{B} = -y^2 + 1$

$\qquad Q = y \quad$ directly in $\text{GF}(3)[y]$

$\qquad R = y \quad$ directly in $\text{GF}(3)[y]$

$\bar{Q} = y$

$\bar{R} = y$

$m' = 1$

$Q = y + (x - 1) \quad (= \text{MPINT}(3, x, 1, 1, 2, Q, \bar{Q}))$

$R = x \cdot y \qquad\quad (= \text{MPINT}(3, x, 1, 1, 2, R, \bar{R}))$

$M = x \cdot (x - 1)$

"$A = B \cdot Q + R$" = true

$\bar{Q} = y + (x - 1)$

$\bar{R} = x \cdot y$

$p' = 1$

$Q = y + (x - 1) \quad (= \text{IPCRA}(1, 3, 1, 2, Q, \bar{Q}))$

$R = x \cdot y \qquad\quad (= \text{IPCRA}(1, 3, 1, 2, R, \bar{R}))$

$P = 3$

"$A = B \cdot Q + R$" = false

$p = 5$

$\bar{A} = (x + 1)y^3 + (x^2 - 2x + 2)y^2 + (2x^2)y + (2x^3 - 2x^2 - 2x - 2)$

$\bar{B} = (x + 1)y^2 + (2x^2 - x)$

$MPQRM(2, 5, \bar{A}, \bar{B}; \bar{Q}, \bar{R})$
$A = \bar{A}$
$B = \bar{B}$
$Q = 0$
$R = 0$
$M = 1$
$m = 0$
$\bar{A} = y^3 + 2y^2 - 2$
$\bar{B} = y^2$

$\qquad MPQRM(1, 5, \bar{A}, \bar{B}; \bar{Q}, \bar{R})$
$\qquad A = \bar{A} = y^3 + 2y^2 - 2$
$\qquad B = \bar{B} = y^2$
$\qquad Q = y + 2 \quad$ directly in $GF(5)[y]$
$\qquad R = -2 \quad$ directly in $GF(5)[y]$

$\bar{Q} = y + 2$
$\bar{R} = -2$
$m' = 1$
$Q = y + 2 \quad (= MPINT(5, 1, 0, 1, 2, Q, \bar{Q}))$
$R = -2 \quad (= MPINT(5, 1, 0, 1, 2, R, \bar{R}))$
$M = x$
"$A = B \cdot Q + R$" = false
$m = 1$
$\bar{A} = 2y^3 + y^2 + 2y + 1$
$\bar{B} = 2y^2 + 1$

$\qquad MPQRM(1, 5, \bar{A}, \bar{B}; \bar{Q}, \bar{R})$
$\qquad A = \bar{A} = 2y^3 + y^2 + 2y + 1$
$\qquad B = \bar{B} = 2y^2 + 1$
$\qquad Q = y - 2 \quad$ directly in $GF(5)[y]$
$\qquad R = y - 2 \quad$ directly in $GF(5)[y]$

$\bar{Q} = y - 2$
$\bar{R} = y - 2$
$m' = 1$
$Q = y + (x + 2) \quad (= MPINT(5, x, 1, 1, 2, Q, \bar{Q}))$
$R = x \cdot y - 2 \quad (= MPINT(5, x, 1, 1, 2, R, \bar{R}))$
$M = x \cdot (x - 1)$
"$A = B \cdot Q + R$" = true

$\bar{Q} = y + (x + 2)$
$\bar{R} = x \cdot y - 2$
$p' = 2$
$Q = y + (x + 2) \quad (= IPCRA(3, 5, 2, 2, Q, \bar{Q}))$
$R = x \cdot y + 3 \quad (= IPCRA(3, 5, 2, 2, R, \bar{R}))$
$P = 15$
"$A = B \cdot Q + R$" = true

Result:

$Q = y + (x + 2)$
$R = x \cdot y + 3. \quad\blacksquare$

2.5 Applications

Modular algorithms can be used for arithmetic in various domains like $\mathbb{Z}$, $\mathbb{Z}[y_1, y_2, \ldots, y_r]$, or the ring of matrices over $\mathbb{Z}[y_1, y_2, \ldots, y_r]$; see [4], p. 268 – 276, for the advantages and disadvantages of modular arithmetic.

Perhaps the most successful application is to the gcd problem in $\mathbb{Z}[y_1, y_2, \ldots, y_r]$, and to the closely related calculation of polynomial remainder sequences and resultants; cf. the corresponding chapter of this volume and the references cited there.

Modular algorithms are used to solve problems in Linear Algebra over $\mathbb{Z}[y_1, y_2, \ldots, y_r]$ such as matrix inversion, determinant calculation, and the solution of matrix equations; cf. [11].

Other applications are the squarefree decomposition of polynomials in $\mathbb{Z}[y_1, y_2, \ldots, y_r]$, cf. [21], and rational function integration [3].

3. Lifting Algorithms

p-adic methods are well known in algebraic and number theoretic computations. They compute ("lift") from a solution of a problem (more precisely: a solution of a system of equations) mod p^t a solution mod p^{t+1}, where p is a prime element of the underlying domain. Such lifting algorithms are obtained by applying Newton's method for the approximation of the root of an equation, formulated in the appropriate algebraic setting; see, for example, [7, 24, 9, 5, 26]. In [6] this approach is formulated in the abstract setting of commutative rings and finitely generated ideals.

3.1 Linear Lifting

We first state – without proof – a lemma about Taylor expansion in

$$R[x_1, x_2, \ldots, x_r],$$

where R is a commutative ring with identity element.

Lemma (Taylor Expansion). *Let* $f \in R[x_1, x_2, \ldots, x_r]$, $r \geq 1$; $y_1, y_2, \ldots, y_r$ *new indeterminates. Then*

$$f(x_1 + y_1, \ldots, x_r + y_r) = f(x_1, \ldots, x_r) + \sum_{j=1}^{r} \frac{\partial f}{\partial x_j} \cdot y_j + F, \qquad (3.1)$$

where

$$F \in R[x_1, \ldots, x_r][y_1, \ldots, y_r] \qquad and \qquad F \equiv 0 \pmod{(y_1, \ldots, y_r)^2}.$$

$[(y_1, \ldots, y_r)^2$ *denotes the square of the ideal generated by* $y_1, \ldots, y_r$ *in* $R[x_1, \ldots, x_r][y_1, \ldots, y_r].]$

Note that $\partial f / \partial x_j \in R[x_1, \ldots, x_r]$.

Theorem (Abstract Linear Lifting). *Let* I *be a finitely generated ideal in* R; $f_1, \ldots, f_n \in R[x_1, \ldots, x_r]$, $r \geq 1$; $a_1, \ldots, a_r \in R$ *with*

$$f_i(a_1, \ldots, a_r) \equiv 0 \pmod{I}, \qquad i = 1, \ldots, n.$$

Let

$$U = (u_{ij}), \quad i = 1, \ldots, n, \quad j = 1, \ldots, r, \quad \text{with} \quad u_{ij} = \frac{\partial f_i}{\partial x_j}(a_1, \ldots, a_r) \in R$$

(i.e. U is the Jacobian matrix of $f_1, \ldots, f_n$, evaluated at $a_1, \ldots, a_r$). Assume that U is invertible mod I.

Then for each positive integer t there exist $a_1^{(t)}, \ldots, a_r^{(t)} \in R$ such that

$$f_i(a_1^{(t)}, \ldots, a_r^{(t)}) \equiv 0 \pmod{I^t}, \quad i = 1, \ldots, n,$$

and

$$a_j^{(t)} \equiv a_j \pmod{I}, \quad j = 1, \ldots, r.$$

Proof. We proceed by induction on t. If $t = 1$, then the proposition is true with $a_j^{(1)} = a_j, j = 1, \ldots, r$. So let $t \geq 1$ and assume that the proposition is true for t. Hence, there exist $a_1^{(t)}, \ldots, a_r^{(t)} \in R$ with

$$f_i(a_1^{(t)}, \ldots, a_r^{(t)}) \equiv 0 \pmod{I^t}, \quad i = 1, \ldots, n; \tag{3.1}$$

and

$$a_j^{(t)} \equiv a_j \pmod{I}, \quad j = 1, \ldots, r. \tag{3.2}$$

Since I is finitely generated, I^t is finitely generated. Let

$$I^t = (q_1, \ldots, q_m). \tag{3.3}$$

By (3.1) and (3.3) there exist $v_{ik} \in R$, $i = 1, \ldots, n$, $k = 1, \ldots, m$, with

$$f_i(a_1^{(t)}, \ldots, a_r^{(t)}) = \sum_{k=1}^{m} v_{ik} q_k. \tag{3.4}$$

We set

$$a_j^{(t+1)} = a_j^{(t)} + B_j, \quad j = 1, \ldots, r, \tag{3.5}$$

with

$$B_j = \sum_{k=1}^{m} b_{jk} q_k \in I^t, \quad b_{jk} \in R, \quad j = 1, \ldots, r, \quad k = 1, \ldots, m, \tag{3.6}$$

where we want to choose the "unknowns" b_{jk} such that the proposition of the theorem holds for $a_j^{(t+1)}, j = 1, \ldots, r$.

Let

$$u_{ij}^{(t)} = \frac{\partial f_i}{\partial x_j}(a_1^{(t)}, \ldots, a_r^{(t)}), \quad i = 1, \ldots, n, \quad j = 1, \ldots, r. \tag{3.7}$$

Then we have, by (3.5)

$$f_i(a_1^{(t+1)}, \ldots, a_r^{(t+1)}) = f_i(a_1^{(t)} + B_1, \ldots, a_r^{(t)} + B_r),$$

(by the lemma and (3.7)) $$\equiv f_i(a_1^{(t)}, \ldots, a_r^{(t)}) + \sum_{j=1}^{r} u_{ij}^{(t)} B_j \pmod{I^{t+1}},$$

(by (3.4) and (3.6))
$$\equiv \sum_{k=1}^{m} v_{ik}q_k + \sum_{j=1}^{r} u_{ij}^{(t)} \sum_{k=1}^{m} b_{jk}q_k$$

$$\equiv \sum_{k=1}^{m} \left(v_{ik} + \sum_{j=1}^{r} u_{ij}^{(t)} b_{jk} \right) q_k \quad (\text{mod } I^{t+1}). \qquad (3.8)$$

Now by (3.3), (3.8) $\equiv 0 \pmod{I^{t+1}}$ if

$$v_{ik} + \sum_{j=1}^{r} u_{ij}^{(t)} b_{jk} \equiv 0 \pmod{I}, \qquad i = 1, \ldots, n, \qquad k = 1, \ldots, m. \qquad (3.9)$$

These are m systems of linear equations, each with coefficient matrix $(u_{ij}^{(t)})$, and we want solutions mod I. By (3.2) and (3.7),

$$u_{ij}^{(t)} \equiv u_{ij} \pmod{I}.$$

So (3.9) is equivalent to

$$v_{ik} + \sum_{j=1}^{r} u_{ij} b_{jk} \equiv 0 \pmod{I}, \qquad i = 1, \ldots, n, \qquad k = 1, \ldots, m. \qquad (3.10)$$

Since we assumed that $U = (u_{ij})$ is invertible mod I, the b_{jk} which satisfy (3.10) do exist, and so, for these b_{jk},

$$f_i(a_1^{(t+1)}, \ldots, a_r^{(t+1)}) \equiv 0 \pmod{I^{t+1}}, \qquad i = 1, \ldots, n.$$

Clearly, by (3.2) and (3.5),

$$a_j^{(t+1)} \equiv a_j \pmod{I}, \qquad j = 1, \ldots, r. \quad \blacksquare$$

As a sample application let us see how we get the famous Hensel Lemma as a specialization of that theorem.

Hensel's Lemma, in its formulation for factoring univariate integral polynomials, reads as follows:

Let $A_1, A_2, C \in \mathbb{Z}[y]$, $p \in \mathbb{Z}$ a prime. Suppose that

$$A_1 A_2 \equiv C \pmod{p},$$

and that A_1 and A_2 are relatively prime mod p. Then, for each positive integer t there exist $A_1^{(t)}, A_2^{(t)} \in \mathbb{Z}[x]$ such that

$$A_1^{(t)} A_2^{(t)} \equiv C \pmod{p^t}$$

and

$$A_i^{(t)} \equiv A_i \pmod{p}, \qquad i = 1, 2.$$

Now, in order to get the so-called Hensel Construction which constitutes a constructive proof of Hensel's Lemma, let — with the terminology of the theorem — $R = \mathbb{Z}[y]$, $I = (p)$, and $f_1 = x_1 x_2 - C \in R[x_1, x_2]$. Let $A_1, A_2 \in R$ with

$$f_1(A_1, A_2) = A_1 A_2 - C \equiv 0 \pmod{I = (p)}.$$

Then

$$u_{11} = \frac{\partial f_1}{\partial x_1}(A_1, A_2) = A_2,$$

$$u_{12} = \frac{\partial f_1}{\partial x_2}(A_1, A_2) = A_1.$$

The 1×2-matrix $U = (A_2, A_1)$ is invertible mod p if and only if A_1, A_2 are relatively prime mod p.

If $A_1^{(t)}, A_2^{(t)}$ satisfy

$$A_1^{(t)} A_2^{(t)} - C = V \cdot p^t \equiv 0 \pmod{(p)^t = (p^t)},$$

we get

$$A_1^{(t+1)} = A_1^{(t)} + B_1 p^t, \qquad A_2^{(t+1)} = A_2^{(t)} + B_2 p^t,$$

by solving (for B_1 and B_2)

$$V + B_1 A_2 + B_2 A_1 \equiv 0 \pmod{p}. \quad \blacksquare$$

The proof of the theorem is constructive and can easily be used to formulate a linear lifting algorithm. We will not do so for the sake of formulating a quadratically "convergent" algorithm.

3.2 Quadratic Lifting

In [25] Zassenhaus proposed a "quadratic Hensel Construction" which can be carried over to the construction in the above theorem.

If we re-examine the proof we see that by the Taylor expansion lemma the congruence (3.8) holds not only mod I^{t+1} but even mod I^{2t}. So we may try to determine the $a_j^{(t+1)}$ such that (3.8) $\equiv 0 \bmod I^{2t}$. By (3.3) this holds if

$$v_{ik} + \sum_{j=1}^{r} u_{ij}^{(t)} b_{jk} \equiv 0 \pmod{I^t}. \tag{3.11}$$

Since in general $u_{ij}^{(t)} \neq u_{ij} \pmod{I^t}$, we cannot pass from $u_{ij}^{(t)}$ to u_{ij} as we did in (3.10). We are done if we can show that the matrix $U^{(t)} = (u_{ij}^{(t)})$ is invertible mod I^t, i.e. that there exists a matrix $W^{(t)}$ such that

$$U^{(t)} W^{(t)} \equiv E \pmod{I^t}, \tag{3.12}$$

where E denotes the $n \times n$ identity matrix over R. But (3.12) is a system of equations, and since $U^{(t)} \equiv U \pmod{I}$ and we assumed U to be invertible mod I, there exists a "starting solution" of (3.12) mod I, i.e. a matrix W such that

$$U^{(t)} W \equiv E \pmod{I}.$$

Moreover, the Jacobian of (3.12) is just U. Hence U (for the f_i) and W (for the a_j) satisfy the conditions of the theorem which now tells us that $W^{(t)}$ satisfying (3.12) exists.

Of course, the algorithm which performs this quadratic lifting lifts the a_j (the "solution") and W (the inverse of the Jacobian) simultaneously, and both quadratically. Since the final $W^{(t)}$ will be needed in another algorithm it belongs to the output. Similarly as in the Chinese Remainder Algorithms of Section 2, the initial matrix W is assumed to be precomputed and belongs to the input. Here is the algorithm:

Algorithm QLA (Quadratic Lifting Algorithm).

Inputs: $P = \{p_1, p_2, \ldots, p_e\} \subset R$ with $I = (p_1, \ldots, p_e)$;
$f_1, f_2, \ldots, f_n \in R[x_1, \ldots, x_r]$, $r \geqslant 1$;
$a_1, a_2, \ldots, a_r \in R$ with $f_i(a_1, \ldots, a_r) \equiv 0 \pmod{I}$, $i = 1, \ldots, n$;
W, a $r \times n$ matrix over R with $UW \equiv E \pmod{I}$, where E is the $n \times n$ identity matrix, $U = (u_{ij})$, and
$u_{ij} = \partial f_i / \partial x_j (a_1, \ldots, a_r)$, $i = 1, \ldots, n$, $j = 1, \ldots, r$;
S, a non-negative integer.

Outputs: $A_1, A_2, \ldots, A_r \in R$ with
$$f_i(A_1, \ldots, A_r) \equiv 0 \pmod{I^{2^S}}, \ i = 1, \ldots, n, \text{ and}$$
$$A_j \equiv a_j \pmod{I}, \ j = 1, \ldots, r;$$
$\bar{W}$, a $r \times n$ matrix over R with
$$\bar{U}\bar{W} \equiv E \pmod{I^{2^S}}, \text{ where } \bar{U} = (\bar{u}_{ij}), \text{ and}$$
$$\bar{u}_{ij} = \partial f_i / \partial x_j (A_1, \ldots, A_r), \ i = 1, \ldots, n, \ j = 1, \ldots, r.$$

(1) [Initialize]
$s := 0$; $a_j^{(s)} := a_j$, $j = 1, \ldots, r$; $W^{(s)} := W$.

(2) [Lift]
while $s < S$ do
begin compute $q_1, q_2, \ldots, q_m \in R$ with $I^{2^s} = (q_1, q_2, \ldots, q_m)$;
[Lift $a_j^{(s)}$]
compute $v_{ik} \in R$, $i = 1, \ldots, n$, $k = 1, \ldots, m$, with
$$f_i(a_1^{(s)}, \ldots, a_r^{(s)}) = \sum_{k=1}^{m} v_{ik} q_k, \ i = 1, \ldots, n;$$
$b_{jk} := - \sum_{i=1}^{n} w_{ji}^{(s)} v_{ik}$, $j = 1, \ldots, r$, $k = 1, \ldots, m$;
$a_j^{(s+1)} := a_j^{(s)} + \sum_{k=1}^{m} b_{jk} q_k$, $j = 1, \ldots, r$.
[Lift $W^{(s)}$]
$u_{ij}^{(s+1)} := \partial f_i / \partial x_j (a_1^{(s+1)}, \ldots, a_r^{(s+1)})$, $i = 1, \ldots, n$, $j = 1, \ldots, r$;
compute D_k, $k = 1, \ldots, m$, where each D_k is a $n \times n$ matrix
over R, and $U^{(s+1)} W^{(s)} = E + \sum_{k=1}^{m} D_k q_k$;
$Z_k := - W^{(s)} D_k$, a $r \times n$ matrix over R, $k = 1, \ldots, m$;
$W^{(s+1)} := W^{(s)} + \sum_{k=1}^{m} Z_k q_k$;
$s := s + 1$
end.

(3) [Finish]
$A_j := a_j^{(s)}$, $j = 1, \ldots, r$; $\bar{W} := W^{(s)}$. ∎

An algorithm for linear lifting can easily be obtained from QLA by essentially omitting the lifting of W and replacing $W^{(s)}$ by W in the computation of the b_{jk}, and by letting $(q_1, \ldots, q_m) = I^t$.

Regarding (3.11) (with $t = 2^s$), a concrete lifting algorithm which is based on QLA will perform the computation of the b_{jk} (and of the Z_k) mod I^{2^s}. However, the linear version of this algorithm will compute the b_{jk} mod I! So the price for quadratic convergence is a more complicated arithmetic, and this price may be high enough to annihilate the gain of the quadratic convergence versus the linear one (cf. [17]). Of course, also a linear lifting algorithm can easily be extended such that it also lifts (linearly) the matrix W so that we do not have to relinquish with $W^{(t)}$ satisfying

$$U^{(t)} W^{(t)} \equiv E \quad (\text{mod } I^t).$$

3.3 Iterated Lifting

Quite similar to the Chinese Remainder Algorithms the lifting algorithms become significantly simpler if the ideal I is principal (cf. Section 2.2). If $I = (p)$, then $I^t = (p^t)$. Thus, the rather vaguely formulated step "compute $q_1, q_2, \ldots, q_m \in R$ with $\ldots$" in algorithm QLA becomes "$q^{(s)} := q^{(s-1)} \cdot q^{(s-1)}$" (with $q^{(0)} := p$), or, for the linear lifting, "$q^{(t)} := q^{(t-1)} \cdot p$". Also the computation of the v_{ik} simplifies to the mere division "$v_i := f_i(\cdots)/q^{(s)}$".

For the case of multivariate polynomial factorization Musser in [14, 15] showed how a factorization mod $I = (m_1, m_2, \ldots, m_e)$ can be lifted to a factorization $\text{mod}(m_1^{n_1}, m_2^{n_2}, \ldots, m_e^{n})$ by iterated lifting with respect to $(m_1), (m_2), \ldots, (m_e)$ and hence with respect to principal ideals. The following theorem shows that this technique can be applied to achieve iterated lifting with respect to finitely generated ideals.

Theorem (Iterated Lifting Theorem). *Let $P_1, \ldots, P_m$ be finite subsets of R and $I = (P_1, P_2, \ldots, P_m)$ (i.e. the ideal generated by $P_1 \cup P_2 \cup \cdots \cup P_m$).*

Let $f_1, \ldots, f_n, a_1, \ldots, a_r$ and U be like in the Abstract Lifting Theorem in Section 3.1. Then for each sequence of positive integers $t_1, t_2, \ldots, t_m$ there exist $A_1, A_2, \ldots, A_r \in R$ such that

$$f_i(A_1, \ldots, A_r) \equiv 0 \quad (\text{mod}((P_1)^{t_1}, \ldots, (P_m)^{t_m})), \qquad i = 1, \ldots, n,$$

and

$$A_j \equiv a_j \text{ mod } I, \qquad j = 1, \ldots, r.$$

A rigorous proof is given in [6]. It proceeds by induction on m, and for the induction step it uses the second isomorphism theorem for rings and the fact that the inverse matrix W of U (mod I) can also be lifted.

We only sketch an algorithm which is based on this theorem. It may be formulated recursively with $m = 1$ and algorithm QLA (or the linear version) as a recursion base. If $m > 1$ it makes a recursive call for the image problem in $R/(P_m)$ and the ideals $(P_1)/(P_m), (P_2)/(P_m), \ldots, (P_{m-1})/(P_m)$. It finally incorporates P_m by applying QLA to the images of the result of the recursive call in $R/((P_1)^{t_1}, \ldots, (P_{m-1})^{t_{m-1}})$ and the ideal $(P_m)/((P_1)^{t_1}, \ldots, (P_{m-1})^{t_{m-1}})$. See [6] for a complete description.

3.4 Lifting in $\mathbb{Z}$ and $\mathbb{Z}[y_1, y_2, \ldots, y_r]$

In case $R = \mathbb{Z}$ the f_i are polynomials with integer coefficients, and we are concerned with their integral zeros, i.e. we have a system of diophantine equations. There is a broad theory on diophantine equations and Lewis in [7] describes p-àdic methods for their solution. In fact our theorem on abstract linear lifting (cf. Section 3.1) is a straightforward generalization of Lewis' Theorem D to commutative rings with identity and finitely generated ideals.

We now give an example. Let

$$f = x_1 x_2 - x_2^2 - 10,$$

$$P = \{3\},$$

$$a_1 = 1, \qquad a_2 = -1.$$

Then

$$f(1, -1) = -12 \equiv 0 \pmod{3}.$$

Furthermore,

$$\frac{\partial f}{\partial x_1} = x_2, \qquad \frac{\partial f}{\partial x_2} = x_1 - 2x_2,$$

hence $U = (u_1 u_2)$ with $u_1 = -1$, $u_2 = 3$. Then

$$W = \begin{pmatrix} w_1 \\ w_2 \end{pmatrix}$$

with $w_1 = -1$, $w_2 = 1$ satisfies $UW \equiv 1 \pmod{3}$.

Giving P, f, a_1, a_2, W as inputs to algorithm QLA (and leaving the termination value S unspecified) the algorithm works as follows: (According to the remarks at the end of Section 3.2 the computation of the b_j and of Z is performed mod 3^{2^s}).

(1) $s = 0$, $a_1^{(0)} = 1$, $a_2^{(0)} = -1$, $w_1^{(0)} = -1$, $w_2^{(0)} = 1$.

(2) $q_1 = 3$

$$v = \frac{f(a_1^{(0)}, a_2^{(0)})}{3} = \frac{-12}{3} = -4 \equiv -1 \pmod{3}$$

$b_1 = -1$, $b_2 = 1$
$\underline{a_1^{(1)} = -2, \ a_2^{(1)} = 2}$

[Now $f(a_1^{(1)}, a_2^{(1)}) = -18 \equiv 0 \pmod{3^{2^1} = 9}$]

$$u_1^{(1)} = \frac{\partial f}{\partial x_1}(a_1^{(1)}, a_2^{(1)}) \equiv -1, \ u_2^{(1)} \equiv 0 \pmod{3}$$

$D = 0$
$Z = 0$
$w_1^{(1)} = -1$, $w_2^{(1)} = 1$

$s = 1$

$q = 9$

$v = \dfrac{f(a_1^{(1)}, a_2^{(1)})}{9} = -2$

$b_1 = -2$

$b_2 = 2$

$a_1^{(2)} = -20, \ a_2^{(2)} = 20$

[Now $f(a_1^{(2)}, a_2^{(2)}) = -810 \equiv 0 \ (\mathrm{mod}\ 3^{2^2} = 81)$]

$u_1^{(2)} \equiv 2, \ u_2^{(2)} \equiv 3 \ (\mathrm{mod}\ 9)$

$D = 0$

$Z = 0$

$w_1^{(2)} = -1, \ w_2^{(2)} = 1$

$s = 2$

$q = 81$

$v = \dfrac{f(a_1^{(2)}, a_2^{(2)})}{81} = -10$

$b_1 = -10$

$b_2 = 10$

$a_1^{(3)} = -830, \ a_2^{(3)} = 830$

[Now $f(a_1^{(3)}, a_2^{(3)}) = 1\,377\,810 \equiv 0 \ (\mathrm{mod}\ 3^{2^3} = 6561)$].

We stop at this point. The equation $x_1 x_2 - x_2^2 - 10 = 0$ has 8 integral solutions $((\pm 11, \pm 1), (\pm 7, \pm 2), (\pm 7, \pm 5), (\pm 11, \pm 10))$ and it is evident that the above lifting process will not produce any of those. It even does not hold that $a_i^{(s)} \equiv A_i$ $(\mathrm{mod}\ 3^{2^s})$, $i = 1, 2$, for $s \geqslant 2$, if (A_1, A_2) is one of the solutions over $\mathbb{Z}$, although the "starting solution" $a_1 = 1$, $a_2 = -1$ is congruent to some of the true solutions mod 3. The reason for this flaw is that in general the solution of (3.10) is not unique. In the above example we had to solve

$$-2 + u_1^{(1)} b_1 + u_2^{(1)} b_2 \equiv 0 \quad (\mathrm{mod}\ 9),$$

where $u_1^{(1)} = 2$, $u_2^{(1)} = -6$. The solution computed by the algorithm was $b_1 = -2$, $b_2 = 2$. But $b_1 = 1$, $b_2 = 0$ is another solution which would have lead to $a_1^{(2)} = 7$, $a_2^{(2)} = 2$ which is one of the solutions (over $\mathbb{Z}$) of the given equation.

The final aim of computation in a domain R is in general a solution over R, not a solution modulo some ideal. It is intuitively clear that this implies that, if $A_1, \dots, A_r$ is a solution over R, and $a_1^{(s)}, \dots, a_r^{(s)}$ is a solution mod I^{2^s} obtained by algorithm QLA,

$$A_j \equiv a_j^{(s)} \quad (\mathrm{mod}\ I^{2^s}), \qquad j = 1, \dots, r. \tag{3.13}$$

For $R = \mathbb{Z}$ it seems difficult to choose a particular solution of (3.10) which keeps the lifted solution such that it satisfies (3.13), and therefore algorithm QLA seems only to be of theoretical interest in case $R = \mathbb{Z}$.

The situation becomes different if $R = \mathbb{Z}[y_1, y_2, \ldots, y_r]$, $r \geqslant 1$, which is the most commonly used domain in which the concrete lifting algorithms work. The general method to keep the lifted solution satisfying (3.13) seems to give degree constraints on the solution of (3.10). Exactly for this reason the description of the Hensel construction which we derived from the abstract lifting theorem in Section 3.1 was not complete. We quoted that, for one lifting step, we have to solve (for B_1 and B_2)

$$V + B_1 A_2 + B_2 A_1 \equiv 0 \pmod{p}$$

(cf. the end of Section 3.1). There are infinitely many such solutions B_1, B_2, but there is a unique solution satisfying

$$\deg(B_1) < \deg(A_1).$$

If C has two irreducible factors over $\mathbb{Z}$ and these B_1, B_2 (with $\deg(B_1) < \deg(A_1)$) are used in the lifting process, then the lifted factors will eventually become the true factors. See the chapter on factorization for further details, especially for the case that C has more than two factors, or as less factors over $\mathbb{Z}$ than mod p.

The most commonly used lifting algorithm in $R = \mathbb{Z}[y_1, y_2, \ldots, y_r]$ is the one for iterated lifting; cf. Section 3.3. As mentioned earlier, the starting ideal $I = (P_1, P_2, \ldots, P_m)$ is chosen such that $R/I \simeq \mathrm{GF}(p)[y_r]$. There are different ways to choose $P_1, \ldots, P_m$ to achieve this. Musser in [14, 15] proposed $P_i = \{y_i - d_i\}$, $d_i \in \mathbb{Z}$, $i = 1, \ldots, r - 1$, $P_r = \{p\}$, $p \in \mathbb{Z}$ prime, which leads to iterated lifting with principal ideals. Yun in [23] and Wang and Rothschild in [19, 20] use $P_1 = \{p\}$, $P_2 = \{y_1 - d_1, \ldots, y_{r-1} - d_{r-1}\}$ which has the advantage that, if linear lifting is used for P_2, all essential calculations are performed in

$$\mathbb{Z}/(p^k)[y_1, \ldots, y_r]/(P_2) \simeq Z/(p^k)[y_r].$$

Note that the basis elements of the ideals (except p) are of the form $y_i - d_i$ and therefore the mod-operation becomes an evaluation. There is a remarkable computational trick in this connection. The computation of the v_{ik} in algorithm QLA is rather difficult if $d_i \neq 0$, even for principal ideals. The inputs are therefore translated such that the evaluation points all become zero by substituting $y_i' + d_i$ for y_i, and this translation is reversed after the lifting.

3.5 A Complete Sample Application: Polynomial Division with Remainder

We take the same problem as in Section 2.4, but now want to employ a lifting algorithm. Like in Section 2.4 we assume that we know for $A, B \in \mathbb{Z}[y_1, y_2, \ldots, y_r]$ there exist Q, R with

$$A = B \cdot Q + R$$

and

$$\deg_{y_r}(R) < \deg_{y_r}(B).$$

In order to apply a lifting algorithm we have to specify Q and R as zeros of (one or more) polynomials over $\mathbb{Z}[y_1, y_2, \ldots, y_r]$. Obviously

$$f = A - Bx_1 - x_2,$$

with the "side condition" $\deg_{y_r}(x_2) < \deg_{y_r}(B)$.

We now have to decide which lifting algorithm we want to use. We have

$$\frac{\partial f}{\partial x_1} = -B, \qquad \frac{\partial f}{\partial x_2} = -1.$$

This means that (3.10) which has to be solved for the "increments" turns out to become a division with remainder problem. Even if we solve this (sub-) problem using the "inverse matrix" $W^{(t)}$ we have to reduce the (degree of the) sub-problem remainder mod B in order to obtain a lifted solution which satisfies (3.13); see also the remarks on the validity of the algorithm below. Hence, we will not gain much over a division in $\mathbb{Z}[y_1, y_2, \dots, y_r]$ unless the sub-problem division is always carried out in $GF(p)[y_r]$. This causes us to use linear, non-iterated lifting, i.e. $I = (p, y_1 - d_1, \dots, y_{r-1} - d_{r-1})$.

$$\text{IPQRL}(r, A, B; Q, R).$$

[Integral polynomial quotient and remainder, lifting algorithm. The specifications are as an IPQRM, cf. Section 2.4.]

```
begin
find "lucky" p, d₁,...,d_{r-1} ∈ Z, p prime;
```
$\bar{A}(y_r) := A(d_1, \dots, d_{r-1}, y_r) \bmod p;$
$\bar{B}(y_r) := B(d_1, \dots, d_{r-1}, y_r) \bmod p;$
compute (directly) $\bar{Q}, \bar{R} \in GF(p)[y_r]$ with
$\qquad \bar{A} - \bar{B}\bar{Q} - \bar{R} = 0$ (in $GF(p)[y_r]$)
$\qquad$ and $\deg(\bar{R}) < \deg(\bar{B})$;
$A' := A(y_1 + d_1, \dots, y_{r-1} + d_{r-1}, y_r)$ [translation]
$B' := B(y_1 + d_1, \dots, y_{r-1} + d_{r-1}, y_r)$ [translation]
$t := 1; \; Q' := \bar{Q}; \; R' := \bar{R};$
$V' := A' - B'Q' - R';$ (in $\mathbb{Z}[y_1, \dots, y_r]$)
while $V' \neq 0$ do
begin [let $V' = \sum_{i=0}^{n_t} v_i' y_r^i, \; v_i' \in \mathbb{Z}[y_1, \dots, y_{r-1}]$]
$\qquad$ for all $q = p^{t_0} y_1^{t_1} y_2^{t_2} \cdots y_{r-1}^{t_{r-1}}$ with $t_0 + t_1 + \cdots + t_{r-1} = t$ do
$\qquad$ begin $\bar{V} := 0;$
$\qquad\qquad$ for $i := 0$ to n_t do
$\qquad\qquad\qquad$ if v_i' contains a term $s \cdot y_1^{t_1} \cdots y_{r-1}^{t_{r-1}}, \; s \in \mathbb{Z}$, then
$\qquad\qquad\qquad$ begin $\bar{v}' := s/p^{t_0};$ (in $\mathbb{Z}$)
$\qquad\qquad\qquad\qquad \bar{v} := \bar{v}' \bmod p;$
$\qquad\qquad\qquad\qquad \bar{V} := \bar{V} + \bar{v} \cdot y_r^i$
$\qquad\qquad$ end;
$\qquad\qquad$ compute (directly) $\bar{Q}, \bar{R} \in GF(p)[y_r]$ with
$\qquad\qquad\qquad \bar{V} - \bar{B}\bar{Q} - \bar{R} = 0$ (in $GF(p)[y_r]$)
$\qquad\qquad\qquad$ and $\deg(\bar{R}) < \deg(\bar{B})$;
$\qquad\qquad Q' := Q' + \bar{Q} \cdot q;$
$\qquad\qquad R' := R' + \bar{R} \cdot q;$
$\qquad$ end;
$\qquad t := t + 1;$
$\qquad V' := A' - B'Q' - R'$ (in $\mathbb{Z}[y_1, \dots, y_r]$)
end;

$Q := Q'(y_1 - d_1, \ldots, y_{r-1} - d_{r-1}, y_r);$ [reverse translation]
$R := R'(y_1 - d_1, \ldots, y_{r-1} - d_{r-1}, y_r)$ [reverse translation]
end. ∎

(The luckiness in the first statement means that $\deg(\bar{A}) = \deg_{y_r}(A)$ and $\deg(\bar{B}) = \deg_{y_r}(B)$.)

For a proof of the termination of IPQRL we have to show that after finitely many steps Q' and R' become the true quotient Q^* and the true remainder R^* of A' and B'. This amounts to showing that, for each t,

$$Q^* \equiv Q', \qquad R^* \equiv R' \pmod{I'},$$

i.e. that Q', R' satisfy (3.13). We have

$$A' \equiv B'Q' + R' \pmod{I'},$$

and, of course,

$$A' \equiv B'Q^* + R^* \pmod{I'},$$

hence

$$(Q^* - Q')B' \equiv R' - R^* \pmod{I'}.$$

Thus B' divides $R' - R^* \bmod I'$ while $\deg(B') > \deg(R' - R^*)$. But this implies that the leading coefficient of B' (an element of $\mathbb{Z}[y_1, \ldots, y_{r-1}]$) is a zero divisor mod I' which contradicts the assumption that I is lucky.

The way $\bar{V}$ is computed suggests to choose y_r as a main variable and to represent the coefficients (which are elements of $\mathbb{Z}[y_1, \ldots, y_{r-1}]$) in distributive form.

As in Section 2.4 we trace a sample run, and we take the same inputs $A, B \in \mathbb{Z}[x, y]$.

IPQRL$(2, A, B; Q, R)$
$A \quad = (x + 1) \cdot y^3 + (x^2 + 3x + 2)y^2 + (2x^2) \cdot y + (2x^3 + 3x^2 - 2x + 3)$
$B \quad = (x + 1) \cdot y^2 + (2x^2 - x)$
$p \quad = 3$
$d \quad = 0$
$\bar{A} \quad = y^3 - y^2$
$\bar{B} \quad = y^2$
$\bar{Q} \quad = y - 1$
$\bar{R} \quad = 0$
$A' \quad = A$
$B' \quad = B$
$t \quad = 1$
$Q' \quad = y - 1$
$R' \quad = 0$
$V' \quad = (x^2 + 4x + 3) \cdot y^2 + x \cdot y + (2x^3 + 5x^2 - 3x + 3)$
"$V' \neq 0$" $=$ true
$q \quad = 3^0 x^1 = x$
$\bar{V} \quad = y^2 + y$
$\bar{Q} \quad = 1$

$$\bar{R} = y$$
$$Q' = (y - 1) + 1 \cdot x = y + (x - 1)$$
$$R' = 0 + y \cdot x \qquad = xy$$
$$q = 3^1 \cdot x^0 = 3$$
$$\bar{V} = y^2 + 1$$
$$\bar{Q} = 1$$
$$\bar{\bar{R}} = 1$$
$$Q' = y + (x - 1) + 1 \cdot 3 = \quad y + (x + 2)$$
$$R' = xy \qquad\qquad + 1 \cdot 3 = xy + 3$$
$$V' = 0$$
$$\text{``}V' \neq 0\text{''} = \text{false}$$
$$Q = y + (x + 2)$$
$$R = xy + 3. \quad \blacksquare$$

3.6 Applications

The field of applications of lifting algorithms is at least as broad as the one for Chinese Remainder Algorithms. One of the best known applications is the complete factorization of multivariate polynomials over the integers; cf. [25, 15, 20, 19]. Extending these ideas to non-complete factorization, lifting algorithms are used to compute the gcd of multivariate polynomials; cf. [16]. Other applications are square-free decomposition [21, 22, 24], partial fraction decomposition [3], and polynomial division with remainder [23]. See also [24] for a survey of these applications. Further, we mention the computation of rational zeros of (univariate) integral polynomials [10] and the manipulation of power series [9, 5].

References

[1] Collins, G. E.: Computer Algebra of Polynomials and Rational Functions. Am. Math. Monthly **80**, 725 – 755 (1973).

[2] Collins, G. E.: The Calculation of Multivariate Polynomial Resultants. J. ACM **18**, 515 – 532 (1971).

[3] Horowitz, E.: Algorithms for Partial Fraction Decomposition and Rational Function Integration. SYMSAM **1971**, 441 – 457.

[4] Knuth, D. E.: The Art of Computer Programming, Vol. II. Seminumerical Algorithms, 2nd ed. Reading, Mass.: Addison-Wesley 1981.

[5] Kung, H. T.: On Computing Reciprocals of Power Series. Numer. Math. **22**, 341 – 348 (1974).

[6] Lauer, M.: Generalized p-adic Constructions. SIAM J. Comp. (to appear).

[7] Lewis, D. J.: Diophantine Equations: p-adic Methods. In: Studies in Number Theory (LeVeque, ed.), pp. 25 – 75 (1969).

[8] Lipson, J. D.: Chinese Remainder and Interpolation Algorithms. SYMSAM **1971**, 372 – 391.

[9] Lipson, J. D.: Newton's Method: A Great Algebraic Algorithm. SYMSAC **1976**, 260 – 270.

[10] Loos, R.: Computing Rational Zeros of Integral Polynomials by p-adic Expansion. Dec. 1981 (to appear in SIAM J. Comp.)

[11] McClellan, M. T.: The Exact Solution of Systems of Linear Equations with Polynomial Coefficients. SYMSAM **1971**, 399 – 414.

[12] Moenck, R. T.: Practical Fast Polynomial Multiplication. SYMSAC **1976**, 136 – 148.

[13] Moenck, R. T.: Another Polynomial Homomorphism. Acta Informatica **6**, 153 – 169 (1976).

[14] Musser, D. R.: Algorithms for Polynomial Factorization. Techn. Report No. 134, Ph.D. Thesis, Comp. Sciences Dept., University of Wisconsin, Madison, 1971.

[15] Musser, D. R.: Multivariate Polynomial Factorization. J. ACM **22**, 291 – 308 (1975).

[16] Moses, J., Yun, D. Y. Y.: The EZGCD Algorithm. Proc. ACM Annual Conference, Atlanta, pp. 159 – 166 (August 1973).

[17] Miola, A., Yun, D. Y. Y.: The Computational Aspects of Hensel-Type Univariate Greatest Common Divisor Algorithms. EUROSAM **1974**, 46 – 54.

[18] Schönhage, A.: Asymptotically Fast Algorithms for the Numerical Multiplication and Division of Polynomials with Complex Coefficients. EUROCAM **1982** (to appear).

[19] Wang, P. S. H.: An Improved Multivariate Polynomial Factoring Algorithm. Math. Comp. **32**, 1215 – 1231 (1978).

[20] Wang, P. S. H., Rothschild, L. P.: Factoring Multivariate Polynomials over the Integers. Math. Comp. **29**, 935 – 950 (1975).

[21] Wang, P. S. H., Trager, B. M.: New Algorithms for Polynomial Square-Free Decomposition over the Integers. M.I.T. (private communication).

[22] Yun, D. Y. Y.: The Hensel Lemma in Algebraic Manipulation. Ph.D. Thesis, Dept. of Math., MAC-TR-38, M.I.T., 1974.

[23] Yun, D. Y. Y.: A p-adic Division with Remainder Algorithm. ACM SIGSAM Bull. **8** (Issue 32), 27 – 32 (1974).

[24] Yun, D. Y. Y.: Algebraic Algorithms Using p-adic Constructions. SYMSAC **1976**, 248 – 259.

[25] Zassenhaus, H.: On Hensel Factorization, I. J. Number Theory **1969**, 291 – 311.

[26] Zippel, R.: Newton's Iteration and the Sparse Hensel Algorithm. SYMSAC **1981**, 68 – 72.

Dipl.-Math. M. Lauer
Institut für Informatik
Universität Karlsruhe
Zirkel 2
D-7500 Karlsruhe
Federal Republic of Germany

Computing in Transcendental Extensions

A. C. Norman, Cambridge

Abstract

Performing the rational operations in a field extended by a transcendental element is equivalent to performing arithmetic in the field of rational functions over the field. The computational difficulty associated with such extensions is in verifying that proposed extensions are transcendental. When the extensions being considered are functions, and where a differentiation operator can be defined for them, structure theorems can be used to determine the character of the extension and to exhibit a relationship between the adjoined element and existing quantities in case the adjoined element is not transcendental.

If F is a field and z is transcendental over it, performing the rational operations in F extended by z is just equivalent to performing arithmetic in the field of rational functions $F(z)$. Multiple extensions do not raise further problems. The computational difficulty associated with such extensions is, then, mainly in verifying that proposed extensions are transcendental. In the case of constant fields this is a difficult problem. For a survey of known results see [3]. Computer algebra has at present nothing new to contribute in this area.

When the extensions being considered are functions, and where a differentiation operator can be defined for them, there are a several so-called structure theorems that can be used to determine the character of the extensions. These theorems provide computable tests that decide if there is some algebraic relationship between the quantities present in a field and a new object that is to be adjoined. If there are no such relationships the new quantity is termed a regular monomial, and extending by it gives rise to no problems. If the new monomial is not regular the structure theorems may sometimes exhibit a relationship between it and existing quantities that can be solved explicitly. When this happens it is easy to reformulate problems so that the quantities in them satisfy a structure theorem and so can be computed with easily. For instance over the field $\mathbb{Q}(e^x)$ the monomial e^{3x} is not regular, but it is easy to discover that it can be replaced by $(e^x)^3$. It must be stressed, however, that transformations of the above type are not always available, and that the primary purpose of a structure theorem does not go beyond its use as a test of field extensions.

Structure Theorem for Elementary Transcendental Functions

Let the series of differential fields F_k be obtained by the extension of a ground field $\mathbb{C}$ by the quantities $z_0, z_1, \ldots, z_k$. Let $Dz_0 = 1$ (i.e. z_0 is the independent variable, with respect to which differentiation is performed). The structure theorem presented here will give conditions that can be applied to the rest of the z_i that will ensure that (a) they model exponentials and logarithms and (b) each z_i is a regular monomial over F_{i-1}, and so elements of F_k can be represented as rational functions in the $k+1$ quantities z_0 to z_k. Part (a) is achieved by defining the action of the differential operator D on each successive z_i. Apart from the initial case $Dz_0 = 1$ each z_i must satisfy one of

$$(i \text{ in } L) \quad Dz_i = Df_i/f_i$$

or

$$(i \text{ in } E) \quad Dz_i = z_i Df_i,$$

where in each case f_i is in F_{i-1}. In the first of these cases z_i models $\log(f_i)$ and in the latter it behaves as $\exp(f_i)$.

For z_i to be a regular monomial over F_{i-1} it is necessary that z_i is algebraically independent of all the previous z_j, and the constant field of F_i is the same as that of F_0. The structure theorem gives restrictions on the forms f_i that can be used to generate z_i. A z_i will fail to be a regular monomial over F_{i-1} if either of the following conditions hold:

(1) i in E and there is some constant k and rational numbers n_j such that

$$f_i = k + \sum_{j \in L} n_j z_j + \sum_{j \in E} n_j f_j.$$

(2) i in L and there is some constant k and rational numbers n_j such that

$$f_i = k \prod_{j \in L} f_j^{n_j} \prod_{j \in E} z_j^{n_j}.$$

These constraints capture the essence of the obvious ways in which the law of logarithms can introduce dependencies among the z_i. If $\log(x)$ and e^x have already been introduced it is easy to see that the above rules show that none of $\log(\mathrm{sqrt}(x))$, $e^{\log(x)+3x}$, $\log(2x)$ and e^{x+1} are regular monomials.

Each of the conditions given in the structure theorem is an expression linear in a collection of quantities k and c_j. Since it lies in F_{i-1} which is known to be built up from $\mathbb{C}$ by a series of regular extensions, it is legitimate to clear denominators and compare coefficients of all possible powers of the z_j. This leads to a set of linear equations in k and the n_j. If these equations have a solution, and if that solution has all the c_j rational then the proposed extension was not regular.

As an example of the application of these results, consider an attempt to set up a model for a differential field containing the expression

$$\log(z \exp(z)) + \exp(\exp(z) + \log(z))$$

(taken from [1]). An attempt is made to generate a tower of field extensions, starting from the rational numbers and containing successively more of the parts of this expression. Working from the left the process starts by identifying $\exp(z)$ as a potentional regular monomial over $\mathbb{Q}(z)$. If it is not regular there will be some constant k with

$$z = k$$

identically, and the fact that this is clearly absurd shows that $z_1 = \exp(z)$ is regular. Next try $z_2 = \log(zz_1)$. If this is not regular there is a constant k and a rational constant n with

$$zz_1 = kz_1^n.$$

Comparing like powers of z shows that this is impossible, so z_2 is regular. Moving to the second term of the original expression a third monomial $z_3 = \log(z)$ is considered. The structure theorem asks about the identity

$$z = k(zz_1)^m z_1^n$$

and this does have solutions. It should be noted that the existence of one readily computable solution to this relationship does not imply that there is a unique way of expressing z_3 in terms of z and z_1: indeed here an informal expression of the state of affairs is that we have

$$z_3 = z_2 - z + k',$$

where k' is some constant known at best modulo $2i\pi$. Having determined this problem the structure theorem can do no more. However, the user is entitled to modify the problem being attacked in the light of difficulties found by applying the structure theorem. Here it may well be reasonable to reformulate the original expression as

$$\log(z \exp(z)) + \exp(\exp(z) + \log(z \exp(z)) - z)$$

so avoiding the ambiguity inherent in the constant k' introduced above. Applying the structure theorem establishes z_1 and z_2 as before, and now investigates

$$z_3 = \exp(z_1 + z_2 - z).$$

This monomial fails to be regular if a constant k and rational constants m and n can be exhibited satisfying

$$z_1 + z_2 - z = k + mz_2 + nz$$

and it is legitimate to compare coefficients in this to obtain a set of simpler identities each of which must hold:

$$k = 0 \quad \text{[constant term]},$$
$$1 = 0 \quad \text{[coefficient of } z_1\text{]},$$
$$1 = m \quad \text{[coefficient of } z_2\text{]},$$
$$-1 = n \quad \text{[coefficient of } z\text{]}.$$

By observing that the second of these cannot be satisfied it is established that z_3 is regular. Now in the differential field that has been constructed the original expression may be written as $z_2 + z_3$.

The construction of a differential field to contain this expression could equally well have started by introducing regular monomials for $\log(z)$, $\exp(z)$ and $\exp(\exp(z) + \log(z))$, and in that case the problem reformulation would have been needed in response to a report that $\log(z \exp(z))$ was not regular over the other monomials.

The basic form of the structure theorem given above has been extended in a number of ways, see for instance [4]. The essential result is not disturbed if two additional modes of extension are admitted:

$$(i \text{ in } X) \quad Dz_i = f_i$$

with the integral non-elementary over F_{i-1}

$$(i \text{ in } A) \quad z_i \text{ satisfies } p(z_i) = 0,$$

where p is a polynomial with coefficients in F_{i-1}.

The Liouvillian extensions X cause no problems at all provided the non-elementary nature of the integral of f_i can be tested. The algebraic case A does not change the statement of the structure theorem but does make some difference to the mechanics of testing the conditions that it imposes. This means that in general it will not be possible to apply the results when several X and A type extensions coexist.

References

[1] Caviness, B. F.: Methods for Symbolic Computation with Transcendental Functions. MAXIMIN 1977, 16–43.
[2] Kaplansky, I.: An Introduction to Differential Algebra. Paris: Hermann 1957.
[3] Lang, S.: Transcendental Numbers and Diophantine Approximations. Bull. AMS 77/5, 635–677.
[4] Rothstein, M., Caviness, B. F.: A Structure Theorem for Exponential and Primitive Functions. SIAM J. Computing 8/3, 357–367 (1979).

Dr. A. C. Norman
Computer Laboratory
University of Cambridge
Corn Exchange Street
Cambridge CB2 3QG
United Kingdom

Computing in Algebraic Extensions

R. Loos, Karlsruhe

Abstract

The aim of this chapter is an introduction to elementary algorithms in algebraic extensions, mainly over $\mathbb{Q}$ and, to some extent, over GF(p). We will talk about arithmetic in $\mathbb{Q}(\alpha)$ and GF(p^n) in Section 1 and some polynomial algorithms with coefficients from these domains in Section 2. Then, we will consider the field K of all algebraic numbers over $\mathbb{Q}$ and show constructively that K indeed is a field, that multiple extensions can be replaced by single ones and that K is algebraically closed, i.e. that zeros of algebraic number polynomials will be elements of K (Section 4 – 6). For this purpose we develop a simple resultant calculus which reduces all operations on algebraic numbers to polynomial arithmetic on long integers together with some auxiliary arithmetic on rational intervals (Section 3). Finally, we present some auxiliary algebraic number algorithms used in other chapters of this volume (Section 7). This chapter does not include any special algorithms of algebraic number theory. For an introduction and survey with an extensive bibliography the reader is referred to Zimmer [15].

0. Introduction

The problem type of this chapter is the task of extending exact operations in a given domain to new domains generated by adjunction of new elements. To solve this problem by algebraic algorithms means to search for exact operations in the extended domains. In most cases the representation of elements in the extension will have a more algebraic part and a more numerical part. Since the full information is always available, the general strategy is to keep the numerical precision as low as possible – just high enough to avoid ambiguities. The process of extension is a powerful method to generate new algebraic structures from given ones; in particular, the process can be iterated leading to telescopic towers of extensions. We hope that the restriction to $\mathbb{Q}$ and GF(p) will not hide the generality of this problem solving strategy.

1. Algorithms in $\mathbb{Q}(\alpha)$ and GF(p^n)

1.1 Representation

The elements of $\mathbb{Q}$ (see the chapter on arithmetic) are represented as 0 or as pairs of integers (a, b) with $b > 0$ and $\gcd(a, b) = 1$. The algebraic number α is defined as a zero of a rational polynomial $M \in \mathbb{Q}[x]$ of positive degree. We will normalize M to be monic and squarefree so that we deal only with separable extensions. In the following we will also assume M to be minimal in order to have a canonical representation of α. From a mathematical point of view things are only slightly more complicated if the minimality assumption is dropped. It would have the computational advantage that only polynomial time algorithms are needed for arithmetic in K, the field of all algebraic numbers over $\mathbb{Q}$. In practice, however, complete factorization algorithms are fast enough that their exponential worst case computing time can be ignored*. Also, the advantage of working with a defining polynomial of minimal degree is computationally very attractive, since in many applications the speed of arithmetic in $\mathbb{Q}(\alpha)$ is the limiting factor. Also, practical considerations suggest the use of a rational polynomial instead of the similar primitive integral polynomial defining α. Otherwise, one would need a mixed integral and rational polynomial arithmetic.

Elements β of $\mathbb{Q}(\alpha)$ are represented with rational coefficients b_i in the integral base $1, \alpha, \ldots, \alpha^{m-1}$ of the vector space $\mathbb{Q}(\alpha)$ of dimension m, which is the degree of the minimal polynomial M of α. We use therefore the isomorphism $\mathbb{Q}(\alpha) \equiv \mathbb{Q}[x]/(M(x))$.

Algebraically, α has not to be distinguished from its conjugates. However, if α is real, for example, the sign of α or its size may be of interest. For this reason we indicate which root of $M(x)$ is α by an isolating interval $(r, s]$ such that $r < \alpha \leqslant s$, $r, s \in \mathbb{Q}$ and $\alpha \notin \mathbb{Q}$ or by the point interval $[r, r]$ if $\alpha = r$. Computationally, the interval endpoints can be required to have only denominators of the form 2^k in order to simplify the gcd-calculations greatly. Also, it is wise to let the degree of α be greater than 1 in order to avoid overhead. This can be easily achieved by real root calculations (see the chapter on this topic) or by special algorithms for rational root finding [11]. The algorithms to follow will also work for $m = 1$. In case α is not real, the isolating interval is an isolating rectangle in the complex plane with (binary) rational endpoints.

Elements in $GF(q)$, $q = p^m$, $m \geqslant 1$, p a rational prime, are m-dimensional vectors over $GF(p)$. In fact, the isomorphism $GF(p^n) \equiv GF(p)[x]/(M(x))$ is used for the arithmetic in $GF(q)$. Here $M(x)$ is any irreducible polynomial of degree m over $GF(p)$. Such defining polynomials can quickly be found by probabilistic methods [3, 13]. For definiteness we assume M to be monic again.

1.2 Arithmetic

The operations of negation, addition and subtraction in $\mathbb{Q}(\alpha)$ are operations in $\mathbb{Q}[x]$; only after multiplication is a reduction mod $M(x)$ sometimes necessary. Let d be the maximal seminorm of the representing polynomials of $\beta, \gamma \in \mathbb{Q}(\alpha)$ and of $M(x)$. Then, the time for $+, -$ is at most $O(mL(d)^2)$ and for multiplication $O(m^3 L(d)^2)$. Similarly, in $GF(q)$ the times are $O(mL(p)^2)$ and $O(m^2 L(p)^2)$ at most; in the most frequently implemented case, $L(p) = 1$ holds.

The inverse β^{-1} is computed in $\mathbb{Q}(\alpha)$ by the extended Euclidean algorithm for the representing polynomial B and M mod M.

$$UB + VM = 1.$$

Therefore

$$U \equiv B^{-1} \bmod M.$$

Using the modular gcd-algorithm this can be done in $O(m^3 L(d) + m^2 L(d)^2)$ steps. Here the first time the assumption M being minimal is crucial.

---

* See "Note added in proof", p. 187.

The same method applied in $GF(p)[x]$ yields a maximal computing time of $O(m^2)$.

1.3 The Sign of a Real Algebraic Number

Given α by the minimal polynomial M and its isolating interval $I = (r, t]$ and given furthermore $\beta \in \mathbb{Q}(\alpha)$ by its representing polynomial B, then the following algorithm computes $s = \text{sign}(\beta)$.

(1) [β rational.] If $B = 0$ then $\{s \leftarrow 0; \text{return}\}$;
    If $\deg B = 0$ then $\{s \leftarrow \text{sign}(\text{lc}(B)); \text{return}\}$.

(2) Compute B^* and b such that $(1/b)B^* = B$, $B^* \in \mathbb{Z}[x]$ and $b \in \mathbb{Q} - \{0\}$.
    Set $\bar{B}$ to the greatest squarefree divisor of B^*.

(3) Obtain from I an isolating interval I^* of α containing no roots of B^* by counting the roots of $\bar{B}$ in I^* and bisection.

$$\text{repeat } \{n \leftarrow \text{ \# of roots of } \bar{B} \text{ in } I^*;$$
$$\text{if } n = 0 \text{ then } \{s \leftarrow \text{sign}(b) * \text{sign}(B^*(t^*)); \text{return}\};$$
$$w \leftarrow (r^* + t^*)/2;$$
$$\text{if } M(r^*)M(w) < 0 \text{ then } t^* \leftarrow w \text{ else } r^* \leftarrow w\}. \quad \blacksquare$$

The correctness of the algorithm depends on the fact that B is reduced mod M and M is minimal. Therefore, B and M are relatively prime and the desired interval I^* for α exists. Then $\text{sign}(\beta) = \text{sign}(B(\alpha)) = \text{sign}(B(t^*))$.

How close can the roots $\alpha_1, \ldots, \alpha_m$ of M and the roots $\gamma_1, \ldots, \gamma_n$, $n < m$, of B lie together? We will see in Section 7 that

$$\min_{\substack{1 \le i \le m \\ 1 \le j \le n}} |\alpha_i - \gamma_j| > \frac{1}{d^m},$$

where d is the maximal seminorm of M and B. Since $\max|\alpha_i| < d$, we have $|I| < d$ at the beginning and the number k of bisections required is such that $d/2^k < d^{-m}$, or $k = O(mL(d))$.

With the results of the chapter on root isolation the analysis can be completed.

2. Polynomial Algorithms

Having available the arithmetic in $\mathbb{Q}(\alpha)$ and $GF(q)$, it is straightforward to realize polynomial arithmetic in $\mathbb{Q}(\alpha)[x_1, \ldots, x_r]$ and $GF(q)[x_1, \ldots, x_r]$. The gcd-algorithm for univariate polynomials can easily be implemented by Euclid's natural remainder sequence. For efficiency reasons it is advisable to use the monic natural p.r.s. Recently, A. K. Lenstra [10] has given a modular gcd algorithm for monic univariate polynomials in $\mathbb{Q}(\alpha)[x]$. He suggests applying the EZGCD algorithm (see chapter on polynomial remainder sequences in this volume) for $\mathbb{Q}(\alpha)[x]$. After this extension, the modified Uspensky algorithm or any other real root isolation algorithm can be extended to polynomials over $\mathbb{Q}(\alpha)$ for real algebraic numbers α. Root finding over $GF(q)$ can be done by interesting probabilistic methods [13, 2].

3. Resultant Calculus

Let $A = \sum_{i=0}^{m} a_i x^i$ and $B = \sum_{i=0}^{n} b_i x^i$ be two polynomials over a commutative ring R with identity. The Sylvester matrix of A and B is the $m + n$ by $m + n$ matrix

$$
M = \begin{pmatrix}
a_m & a_{m-1} & \cdots & & a_0 & & & \\
 & a_m & a_{m-1} & \cdots & & a_0 & & \\
 & & \cdots\cdots\cdots\cdots\cdots & & & & & \\
 & & & a_m & a_{m-1} & \cdots & a_0 & \\
b_n & b_{n-1} & \cdots & b_1 & b_0 & & & \\
 & b_n & \cdots & & & & b_0 & \\
 & & \cdots\cdots\cdots\cdots\cdots & & & & & \\
 & & & b_n & \cdots & & & b_0
\end{pmatrix}
$$

The upper part of M consists of n rows of elements of A, the lower part of m rows of elements of B, where all entries not shown are zero. The *resultant* of A and B is defined by

$$
\operatorname{res}(A, B) = \det(M).
$$

Clearly the resultant is an element of R and we have

$$
\operatorname{res}(A, B) = (-1)^{mn} \operatorname{res}(B, A), \tag{1}
$$

$$
\operatorname{res}(aA, B) = a^n \operatorname{res}(A, B), \qquad a \in R. \tag{2}
$$

By definition

$$
\operatorname{res}(a, B) = a^n, \qquad a \in R,
$$

$$
\operatorname{res}(a, b) = 1, \qquad a, b \in R. \tag{3}
$$

With m indeterminates α_i, $1 \leqslant i \leqslant m$, we construct

$$
A_m(x) = \prod_{i=1}^{m} (x - \alpha_i) = \sum_{i=0}^{m} a_i^{(m)} x^i.
$$

Clearly, we will be interested mainly in the case where the roots of $A_m(x)$ are substituted for the indeterminates α_i. But all resultant relations in this section will be derived without the assumption of the existence of roots, thus with the weaker assumption that the α_i are indeterminates. The coefficients a_i are related to the indeterminates α_i by

$$
a_m^{(m)} = s_m = 1,
$$

$$
-a_{m-1}^{(m)} = s_{m-1} = \alpha_1 + \cdots + \alpha_m,
$$

$$
a_{m-2}^{(m)} = s_{m-2} = \alpha_1 \alpha_2 + \alpha_1 \alpha_3 + \cdots + \alpha_{m-1} \alpha_m,
$$

$$
\cdots
$$

$$
(-1)^m a_0^{(m)} = s_0 = \alpha_1 \alpha_2 \cdots \alpha_m,
$$

where the s_i are the elementary symmetrical polynomials.

The coefficients $a_i^{(m)}$ are linear in α_m. Let us define $A_{m-1}(x) = A_m(x)/(x - \alpha_m)$. Between the coefficients of A_m and A_{m-1} considered as polynomials in the α_i's the relation

$$a_{i-1}^{(m-1)}(\alpha_1, \ldots, \alpha_{m-1}) = a_i^{(m)}(\alpha_1, \ldots, \alpha_{m-1}, 0), \qquad 1 \leqslant i \leqslant m \tag{4}$$

holds.

We are now ready to prove the

Lemma. *Let $B(x)$ be a polynomial over an integral domain R, $\deg(B) > 0$, and let $m > 1$ be an integer. With m indeterminates α_i let $A_m(x) = \prod_{i=1}^{m} (x - \alpha_i)$ and $A_{m-1}(x) = A_m(x)/(x - \alpha_m)$. Then*

$$\operatorname{res}(A_m, B) = B(\alpha_m) \operatorname{res}(A_{m-1}, B).$$

Proof. For $1 \leqslant i < m + n$ add to the last column of the Sylvester matrix M of A_m and B α_m^{m+n-i} times the ith column. For the resulting matrix M_1 we have $\det(M_1) = \det(M)$ and the elements of the last column from top to bottom are $\alpha_m^{n-1} A_m(\alpha_m), \ldots, \alpha_m^{0} A_m(\alpha_m), \alpha_m^{m-1} B(\alpha_m), \ldots, \alpha_m^{0} B(\alpha_m)$. Since $A_m(\alpha_m) = 0$ we take the factor $B(\alpha_m)$ out of the last column resulting in a matrix M_2 with the last column $0, \ldots, 0, \alpha_m^{m-1}, \ldots, \alpha_m^{0}$ and

$$\operatorname{res}(A_m, B) = \det(M) = \det(M_1) = B(\alpha_m) \det(M_2). \tag{5}$$

Let us consider both sides of (5) as polynomials in α_m. Since M has n rows of coefficients of $A(\alpha_m)$, which are at most linear in α_m, the left-hand side is of degree n or less in α_m. On the right-hand side the factor $B(\alpha_m)$ is already of degree n. Since R is an integral domain $\det(M_2)$ is of degree 0 in α_m. Taking $\det(M_2)$ at $\alpha_m = 0$ the last column becomes now $0, \ldots, 0, 0, \ldots, 1$ and the coefficients of A_m are transformed into the coefficients of A_{m-1} according to (4). Expansion of $\det(M_2)|_{\alpha_m=0}$ with respect to the last column results in the $m + n - 1$ by $m + n - 1$ matrix with $\det(M_3) = \det(M_2) = \operatorname{res}(A_{m-1}, B)$ which together with (2) proves the lemma. ∎

Theorem 1 immediately follows, which represents the resultant as symmetrical polynomial in the indeterminates α_i [14].

Theorem 1. *Let $A(x) = a_m \prod_{i=1}^{m} (x - \alpha_i)$ and $B(x) = b_n \prod_{i=1}^{n} (x - \beta_i)$ be polynomials over an integral domain R with indeterminates $\alpha_1, \ldots, \alpha_m$ and $\beta_1, \ldots, \beta_n$. Then*

$$\operatorname{res}(A, B) = (-1)^{mn} b_n^{m} \prod_{i=1}^{n} A(\beta_i), \tag{6}$$

$$\operatorname{res}(A, B) = a_m^{n} \prod_{i=1}^{m} B(\alpha_i), \tag{7}$$

$$\operatorname{res}(A, B) = a_m^{n} b_n^{m} \prod_{i=1}^{m} \prod_{j=1}^{n} (\alpha_i - \beta_j). \tag{8}$$

Proof. The theorem holds for $m = 0$ or $n = 0$ with the convention $\prod_{i=k}^{l} f_i = 1$, for $l < k$. Eq. (6) follows from (7) by (1), also (8) follows from (7) immediately. We prove (7) by induction on m. $\operatorname{res}(A_1, B) = B(\alpha_1)$, where $A_1(x) = x - \alpha_1$, follows from the expansion of the determinant with respect to the last row. Now by (2),

$\operatorname{res}(A, B) = a_m^n \operatorname{res}(A_m, B)$ and an inductive application of the Lemma results in (7). ∎

We state now some resultant relations in which the indeterminates α_i do not occur.

Theorem 2. *Let $A(x)$ and $B(x)$ be polynomials of positive degree over a commutative ring R with identity. Then there exist polynomials $S(x)$ and $T(x)$ over R with $\deg(S) < \deg(B)$ and $\deg(T) < \deg(A)$ such that*

$$AS + BT = \operatorname{res}(A, B). \tag{9}$$

Theorem 2 is a special instance of (1) in the chapter on polynomial remainder sequences.

Theorem 3. *Let A, B_1 and B_2 be polynomials over an integral domain. Then*

$$\operatorname{res}(A, B_1 B_2) = \operatorname{res}(A, B_1) \operatorname{res}(A, B_2). \tag{10}$$

Proof.

$$\operatorname{res}(A, B_1 B_2) = a_m^{n_1 + n_2} \prod_{i=1}^{m} (B_1(\alpha_i) B_2(\alpha_i)) = \operatorname{res}(A, B_1) \operatorname{res}(A, B_2). \quad \blacksquare$$

Theorem 4. *Let A, B, Q be polynomials over an integral domain and let $\deg(A) = m$, $\operatorname{lc}(A) = a_m$, $\deg(B) = n$, $\deg(AQ + B) = l$. Then*

$$\operatorname{res}(A, AQ + B) = a_m^{l-n} \operatorname{res}(A, B). \tag{11}$$

Proof. Again we use (7).

$$\operatorname{res}(A, B) = a_m^n \prod_{i=1}^{m} B(\alpha_i) = a_m^n \prod_{i=1}^{m} (A(\alpha_i) Q(\alpha_i) + B(\alpha_i)) = a_m^{n-l} \operatorname{res}(A, AQ + B). \quad \blacksquare$$

Theorem 3 may be used to increase the efficiency of the resultant calculation whenever a factorization of one of the polynomials is known. For example by (10) $\operatorname{res}(A, x^k B) = \operatorname{res}(A, B) \prod_{i=1}^{k} \operatorname{res}(A, x)$ and by (6)

$$\operatorname{res}(A, x) = (-1)^m A(0) = (-1)^m a_0.$$

Therefore

$$\operatorname{res}(A, x^k B) = (-1)^{mk} a_0^k \operatorname{res}(A, B). \tag{12}$$

Let $\deg(A) - \deg(B) = k \geqslant 0$. Then Eq. (12) together with (1) shows also that there is no loss of generality in the assumption that the polynomials of the resultant are of a specific degree as we stated in the chapter on polynomial remainder sequences.

Theorem 4 suggests an alternative way to calculate the value of the resultant. Moreover, it provides a proof of the next theorem, sometimes called the standard theorem on resultants [9], which follows immediately from (8), without any reference to the indeterminates α_i and β_i in (8).

Theorem 5. *Let A and B be non-zero polynomials over an integral domain R. Then $\operatorname{res}(A, B) = 0$ if and only if $\deg(\gcd(A, B)) > 0$.*

Proof. The theorem holds if A or B is constant. Assume $\deg(A) \geqslant \deg(B) > 0$. Working over the quotient field Q of R let $A = P_1$, $B = P_2$, $P_i = Q_i P_{i+1} + P_{i+2}$, $1 \leqslant i \leqslant k - 2$, $k \geqslant 3$, be a polynomial remainder sequence, thus

$$\deg(P_1) \geqslant \deg(P_2) > \cdots > \deg(P_k) \geqslant 0, \qquad P_{k+1} = 0.$$

Let $n_i = \deg(P_i)$ and set $A = P_{i+1}$, $B = P_{i+2}$ and $Q = Q_i$ in (11). Using also (1) we obtain

$$\operatorname{res}(P_i, P_{i+1}) = (-1)^{n_i n_{i+1}} \operatorname{lc}(P_{i+1})^{n_i - n_{i+2}} \operatorname{res}(P_{i+1}, P_{i+2}), \tag{13}$$

or

$$\operatorname{res}(P_1, P_2) = \operatorname{res}(P_{k-1}, P_k) \prod_{i=i}^{k-2} (-1)^{n_i n_{i+1}} \operatorname{lc}(P_{i+1})^{n_i - n_{i+2}}, \tag{14}$$

where lc denotes the leading coefficient.

If $\deg(P_k) = \deg(\gcd(A, B)) = 0$ then $\operatorname{res}(P_{k-1}, P_k) = \operatorname{lc}(P_k)^{n_{k-1}} \neq 0$ by (3). Otherwise we apply (11) again and since $P_{k+1} = 0$ the resultant vanishes. ∎

In [5] efficient algorithms for resultant calculation are given which are finally based on Eq. (14). They are superior to an evaluation of the determinant of the Sylvester matrix. In fact, the maximum computing time to calculate the resultant of two r-variate polynomials of maximal degree n and maximal seminorm d is $O(n^{2r+1} L(d) + n^{2r} L(d)^2)$.

4. Arithmetic in the Field K of All Algebraic Numbers over $\mathbb{Q}$

First we consider arithmetical operations on algebraic numbers. The following theorem gives the arithmetic in the field K of all algebraic numbers over $\mathbb{Q}$:

Theorem 6 (Loos 1973). *Let $A(x) = a_m \prod_{i=1}^{m} (x - \alpha_i)$ and $B(x) = b_n \prod_{j=1}^{n} (x - \beta_j)$ be polynomials of positive degree over an integral domain R with roots $\alpha_1, \ldots, \alpha_m$ and $\beta_1, \ldots, \beta_n$ respectively. Then the polynomial*

$$r(x) = (-1)^{mn} g a_m^n b_n^m \prod_{i=1}^{m} \prod_{j=1}^{n} (x - \gamma_{ij})$$

has the $m \cdot n$ roots, not necessarily distinct, such that

(a) $\qquad r(x) = \operatorname{res}(A(x - y), B(y)), \qquad \gamma_{ij} = \alpha_i + \beta_j, \quad g = 1,$

(b) $\qquad r(x) = \operatorname{res}(A(x + y), B(y)), \qquad \gamma_{ij} = \alpha_i - \beta_j, \quad g = 1,$

(c) $\qquad r(x) = \operatorname{res}(y^m A(x/y), B(y)), \qquad \gamma_{ij} = \alpha_i \beta_j, \quad g = 1,$

(d) $\qquad r(x) = \operatorname{res}(A(xy), B(y)), \qquad \gamma_{ij} = \alpha_i / \beta_j, \quad B(0) \neq 0,$

$$g = (-1)^{mn} B(0)^m / b_n^m.$$

Proof. The proof is based on relation (6) in all four cases.

(a) $\qquad \operatorname{res}(A(x - y), B(y)) = (-1)^{mn} b_n^m \prod_{j=1}^{n} A(x - \beta_j)$

$$= (-1)^{mn} a_m^n b_n^m \prod_{i=1}^{m} \prod_{j=1}^{n} (x - (\alpha_i + \beta_j)).$$

(b) $\operatorname{res}(A(x + y), B(y)) = (-1)^{mn}b_n^m \prod_{j=1}^{n} A(x + \beta_j)$

$$= (-1)^{mn}a_m^n b_n^m \prod_{i=1}^{m} \prod_{j=1}^{n} (x - (\alpha_i - \beta_j)).$$

(c) $\operatorname{res}(y^m A(x(y), B(y)) = (-1)^{mn}b_n^m \prod_{j=1}^{n} \beta_j^m A(x/\beta_j)$

$$= (-1)^{mn}a_m^n b_n^m \prod_{i=1}^{m} \prod_{j=1}^{n} (x - \alpha_i \beta_j).$$

(d) $\operatorname{res}(A(xy), B(y)) = (-1)^{mn}b_n^m \prod_{j=1}^{n} A(x\beta_j)$

$$= (-1)^{mn}a_m^n b_n^m \prod_{i=1}^{m} \prod_{j=1}^{n} (x\beta_j - \alpha_i)$$

$$= (-1)^{mn}a_m^n \prod_{i=1}^{m} b_n \left(\prod_{j=1}^{n} \beta_j \right) \prod_{j=1}^{n} (x - \alpha_i/\beta_j)$$

$$= (-1)^{mn}a_m^n b_0^m \prod_{i=1}^{m} \prod_{j=1}^{n} (x - \alpha_i/\beta_j) \quad \text{mit} \quad b_0 \neq 0. \quad \blacksquare$$

Theorem 6 constructs explicit polynomials and we see that, except in case (d) the polynomial $r(x)$, to within a sign, is monic if A and B are. We have therefore

Corollary 1. *All algebraic integers over R form a ring.*

We denote the ring by R_∞.

Corollary 2. *All algebraic numbers over R form a field.*

We denote the field by K, where Q is the quotient field of R.

Since the degree of $r(x)$ is $m \cdot n$, the resultants are linear in the particular case the given polynomials are. We conclude that the rational numbers over R form a subfield Q of K and that R forms a subring of R_∞. We convince ourselves that R_∞ is an integral domain by considering Theorem 6, case (c) with $A(0) = 0$, $r(0) = \operatorname{res}(y^m A(0), B(y))$. By (1), (2), (3) and (12), we find $r(0) = A(0)^n b_n^m$. Since R has no zero divisors the same holds for R_∞.

Theorem 6 is the base of Algorithm 1. Obviously, it is sufficient to consider only addition and multiplication of algebraic numbers. For if the number α is defined by the polynomial $A(x)$, then the polynomial $A(-x)$ defines $-\alpha$ and $x^m A(1/x)$, $m = \deg(A)$, defines $1/\alpha$ if $\alpha \neq 0$.

Algorithm 1 (Algebraic number arithmetic).

Input: Let R be an Archimedian ordered integral domain. α, β are algebraic numbers represented by two isolating intervals I, J having endpoints in Q and by two defining polynomials A and B of positive degree over R.

Output: An isolating interval K and a defining primitive squarefree polynomial $C(x)$ representing $\gamma = \alpha + \beta$ (or $\gamma = \alpha * \beta$ for multiplication).

(1) [Resultant]
   $r(x) = \text{res}(A(x - y), B(y))$
   $(r(x) = \text{res}(y^m A(x/y), B(y))$ for multiplication).

(2) [Squarefree factorization] $r(x) = D_1(x)D_2(x)^2 \cdots D_f(x)^f$.

(3) [Root isolation] Generate isolating intervals or rectangles $I_{11}, \ldots, I_{1g_1}, \ldots, I_{fg_f}$ such that every root of D_i is contained in exactly one I_{ij} and $I_{ij} \cap I_{kl} = \emptyset$, $1 \leqslant i$, $k \leqslant f$, $1 \leqslant j \leqslant g_i$, $1 \leqslant l \leqslant g_k$, $(i,j) \neq (k,l)$.

(4) [Interval arithmetic] Set $K = I + J$ ($I * J$ for multiplication) using exact interval arithmetic over Q.

(5) [Refinement] If there is more than one I_{ij} such that $K \cap I_{ij} \neq 0$, bisect I and J and go back to step (4). Otherwise, return K and $C(x) = D_i(x)$. ∎

Note that the computing time of the algorithm is a polynomial function of the degrees and seminorms of α and β. In practical implementation it may be preferable to replace in step 2 the squarefree factorization by a complete factorization, which would give the algorithm an exponential maximum computing time.*

The effectiveness of step (3) depends essentially on a non-algebraic property of the underlying ring R, its Archimedean order. A ring is *Archimedean ordered* if there exists for every element A a natural number (i.e. a multiple of the identity of the ring) N such that $N - A > 0$. Let us for example take the non-Archimedean ordered ring of polynomials over the rationals, $\mathbb{Q}[x]$, where an element is called positive if the leading coefficient is positive. Thus the element x is greater than any rational number and there is no assurance that an interval or rectangle containing x can be made arbitrarily small by bisection. It is still possible to count the zeros in intervals over non-Archimedean rings by Sturm's theorem, but root isolation requires Archimedean order.

The loop in step (4) and step (5) is only executed a finite number of times, since by Theorem 6 exactly one isolating interval contains the sum $\alpha + \beta$ (or the product $\alpha * \beta$) and bisection decreases the length of the interval K, computed by exact interval arithmetic, under any bound. Therefore, the input assumption of the Archimedean order enforces the termination of the algorithm.

The proof of the theorem is based on the relation (6) which in turn follows immediately from Lemma 1. We will show, using the equivalent relation (7), how similar constructions of defining polynomials for algebraic numbers can be established. If α is defined by A, we consider the norm $N_\alpha = \text{res}_\alpha(A(\alpha), \cdot)$ as a polynomial operator with indeterminate α. In order to compute any function $g(\alpha)$ composed finally by ring operations on α only, we have to apply the operator N_α to $x = g(\alpha)$ yielding

$$N_\alpha(x - g(\alpha)) = \text{res}_\alpha(A(\alpha), x - g(\alpha)) = a_m^n \prod_{i=1}^m (x - g(\alpha_i))$$

which shows that $N_\alpha(x - g(\alpha))$ is a polynomial having $g(\alpha)$ as root. By iteration, the method can be extended to any polynomial function of several algebraic numbers. Let α, β be defined by A and B respectively. In order to compute, say $f(\alpha, \beta) = \alpha + \beta$, we form

$$N_\alpha(N_\beta(x - f(\alpha, \beta))) = \mathrm{res}_\alpha(A(\alpha), \mathrm{res}_\beta(B(\beta), x - (\alpha + \beta)))$$

$$= \mathrm{res}_\alpha\left(A(\alpha), b_n \prod_{i=1}^{n} (x - (\alpha + \beta_j)) \right)$$

$$= a_m^n b_n^m \prod_{i=1}^{m} \prod_{j=1}^{n} (x - (\alpha_i + \beta_j))$$

which is up to a sign the defining polynomial of Theorem 6, (a). In fact, the method can still be further generalized. All that is required is that the relation $x = f(\alpha, \beta)$ may be transformed into a polynomial relation, say $F(x, \alpha, \beta)$. The following theorem gives an application. Let us consider fractional powers of algebraic numbers. Since the reciprocal of an algebraic number can be computed trivially we restrict ourselves to positive exponents.

Theorem 7 (Fractional powers of algebraic numbers). *Let A be a polynomial of positive degree m over an integral domain R with roots $\alpha_1, \ldots, \alpha_m$. Let p, q be positive integers. Then*

$$r(x) = \mathrm{res}(A(y), x^q - y^p)$$

has the roots $\alpha_i^{p/q}$, $i = 1, \ldots, m$.

Proof.

$$r(x) = a_m^p \prod_{i=1}^{m} (x^q - \alpha_i^p). \quad \blacksquare$$

We can base on Theorem 7 an algorithm for the computation of fractional powers of algebraic numbers, which would be quite similar to Algorithm 1.

Another application of our algebraic number calculus allows a transformation of the algebraic number representation in $\mathbb{Q}(\alpha)$ as a polynomial $\beta = B(\alpha)$ to a defining polynomial for β. We get $N_\alpha(x - B(\alpha))$ as defining polynomial for β.

Theorem 8. *Let α be an algebraic number over R and A its defining polynomial of degree $m > 0$. Let $\beta = \sum_{x=0}^{m-1} b_i \alpha^i = B(\alpha)$, where $b_i \in Q$, $\deg(B) = n < m$. Then β is algebraic over R and a root of*

$$r(x) = \mathrm{res}(x - B(y), A(y)).$$

If α is an algebraic integer then so is β, provided $b_i \in R$.

Proof. By Theorem 1, Eq. (6), $r(x) = (-1)^{mn} a_m^n \prod_{i=1}^{m} (x - B(\alpha_i))$. $\quad \blacksquare$

Corollary 3. *If A of Theorem 8 is the minimal polynomial of α then B is uniquely determined.*

Proof. Suppose $\beta = B^*(\alpha)$, $\deg(B^*) < m$ and $B^* \neq B$. $B(\alpha) - B^*(\alpha) = \beta - \beta = 0$.

This is a polynomial, not identically vanishing, of degree $< m$, a contradiction to the minimal property $\deg(A) = m$. ∎

Theorem 8 can be used to compute the sign of a real algebraic number differently from the approach in Section 1. Given $\beta = B(\alpha)$, we construct $r(x)$ from the theorem and compute $I_\beta = B(I_\alpha)$ by interval arithmetic such that I_β is isolating with respect to $r(x)$. The position of I_β in relation to 0 gives the sign of β. The position in relation to I_α, made disjoint from I_β, gives an algorithm for real algebraic number comparison.

Historical Note. The resultant $\mathrm{res}_y(A((x-y)/2), A((x+y)/2))$ was considered by Householder in [8] and stimulated our interest in resultants.

A special case of Theorem 6 is the resultant $\mathrm{res}_y(A(x+y), A(y))$, a polynomial having the roots $\alpha_i - \alpha_j$. The task of obtaining lower bounds on $\min_{i,j} |\alpha_i - \alpha_j|$ is reduced by it to the problem of a lower bound for the roots of $r(x)/x^m$. By this approach Collins [6] improved Cauchy's [4] lower bound for the minimum root separation.

5. Constructing Primitive Elements

The representation $\beta = B(\alpha)$ allows the construction of extensions of R and Q as shown by the next two theorems.

Theorem 9. *Let α be an algebraic integer over R and A its defining polynomial of degree $m > 1$. Then the set of all algebraic numbers represented by $\beta = \sum_{i=0}^{m-1} b_i \alpha^i = B(\alpha)$, where $b_i \in R$, forms a ring of algebraic integers.*

We call the ring a *simple extension* of R and denote it by $R[\alpha]$.

Proof. Since $C(x) = Q(x)A(x) + B(x)$, where $B = 0$ or $\deg(B) < m$, and $A(\alpha) = 0$ we have $B(\alpha) = C(\alpha)$. Hence, there is an isomorphism between $R[\alpha]$ and $R[x]/(A(x))$, the ring of residue classes of A. ∎

Theorem 10. *Let α be an algebraic number over a field Q and A its defining polynomial of degree $m > 0$. Then the set of all algebraic numbers represented by $\beta = \sum_{i=0}^{m-1} b_i \alpha^i = B(\alpha)$, where $b_i \in Q$, forms a field.*

We call the field a *simple algebraic extension* of Q and denote it by $Q(\alpha)$.

Proof. We have to show that every non-zero element β of $Q(\alpha)$ has a multiplicative inverse. First, assume $\deg(\gcd(A(x), B(x))) = 0$. By Theorem 5, $\mathrm{res}(A, B) \neq 0$. Theorem 2 gives, for $x = \alpha$, $B(\alpha)T(\alpha) = \mathrm{res}(A, B)$ with $\deg(T) < m$. Therefore, $T(\alpha)/\mathrm{res}(A, B)$ is the inverse of $B(\alpha)$. Now, let $C(x) = \gcd(A, B)$, $\deg(C) > 0$, and $A = CA^*$. Clearly, $C(\alpha) \neq 0$, otherwise $\beta = B(\alpha) = 0$. Therefore, $A^*(\alpha) = 0$. Replace A by A^* and apply the first argument of the proof, observing that $\deg(\gcd(A^*, B)) = 0$. ∎

Extensions $Q(\alpha)$ are called *separable*, if the defining polynomial for α is squarefree.

Clearly, $R \subseteq R[\alpha] \subset R_\infty$. Since R_∞ was shown to be an integral domain the same holds for $R[\alpha]$. Also $Q \subseteq Q(\alpha) \subset K$. All previous results remain valid with $R[\alpha]$ and $Q(\alpha)$ in place of R and Q respectively. In particular $R[\alpha][\beta]$ is a ring and $Q(\alpha)(\beta)$ a field again. We call $Q(\alpha)(\beta) = Q(\alpha, \beta)$ a double extension of Q. Of central importance is

Theorem 11. *Every separable multiple extension is a simple extension.*

We give the proof constructively by an algorithm:

Algorithm 2 (SIMPLE).

Inputs: $A(x)$ and $B(x)$, two primitive squarefree polynomials of positive degree over R, an Archimedean ordered integral domain, I and J two isolating intervals over Q such that α is represented by I and A and β by J and B.

Outputs: An isolating interval K, a defining primitive squarefree polynomial $C_0(x)$ over R, representing γ, and two polynomials over Q, $C_1(x)$ and $C_2(x)$, such that $\alpha = C_1(\gamma)$, $\beta = C_2(\gamma)$.

(1) [Resultant] $r(x, t) = \mathrm{res}(A(x - ty), B(y))$.
   [Here, by (6), $r(x, t)$ has root $\gamma_{ij} = \alpha_i + t\beta_j$.]

(2) [Squarefree] Compute the smallest positive integer t_1 such that $\deg(\gcd(r(x, t_1), r'(x, t_1))) = 0$. Set $C_0(x) = r(x, t_1)$.
   [This implies that all γ_{ij}, $1 \leqslant i \leqslant m$, $1 \leqslant j \leqslant n$, are different, so by (1)

$$\alpha_i - \alpha_k \neq t_1(\beta_j - \beta_l)$$

   for all pairs (i, j) and (k, l) with $i \neq k$ and $j \neq l$. Since there are only finitely many pairs but infinitely many positive integers, a t_1 can always be found.]

(3) [Interval arithmetic] By repeated bisection of I and J construct K such that $K = I + t_1 J$ and K is an isolating interval for $C_0(x)$. [Obviously $Q(\gamma) \subseteq Q(\alpha, \beta)$ for $\gamma = \gamma_{11}$.]

(4) [gcd] Using arithmetic in $Q(\gamma)$ compute $B^*(x) = \gcd(A(\gamma - t_1 x), B(x))$. [By the construction β_j is a root of $B^*(x)$ and there is only one such β according to (2). Therefore $\deg(B^*) = 1$ and $B^*(x) = x - \beta$, where B^* is monic by convention.]

(5) [Exit] Set $C_2(x) = -$ trailing coefficient of $B^*(x)$, $C_1(x) = x - t_1 C_2(x)$. [We obtain $\beta = C(Y)$ and $\alpha = \beta - t_1 C(Y)$.]   ∎

We have used a modification of Theorem 6 (a) in step (1). Similarly, (b) – (d) of Theorem 6 may be used for constructing primitive elements γ. The given algorithm occurs under the name SIMPLE in Collin's quantifier elimination algorithm (see the chapter of this name in this volume). As algorithm 1, this algorithm has also polynomial computing time. If n is the maximal degree of α and β and d is its maximal seminorm then the seminorm of C is $O(d^{2n})$. Empirically, it turns out that the most expensive operation of SIMPLE is the division in step (4) to make B^* monic; up to 80% of the total time can be saved if α and β need not to be represented in $Q(\gamma)$. Note that the degree of γ is in general n^2, which indicates the computational limitations of this approach.

6. The Algorithm NORMAL

A last application of the algebraic number calculus allows to represent roots of algebraic number polynomials as roots of integral polynomials. It gives a constructive proof that K is algebraically closed.

Using Theorem 11 we can show that the roots of a polynomial with algebraic number coefficients are algebraic numbers. Let $B^*(x) = \sum_{j=0}^{n} \beta_j x^j$ be such a polynomial. First we compute $Q(\alpha) = Q(\beta_0, \ldots, \beta_n)$ and express $\beta_j = B_j(\alpha)$, i.e. by polynomials over Q according to the construction of the last proof. By the next theorem we obtain a construction for a polynomial $r(x)$ over R having among its roots the roots of $B^*(x)$.

Theorem 12. *Let $A(y) = \sum_{i=0}^{m} a_i y^i = a_m \prod_{i=1}^{m} (y - \alpha_i)$ be a primitive squarefree polynomial over an integral domain R.*
Let $B(y, x) = \sum_{j=0}^{n} B_j(y) x^j$ be a bivariate polynomial over R such that $\deg(\gcd(A(y), B(y))) = 0$. Let $k = \deg(B(y, x))$ and $r(x) = \mathrm{res}(A(y), B(y, x))$. Then $r(x) = a_k^k \prod_{i=1}^{m} B_n(\alpha_i) \prod_{j=1}^{n} (x - \beta_{ij})$, where the β_{ij} are defined by

$$B(\alpha_i, x) = B_n(\alpha_i) \prod_{j=1}^{n} (x - \beta_{ij}), \qquad 1 \leq i \leq m.$$

Proof. By Theorem 1 (7),

$$r(x) = a_m^k \prod_{i=1}^{m} B(\alpha_i, x) = a_m^k \prod_{i=1}^{m} B_n(\alpha_i) \prod_{j=1}^{n} (x - \beta_{ij}). \qquad \blacksquare$$

Corollary 4. *The roots of a polynomial with algebraic number coefficients are algebraic numbers.*

The algorithm NORMAL that occurs with SIMPLE in Collins' quantifier elimination algorithm is based on this theorem.

Algorithm 3 (NORMAL).

Input: A polynomial $B^*(x)$ of degree n over $Q(\beta_0, \ldots, \beta_n)$, where the β_i are given by defining primitive squarefree polynomials $B_i^*(x)$ over R, an integral Archimedean ordered domain, and isolating intervals $I(\beta_i)$, $\beta \leq i \leq n$.

Output: A polynomial $C(x)$ over R having among its roots the roots of $B^*(x)$. By an obvious modification, isolating intervals of $B^*(x)$ over Q may also be computed.

(1) $[Q(\alpha) = Q(\beta_0, \ldots, \beta_n)]$ By repeated application of Algorithm 2 compute a primitive squarefree polynomial A having the root α such that $Q(\alpha) = Q(\beta_0, \ldots, \beta_n)$ and rational polynomials $\bar{B}_j(y)$ such that $\beta_j = \bar{B}_j(\alpha)$ and $\deg(\bar{B}_j) < m = \deg(A)$.

(2) Compute $d \in R$ such that $B_j = d\bar{B}_j \in R[x]$, $0 \leq j \leq n$, and set $B(y, x) = \sum_{j=0}^{n} B_j(y) x^j$.

(3) [gcd] Set $D(y) = \gcd(A(y), B_n(y))$. Here $\deg(D) \leq \deg(B_n) < m$.

(4) [Reduction?] If $\deg(D) > 0$, go to step (6).

(5) Set $C(x) = \text{res}(A(y), B(y, x))$ and exit. By the preceeding Theorem 12 $C(x) \neq 0$ and every root of B^* is a root of $C(x)$.

(6) [Reduce] Set $\bar{A} = A/D$. Now $\bar{A}(\alpha) = 0$, since $B_n(\alpha) \neq 0$ and hence $D(\alpha) \neq 0$. Set $B_j \equiv b_j \bmod \bar{A}$, for $0 \leqslant j \leqslant n$. Compute $d \in R$ such that $B_j = d\bar{B}_j \in R[x]$, for $0 \leqslant j \leqslant n$. Now $\deg(\gcd(\bar{A}, B_n)) = 0$ since A is squarefree. Go back to step (5). ■

Again, the computing time is a polynomial function.

7. Some Applications

Suppose we know that two polynomials are relatively prime. How close can the roots α_i and β_j be? We look for $\min|\alpha_i - \beta_j|$. Using our approach, we construct a polynomial $C(x)$ having all $\alpha_i - \beta_j$ as roots and determine the radius of the circle around the origin, in which no roots of $C(x)$ are located. C is given by $\text{res}_y(A(x + y), B(y))$ according to Theorem 6 (b) and has seminorm $\leqslant d^{2n}$ by Hadamard's inequality (see the chapter on useful bounds). Then we get $\min|\alpha_i - \beta_j| > d^{-2n}$. We need this result to prove termination of algorithms isolating the roots of two relatively prime polynomials "away" from each other.

Next, suppose that a polynomial $A(x)$ is squarefree. How small can $A(\beta_i)$ be at any zero β_i of the derivative? The answer is given by $\text{res}_\beta(A'(\beta), x - A(\beta)) = C(x)$. The lower root bound of C is of the same order as the previous one.

If we do not require A to be squarefree, then the polynomial $C(x)$ is of order $k > 0$, where k gives the degree of $\gcd(A, A')$. Since $c_0, c_1, \ldots, c_{k-1}$ can be expressed by the coefficients of A, the fact that $\deg \gcd(A, A') = k$ is expressible as $c_0 = 0, \ldots, c_{k-1} = 0$ in the coefficients of A only. This is an alternative to Collins' psc-theorem used for quantifier elimination.

The last two observations give termination bounds in the Collins-Loos real root isolation algorithm (see the chapter on root isolation in this volume).

Let $A(x)$ and $B(x)$ be integral polynomials of the form $A = A_1 A_2$ and $B = B_1 B_2$ such that there exists $k \in Q$ with $B_1(x) = A_1(x - k)$, which we call shifted factors. In order to detect such shifted factors we use again Theorem 6 (b) and compute $C(x) = \text{res}_y(A(x + y), B(y)) = c_{n+m}\prod_j(x - (\alpha_i - \beta_j))$. If A and B have a shifted factor k then there is some root pair α_i, β_j with $\alpha_i - \beta_j = k$ which can be detected as rational root of the polynomial $C(x)$. This idea is used in step 3 of Gosper's algorithm (see the chapter on summation). In a similar manner "rotated" factors $B_1(x) = A_1(x \cdot k)$ can be detected using part (c) of Theorem 6.

With a rational root finder all pairs $\alpha_i + \alpha_j = c_{ij}$ and $\alpha_i\alpha_j = d_{ij}$ can be formed from the resultants (a) and (c) of Theorem 6 with $B = A$. Then, one can form trial divisors $x^2 - c_{ij}x + d_{kl}$ for finding quadratic factors of A [11].

The construction of powers of algebraic numbers used in root-squaring algorithms can be expressed by Theorem 7. Hence, the polynomials entering Graeffe's method are resultants.

Note added in proof: In the meantime A. K. Lenstra, H. W. Lenstra and L. Lovacs discovered a polynomial time bounded factorization algorithm for integral polynomials. Therefore, there is no computational objection against the use of minimal polynomials anymore.

References

[1] Arnon, D. S.: Algorithms for the Geometry of Semi-Algebraic Sets., Ph.D. Thesis, Univ. of Wisconsin, Madison, 1981.

[2] Calmet, J.: A SAC-2 Implementation of Arithmetic and Root Finding over Large Finite Fields. Interner Bericht, Universität Karlsruhe 1981 (forthcoming).

[3] Calmet, J., Loos, R.: An Improvement of Robin's Probabilistic Algorithm for Generating Irreducible Polynomials over GF(p). Information Processing Letters **11**, 94 – 95 (1980).

[4] Cauchy, A.: Analyse Algébrique. Oeuvres complètes, II série, tome III, p. 398, formula (48). Paris: 1847.

[5] Collins, G. E.: The Calculation of Multivariate Polynomial Resultants. J. ACM **18**, 515 – 532 (1971).

[6] Collins, G. E., Horowitz, E.: The Minimum Root Separation of a Polynomial. Math. Comp. **28**, 589 – 597 (1974).

[7] Heindel, G. E.: Integer Arithmetic Algorithms for Polynomial Real Zero Determination. J. ACM **18**, 533 – 548 (1971).

[8] Householder, A. S.: Bigradients and the Problem of Routh and Hurwitz. SIAM Review **10**, 57 – 78 (1968).

[9] Householder, A. S., Stuart III, G. W.: Bigradients, Hankel Determinants, and the Padé Table. Constructive Aspects of the Fundamental Theorem of Algebra (Dejon, B., Henrici, P., eds.). London: 1963.

[10] Lenstra, A. K.: Factorisatie van Polynomen. Studieweek Getaltheorie en Computers, Mathematisch Centrum, Amsterdam, 95 – 134 (1980).

[11] Loos, R.: Computing Rational Zeros of Integral Polynomials by p-adic Expansion. Universität Karlsruhe (1981) (to appear in SIAM J. Comp.).

[12] Pinkert, J. R.: Algebraic Algorithms for Computing the Complex Zeros of Gaussian Polynomials. Ph.D. Thesis, Comp. Sci. Dept., Univ. of Wisconsin, May 1973.

[13] Rabin, M. O.: Probabilistic Algorithms in Finite Fields. SIAM J. Comp. **9**, 273 – 280 (1980).

[14] van der Waerden, B. L.: Algebra I. Berlin-Heidelberg-New York: Springer 1971.

[15] Zimmer, H.: Computational Problems, Methods, and Results in Algebraic Number Theory. Lecture Notes in Mathematics, Vol. 262. Berlin-Heidelberg-New York: Springer 1972.

Prof. Dr. R. Loos
Institut für Informatik I
Universität Karlsruhe
Zirkel 2
D-7500 Karlsruhe
Federal Republic of Germany

Arithmetic in Basic Algebraic Domains

G. E. Collins, Madison, **M. Mignotte**, Strasbourg, and **F. Winkler**, Linz

Abstract

This chapter is devoted to the arithmetic operations, essentially addition, multiplication, exponentiation, division, gcd calculation and evaluation, on the basic algebraic domains. The algorithms for these basic domains are those most frequently used in any computer algebra system. Therefore the best known algorithms, from a computational point of view, are presented. The basic domains considered here are the rational integers, the rational numbers, integers modulo m, Gaussian integers, polynomials, rational functions, power series, finite fields and p-adic numbers. Bounds on the maximum, minimum and average computing time (t^+, t^-, t^*) for the various algorithms are given.

The Rational Integers

Most of the material for this section has been taken from [11]. There the reader may also find the details for the following computing time analyses.

Representation

For specifying the algorithms for the various operations on $\mathbb{Z}$ we represent the elements of $\mathbb{Z}$ in a positional number system. In order to do so we choose a natural number $\beta \geqslant 2$ as the base or radix of the positional system. Let $\{-(\beta-1), \ldots, -1, 0, 1, \ldots, (\beta-1)\}$ be the set of β-integers. A positive integer a is then represented (in the positional number system with radix β) by the unique sequence $a_{(\beta)} = (a_0, \ldots, a_{n-1})$ of nonnegative β-integers, such that $a_{n-1} > 0$ and $a = \sum_{k=0}^{n-1} a_i \cdot \beta^i$. Similarly we represent a negative number a by the unique sequence $(a_0, \ldots, a_{n-1})$ of nonpositive β-integers, such that $a_{n-1} < 0$. Finally, we let the empty sequence ( ) represent the integer zero. For practical implementation most systems represent only the integers a with $|a| \geqslant \beta$ as lists, whereas small integers are not represented as lists in order to avoid function call overhead. With this convention 0 is represented by the integer 0 and not by ( ).

We want to be a little bit more specific, at this point, about the data structure of this representation. Since almost never the lengths of the integers occurring during the execution of an algorithm are known in advance, it would be extremely wasteful to allocate the storage statically at the beginning of an algorithm. A means for avoiding this inefficiency is list representation. Thus we represent an element $a \in \mathbb{Z}$ as the list of β-digits of a, the least significant digit a_0 being the first element in the list. Relative to the alternative representation $(a_{n-1}, \ldots, a_0)$ this has both advantages and disadvantages. For addition and multiplication we have the advantage that carries propagate in the direction that the lists are scanned. But there is the disadvantage that the determination of the sign is not immediate if some of the least significant digits are zero, and there is the further disadvantage that in division the most significant digits are required first. All in all, however, the advantages appear to significantly outweigh the disadvantages.

For $a \neq 0$, represented as $(a_0, \ldots, a_{n-1})$, we call n the β-length of a, denoted by $L_\beta(a)$, and we let $L_\beta(0) = 1$. The β-length of a is essentially the β-logarithm of $|a|$, precisely $L_\beta(a) = \lfloor \log_\beta |a| \rfloor + 1$ for $a \neq 0$, where $\lfloor \ \rfloor$ denotes the floor function. If γ is another basis, then $L_\beta(a)$ and $L_\gamma(a)$ are codominant. So we often just speak of $L(a)$, the length of a.

Though list representation is extremely helpful for minimizing the storage needed for an algorithm, it also has the disadvantage that mathematically seemingly trivial operations become nontrivial.

Conversion and Input/Output

A problem frequently occurring is that of conversion from one radix to another. Let us assume that we are converting the integer a from radix β to radix γ. Basically there are two algorithms for this purpose.

The first algorithm, CONVM, involves multiplication by β using only radix γ arithmetic. It amounts to evaluating the polynomial $a(x) = \sum_{i=0}^{n-1} a_i \cdot x^i$ at $x = \beta$ using Horner's method, where $a_{(\beta)} = (a_0, \ldots, a_{n-1})$.

The second algorithm, CONVD, uses only radix β arithmetic and is essentially a repeated division by γ (the remainders being the digits of $a_{(\gamma)}$).

The computing times for both methods are proportional to $L(a)^2$.

One of the most important usages of conversion is for input and output. Here the problem is to convert from decimal to radix β representation (where β usually is a constant whose length differs only slightly from the word size of the machine) and vice versa. Since arithmetic operations will be developed for radix β representation, it is natural to employ algorithm CONVM for input and CONVD for output.

Instead of converting from decimal to radix β representation at once, it is preferable (see [23], Section 4.4) to convert first to radix 10^η ($\eta = \lfloor \log_{10} \beta \rfloor$) and to compute the radix β representation from there. Each digit in the radix 10^η representation is completely determined by η consecutive decimal digits, so the conversion of a from decimal to radix 10^η takes time proportional to $L(a)$. The conversion from radix 10^η to radix β will take time proportional to $L(a)^2$, just as would direct conversion from decimal to radix β, but the constant of proportionality for direct conversion is about η times as large. Since η will typically be approximately 10, a substantial saving is realized. A similar remark holds for conversion from radix β to decimal.

Arithmetic Operations

The major operations we want to perform are comparison, addition, subtraction, multiplication, exponentiation, division and gcd-computation.

In many algorithms we need the computation of the sign of an integer. Because of our data structure the algorithm ISIGN is not totally trivial, but still quite straightforward:

$s \leftarrow$ ISIGN(a)
[integer sign. a is an integer. $s = \text{sign}(a)$, a β-integer]
    Successively check $a_0, a_1, \ldots$ until the first non-zero digit a_k is found, from which the sign s can be determined. ∎

Clearly, $t_{\text{ISIGN}}^+ \sim L(a)$, $t_{\text{ISIGN}}^- \sim 1$. Surprisingly, $t_{\text{ISIGN}}^* \sim 1$ (actually $< (1 - \beta^{-1})^{-2}$). This can be seen from the following consideration: the number of integers of length n $(n \geqslant 2)$, for which k digits have to be checked, is $2(\beta - 1)$ for $k = n$ and $2(\beta - 1)^2 \beta^{n-k-1}$ for $k < n$. Since there are $2(\beta - 1)\beta^{n-1}$ integers of length n $(n \geqslant 2)$, the average number of digits to be checked for integers of length n is

$$N_n = (\beta - 1) \sum_{k=1}^{n-1} k\beta^{-k} + n\beta^{-n+1} < \sum_{k=1}^{\infty} k\beta^{-k+1} = (1 - \beta^{-1})^{-2}.$$

Algorithms for negation, INEG, and computation of the absolute value, IABS, are straightforward. Their computing times are proportional to the length of the input integer a.

Next we give an algorithm for comparing two integers a and b. Let us consider the following example: $\beta = 10$ and we want to compare the two integers $a = -102$ and $b = 3810$. a and b are represented by the lists $(-2, 0, -1)$ and $(0, 1, 8, 3)$, respectively. First we look at the first digits of a and b. Since $b_0 = 0$, we do not know the sign of b. Thus the decision has to be delayed, but we remember that the sign of a is -1. Inspecting the next two digits, we find that the sign of b is 1. Hence, $a < b$. As a second example consider $c = -12$ and $d = -800$, represented as the lists $(-2, -1)$ and $(0, 0, -8)$. After inspecting the first two pairs of digits a decision cannot be made. But since the end of the first list is reached we know that the absolute value of b is greater than that of a. Thus we can make the decision if we determine the sign of b. These ideas are incorporated in the following algorithm:

$s \leftarrow$ ICOMP(a, b)
[integer comparison. a and b are integers. $s = \text{sign}(a - b)$]

(1)   Set u and v to zero and inspect successive digits $a_0, b_0, a_1, b_1, \ldots$ until the end of one of the lists is reached. At the kth step $(k = 0, 1, \ldots)$ set $u \leftarrow \text{sign}(a_k)$ if $a_k \neq 0$, $v \leftarrow \text{sign}(b_k)$ if $b_k \neq 0$ and check whether $u \cdot v = -1$. If so, return $s \leftarrow u$. Otherwise set $s \leftarrow \text{sign}(a_k - b_k)$.

(2)   If both lists are finished, then return s.

(3)   If the list for a is finished then return $s \leftarrow$ ISIGN$((b_{k+1}, \ldots))$ otherwise $s \leftarrow$ ISIGN$((a_{k+1}, \ldots))$. ∎

The average computing time for ICOMP is proportional to $\min(L(a), L(b))$.

Our next problem is summation of two integers a and b. We divide the problem into the following subproblems:

either a or b is zero: in this case we simply return b or a as the sum,

neither a nor b is zero and $\mathrm{ISIGN}(a) = \mathrm{ISIGN}(b)$,

neither a nor b is zero and $\mathrm{ISIGN}(a) \neq \mathrm{ISIGN}(b)$.

The usual classical algorithm for the second subproblem ([23], p. 251) takes time proportional to $\max(L(a), L(b))$. If, however, a and b are not of the same length, the carry usually will not propagate far beyond the most significant digit of the shorter input. This gives us an algorithm with computing time proportional to $\min(L(a), L(b))$:

$c \leftarrow \mathrm{ISUM1}(a, b)$
[integer sum, first algorithm. a and b are integers with $\mathrm{ISIGN}(a) = \mathrm{ISIGN}(b)$. $c = a + b$]

(1)   Successively add the digits of a and b with carry until one of the two lists (say b) is finished, getting $c_0, \ldots, c_k$ and a carry d.

(2)   Add the carry to the successive digits of the longer input, getting $c_{k+1}, \ldots, c_l$, until either the carry disappears, in which case the concatenation of the two lists $(c_0, \ldots, c_l)$ and $(a_{l+1}, \ldots)$ is returned, or the longer list is also finished, in which case $(c_0, \ldots, c_l, d)$ is returned.   ∎

Now we would like to have an algorithm for the summation of two integers with different signs, which also makes use of the above observation in order to achieve a computing time proportional to $\min(L(a), L(b))$. Let us first consider an example: $\beta = 10$, $a = 30578$ and $b = -2095$. If we just look at the least significant digits in the representations $(8, 7, 5, 0, 3)$ and $(-5, -9, 0, -2)$ of a and b we cannot know the sign of the result c in advance. So we simply add the successive digits of a and b without carry (the partial results are guaranteed to be β-digits), getting $(3, -2, 5, -2)$ until one of the lists, in this case b, is finished. Now we look at the remaining digit of a and see that the result has to be positive. So we go back and change all digits to positive ones, getting $(3, 8, 4, 8, 2)$.

$c \leftarrow \mathrm{ISUM2}(a, b)$
[integer sum, second algorithm. a and b are integers with $\mathrm{ISIGN}(a) = \mathrm{ISIGN}(b)$. $c = a + b$]

(1)   Set $u \leftarrow 0$ and successively add the digits of a and b without carry until one of the two lists (say b) is finished, getting $c_0, \ldots, c_k$, and at each step set $u \leftarrow \mathrm{sign}(c_i)$ if $c_i \neq 0$.

(2)   If both lists are finished and all the c_i $(0 \leqslant i \leqslant k)$ are zero, then return $c \leftarrow (\ )$.

(3)   If $u \neq \mathrm{sign}(a)$ then successively set $c_{k+1} \leftarrow a_{k+1}, \ldots, c_l \leftarrow a_l$ until $a_l \neq 0$.

(4)   Now $\mathrm{sign}(c) = \mathrm{sign}(c_l)$. Go back to c_0 and propagate carries, getting new $c_0, \ldots, c_l$.

(5)   Concatenate the two lists $(c_0, \ldots, c_l)$ and $(a_{k+1}, \ldots)$, getting the result c.   ∎

Note that an average computing time proportional to $\min(L(a), L(b))$ can only be achieved by overlapping of lists, a feature not available for integers represented by arrays.

The difference of two integers a and b can be computed as $\mathrm{ISUM}(a, \mathrm{INEG}(b))$, the average computing time being proportional to $L(b)$.

The next problem to attack is multiplication. The "classical" algorithm ([23], p. 253) for multiplying two integers a and b has a computing time proportional to $L(a) \cdot L(b)$. For very long integers a and b, however, we can do much better than that by using an algorithm attributable to A. Karatsuba [22]. The idea is to bisect the two integers a and b into two parts each, getting $a = A_1\beta^k + A_0$, $b = B_1\beta^k + B_0$ where $k = \max(m, n)/2$, $m = L(a)$, $n = L(b)$ and $A_0 = (a_0, \ldots, a_{k-1})$, $A_1 = (a_k, \ldots, a_{m-1})$, $B_0 = (b_0, \ldots, b_{k-1})$, $B_1 = (b_k, \ldots, b_{n-1})$. We might view this as representing a and b in radix β^k representation. The product $c = a \cdot b$ can then be computed with only three multiplications of integers of lengths k or less plus some shiftings and additions using the formula

$$(*) \quad c = a \cdot b = (A_1 \cdot B_1)\beta^{2k} + (A_1 B_0 + A_0 B_1)\beta^k + (A_0 B_0)$$
$$\text{where } A_1 B_0 + A_0 B_1 = (A_1 + A_0)(B_1 + B_0) - A_1 B_1 - A_0 B_0.$$

If k itself is still large then the same method can be reapplied for computing the three smaller products. Thus we get a recursive integer multiplication algorithm. The time needed to multiply two integers of lengths n or less is three times the time needed for multiplying three integers of lengths $n/2$ plus some linear term in n. The solution of the equation

$$M(n) = 3 \cdot M(n/2) + d \cdot n$$

is $M(n) = n^{\log_2 3}$ (see [1], p. 64). Thus the cost of multiplication has been reduced from $L(a) \cdot L(b)$ to $\max(L(a), L(b))^{\log_2 3}$.

The critical values for $L_\beta(a)$ and $L_\beta(b)$, for which the Karatsuba method becomes faster than the classical method, the extra additions taking less time than the saved multiplication (the so-called trade-off point), depend heavily on the machine and on the basic operations for list processing. Let us assume, for the time being, that this trade-off point is reached for $\min(L_\beta(a), L_\beta(b)) \geq N$ (a realistic assumption is that a and b have more than 200 decimal digits). Then we can describe an algorithm based on the Karatsuba method as follows:

$$c \leftarrow \mathrm{IPRODK}(a, b)$$
[integer product, Karatsuba method. a and b are integers. $c = a \cdot b$]

(1)   For short integers (i.e. $\max(L_\beta(a), L_\beta(b)) < N$) apply the "classical" multiplication algorithm and return the result as the value of c.

(2)   Compute the partial results $A_1 B_1$, $A_1 B_0 + A_0 B_1$, $A_0 B_0$ using $(*)$ and applying IPRODK recursively.

(3)   Combine the partial results by shifting and adding, getting the result c.   ∎

Algorithm IPRODK is just a slight modification of the first of an infinite sequence of algorithms M_1, M_2, M_3, $\ldots$. In algorithm M_r we cut a and b into $r + 1$ pieces such that

$$a = \sum_{i=0}^{r} A_i \cdot \beta^{ik}, \qquad b = \sum_{i=0}^{r} B_i \cdot \beta^{ik},$$

where $k = \lceil \max(L_\beta(a), L_\beta(b))/(r+1) \rceil$. a and b are just the values of the two polynomials

$$a(x) = \sum_{i=0}^{r} A_i x^i, \qquad b(x) = \sum_{i=0}^{r} B_i x^i$$

at the point $x = \beta^k$. Hence, if we set $c(x) = a(x) \cdot b(x)$, then $c = a \cdot b = c(\beta^k)$. $c(x)$ is a polynomial of degree $2r$ or less. Therefore the coefficients of $c(x)$ can be computed from the values at $2r + 1$ distinct integer points, say $0, \mp 1, \mp 2, \ldots, \mp r$, using interpolation. In fact, if

$$c(x) = \sum_{i=0}^{2r} c_i x^i$$

then c_i is a linear combination

$$c_i = \sum_{i=-r}^{r} d_{ij} \cdot c(j)$$

in which the d_{ij} are certain rational numbers independent of a and b. The computing time of M_r is dominated by $\max(L(a), L(b))^{\alpha_r}$, where $\alpha_r = \log_{r+1}(2r+1) < 1 + \log_{r+1} 2$. Since $\log_{r+1} 2$ converges to 0 for $r \to \infty$, there is a multiplication algorithm with computing time dominated by $\max(L(a), L(b))^{1+\varepsilon}$ for every $\varepsilon > 0$. (Of course, the factor of proportionality grows accordingly).

Other algorithms are known (see [23], Section 4.3.3) which are faster than every algorithm M_r. The fastest currently known algorithm, due to A. Schönhage and V. Strassen [36] has a computing time dominated by $m \cdot L(m) \cdot L(L(m))$ where $m = \max(L(a), L(b))$. These results are of theoretical interest but have not yet an immediate practical significance for computer algebra because these "fast" algorithms are faster only for enormous numbers.

When a multiplication algorithm is given whose computing time for multiplying two integers of length m or less is dominated by $M(m)$, one can always design a multiplication algorithm whose computing time for multiplying two integers a and b (with $m = L(a)$, $n = L(b)$) is dominated by the function

$$M'(m, n) = \begin{cases} m \cdot n^{-1} \cdot M(n) & \text{if} \quad m \geqslant n \\ n \cdot m^{-1} \cdot M(m) & \text{if} \quad m < n \end{cases}.$$

This can be achieved by breaking a up (assuming without loss of generality that $m \geqslant n$) into pieces each of length n or less and then multiplying each piece by b.

These fast multiplication algorithms can be used to improve the computing times of many other algorithms. One important application is conversion. R. P. Brent [4] has shown how to make use of fast multiplication for efficiently evaluating functions such as log, exp and arctan (an evaluation to n significant digits can be achieved in time dominated by $M(n) \cdot \log(n)$ if the time needed for multiplying two n-digit numbers is dominated by $M(n)$).

A problem related to multiplication is exponentiation. The most obvious algorithm for computing a^n, $n > 1$, uses repeated multiplications by a, computing a, a^2, $a^3, \ldots, a^n$. But we also consider the so-called left-to-right and right-to-left binary exponentiation methods.

Let $n_{(2)} = \sum_{i=0}^{k-1} b_i 2^i$ be the binary representation of n. In the left-to-right binary exponentiation method we compute successively $a^{n_1}, a^{n_2}, \ldots, a^{n_k}$ for $n_i = \lfloor n/2^{k-1} \rfloor$ (note that $n_k = n$) by the formula

$$a^{n_{i+1}} = \begin{cases} (a^{n_i})^2, & \text{if } b_{k-i-1} = 0 \\ (a^{n_i})^2 a, & \text{if } b_{k-i-1} = 1 \end{cases}.$$

In the right-to-left method we compute two sequences $a_0, \ldots, a_k$ and $c_0, \ldots, c_{k-1}$ in which

$$c_0 = a, \quad c_{i+1} = c_i^2 \quad \text{and} \quad a_0 = 1, \quad a_{i+1} = \begin{cases} a_i, & \text{if } b_i = 0 \\ a_i \cdot c_i, & \text{if } b_i = 1 \end{cases}.$$

Finally $a_k = a^n$.

It turns out that the computing times of the three methods are mutually codominant. If $M(n)$ is the cost of multiplying two numbers of length n or less, then the time needed for computing a^n is proportional to $M(n \cdot L(a))$. Although fewer multiplications are needed in the binary methods, the integers to be multiplied are usually greater than in the conventional method. This remark no longer holds when we are doing modular arithmetic. In this case the binary methods are actually superior.

Let us now consider division of rational integers. For $a, b \in \mathbb{Z}$, $b \neq 0$, there exist unique integers $\text{quot}(a, b)$ and $\text{rem}(a, b)$ such that $a = \text{quot}(a, b) \cdot b + \text{rem}(a, b)$ and $0 \leqslant \text{rem}(a, b) < |b|$ if $a \geqslant 0$, $-|b| < \text{rem}(a, b) \leqslant 0$ if $a < 0$. In order to compute the quotient and remainder of two positive integers a and b (with $m = L_\beta(a) > L_\beta(b) = n$) we make a guess about the first quotient digit, say q_{k-1}. Having determined q_{k-1} we replace a by $a - b \cdot q_{k-1} \cdot \beta^{k-1}$ and determine q_{k-2} in the same manner, using the new a. In determining q_j we have $0 \leqslant a < \beta^{j+1}$. Thus $q_j = \lfloor a/b\beta^j \rfloor$ and

$$q_j^* = \lfloor \lfloor a/\beta^{n+j-1} \rfloor / \lfloor b\beta^j/\beta^{n+j-1} \rfloor \rfloor = \lfloor (a_{n+j}\beta + a_{n+j-1})/b_{n-1} \rfloor$$

should be an accurate approximation to q_j. We will have $q_j^* \geqslant \beta$ just in case $a_{n+j} \geqslant b_{n-1}$; in this case we may set $q_j^* = \beta - 1$. It turns out that $q_j \leqslant q_j^* \leqslant q_j + 2$ if $b_{n-1} \geqslant \beta/2$. This fact was first observed by D. A. Pope and M. L. Stein in [34]. G. E. Collins and D. R. Musser [12] later showed that the respective probabilities of $q_j^* = q_j + i$, $i = 0, 1, 2$, are approximately 0.67, 0.32, 0.01. The requirement $b_{n-1} \geqslant \beta/2$ can be met by normalizing b to $b' = b \cdot d$ and a to $a' = a \cdot d$ where $d = \lfloor \beta/(b_{n-1} + 1) \rfloor$. This does not change the value of the quotient q and $\text{rem}(a, b) = \text{rem}(a', b')/d$.

For describing the algorithm for division we assume that we already have an algorithm DQR for dividing an integer of β-length 2 by a β-digit. This could be achieved using the idea above. Most computers, however, have an instruction which divides a double precision integer by a single precision integer, producing a single precision quotient and a single precision remainder. Such an instruction essentially realizes DQR. Based on DQR an algorithm IDQR for dividing an integer by a digit can easily be programmed.

$(q, r) \leftarrow \text{IQR}(a, b)$

[integer quotient-remainder algorithm. a and b are integers, $L_\beta(a) \geq L_\beta(b) \geq 2$. $q = \text{quot}(a, b)$, $r = \text{rem}(a, b)$]

(1)   Compute the signs s and t of a and b, respectively, and the normalization factor $d = \beta/(|b_{n-1}| + 1)$, where $n = L_\beta(b)$.
        Normalize a and b: $a' \leftarrow \text{IPROD}(a, s \cdot d)$, $b' \leftarrow \text{IPROD}(b, t \cdot d)$.

(2)   For $j = 0$ to $m - n$ ($m = L_\beta(a)$) do steps (3)–(5).

(3)   Calculate $q^* \leftarrow \min(\beta - 1, (a'_{m-j-1} + \beta a'_{m-j})/b'_{n-1})$
        where we let $a'_m = 0$ if $L_\beta(a') = m$.

(4)   Adjust the quotient by reducing q^* by 1 as long as
        $\text{IDIF}((a'_{m-n-j}, \ldots, a'_{m-j}), q^* b')$ is negative.

(5)   Set $q_{m-n-j} \leftarrow q^*$, $a' \leftarrow \text{IDIF}(a', q^* \cdot \beta^{m-n-j} \cdot b')$.

(6)   If $q_{m-n} = 0$ then set $q \leftarrow (q_0, \ldots, q_{m-n-1})$.
        If $s \cdot t = -1$ then set $q \leftarrow \text{INEG}(q)$.
        $r \leftarrow a'/s \cdot d$. ∎

It is clear from the above remarks that step (4) has to be done at most twice in each cycle. If we replace step (4) by

(4′)   As long as $b'_{n-2} \cdot q^* > (a'_{m-j}\beta + a'_{m-j-1} - q^* b'_{n-1})\beta + a'_{m-j-2}$
        reduce q^* by 1.
        If $\text{IDIF}((a'_{m-n-j}, \ldots, a'_{m-j}), q^* b')$ is negative reduce q^* by 1

we have to use IDIF only once. This correction has been suggested by D. E. Knuth in [23], Section 4.3.1.

The execution time of IQR is dominated by $L(b) \cdot (L(a) - L(b) + 1)$, which is the same as for computing $b \cdot \text{quot}(a, b)$.

Instead of computing a/b directly, we could first calculate b^{-1} and then multiply by a. In order to use this method effectively we must be able to approximate b^{-1}, i.e. the zero of the function $f(x) = 1/x - b$. This, however, can be achieved by Newton's method. If b_i is already an approximation to b^{-1} then $b_{i+1} = b_i \cdot (2 - b \cdot b_i)$ is a better one with $|b_{i+1} - b^{-1}| < \varepsilon^2$ if $\varepsilon = |b_i - b^{-1}|$. Thus we need a subalgorithm IRECIP, which takes as input an integer a and a positive β-integer n and computes a positive β-integer b such that $|b \cdot 2^{-n-1} - a^{-1} \cdot 2^m| \leq 2^{-n}$ and $b \leq 2^{n+2}$, where $m = L_2(a)$. IRECIP can now be used to obtain a fast quotient-remainder algorithm IQRN corresponding to the fast product algorithm IPRODK. To divide a by b we approximate b^{-1} with b', using IRECIP, to $m - n + 3$ places, where $m = L_2(a)$ and $n = L_2(b)$. We approximate a by a' to $m - n + 3$ places. Multiplying a' by b' and truncating, we obtain an approximation q' to a/b and hence to $q = \lfloor a/b \rfloor$. The following theorem (taken from [11]) shows that $q' = q$ or $q' = q + 1$, so $q = q' - 1$ just in case $a - b \cdot q' < 0$.

Theorem. *Let a and b be integers,*

$$m = L_2(a), \qquad n = L_2(b), \qquad k = m - n + 1 \geqslant 1,$$

$$|b' \cdot 2^{-k-3} - b^{-1} \cdot 2^n| \leqslant 2^{-k-2}, \qquad a' = \lfloor a/2^{n-3} \rfloor,$$

$$q^* = \lfloor a' \cdot b'/2^{k+4} \rfloor, \qquad q' = \lfloor (q^* + 3)/4 \rfloor, \qquad q = \lfloor a/b \rfloor.$$

Then $q' = q$ or $q' = q + 1$.

The computing time of IQRN is dominated by $M(L(q)) + M'(L(b), L(q))$ for $a \geqslant b$, where $q = \lfloor a/b \rfloor$.

Finally, let us just state that one could design a quotient-remainder algorithm with a computing time bound comparable with that for fast multiplication. Let $m = L_2(a)$, $n = L_2(b), k = m - n + 1$. If either n or k is less than some critical value then IQR is applied. Otherwise, a is expressed in the radix β^n,

$$a = \sum_{i=0}^{h-1} a_i \beta^{ni}, \qquad h = \lceil m/n \rceil.$$

The β^n-digits q_i in $q = \sum_{i=0}^{h-1} q_i \beta^{ni}$ are computed, and q is computed from the q_is by shifting and adding. The computing time for this algorithm is dominated by $M'(L(b), L(q))$ for $|a| \geqslant |b|$, where $q = \lfloor a/b \rfloor$.

Finally we want to compute greatest common divisors of integers. Let a and b be integers not both equal to zero. Then we let $\gcd(a, b)$ be the greatest integer that evenly divides both a and b; $\gcd(0, 0)$ is defined to be 0. If we divide a by b, getting a quotient q and a remainder c, it is immediate from the equation $a = bq + c$ that a and b have the same common divisors as b and c. If $c > 0$ the process can be repeated with (b, c) in place of (a, b). Continuing until a zero remainder is reached, we obtain the so-called Euclidean remainder sequence $a_1, a_2, \ldots, a_r, a_{r+1}$ with $a_1 = a, a_2 = b, a_{r+1} = 0$, and a_r is the greatest common divisor of a and b. From these considerations we obtain immediately the classical Euclidean greatest common divisor algorithm IGCDE. The computing time of IGCDE is proportional to $n \cdot (m - k + 1)$, where $m = \max(L(a), L(b))$, $n = \min(L(a), L(b))$, $k = L(\gcd(a, b))$, for $a \neq 0, b \neq 0$. The analysis of the computing time of IGCDE, depending only on the inputs a, b, is an interesting subject in its own right. Let us just mention a theorem of G. Lamé (1845), which exhibits the relation between Euclid's algorithm and the Fibonacci numbers:

Theorem. *For $r \geqslant 1$, let a and b be integers with $0 < b < a$ such that Euclid's algorithm applied to a and b requires exactly r division steps, and such that a is as small as possible satisfying these conditions. Then $a = F_{r+2}$ and $b = F_{r+1}$.*

As a consequence of the theorem of Lamé we get $\lceil \log_\phi(\sqrt{5} N) \rceil - 2$ as an upper bound for the number of division steps in IGCDE, where $N = \max(a, b)$ and

$\phi = (1 + \sqrt{5})/2$. G. E. Collins [11] gives another bound for the number of division steps:

Theorem. *Let $0 < b < a$, then the number of division steps in IGCDE for computing* $\gcd(a, b)$ *is bounded by* $(\log_\phi 2) \cdot L(b) + 2$.

For lack of space, we cannot go into more details, here. The reader may find an extensive treatment in [23], Section 4.5.3 and [11].

One can extend the Euclidean algorithm in the following way: in addition to computing $\gcd(a, b)$ we also compute integers u and v such that

$$a \cdot u + b \cdot v = \gcd(a, b).$$

$(c, u, v) \leftarrow \text{IGCDEXE}(a, b)$

[integer greatest common divisor algorithm, extended Euclidean. a and b are integers, $c = \gcd(a, b)$, $a \cdot u + b \cdot v = c$]

(1)　　　Initialize the vectors u and v:
$(u_1, u_2, u_3) \leftarrow (\text{ISIGN}(a), 0, \text{IABS}(a))$,
$(v_1, v_2, v_3) \leftarrow (0, \text{ISIGN}(b), \text{IABS}(b))$.

(2)　　　Compute remainder sequence and cosequences until zero remainder is reached: as long as $v_3 \neq 0$ do the following:
$(q, c') \leftarrow \text{IQR}(u_3, v_3)$
$(t_1, t_2, t_3) \leftarrow (u_1, u_2, u_3) - (v_1, v_2, v_3) \cdot q$
$(u_1, u_2, u_3) \leftarrow (v_1, v_2, v_3), (v_1, v_2, v_3) \leftarrow (t_1, t_2, t_3)$.

(3)　　　Return $c \leftarrow u_3$, $u \leftarrow u_1$, $v \leftarrow u_2$. ∎

In algorithm IGCDE (and IGCDEXE) a lot of divisions of long integers have to be performed. Since this is quite time consuming, we want to investigate whether there is a faster way of computing the gcd of two integers.

D. H. Lehmer in 1938 [25] has devised an algorithm, which uses only the most significant digits of a and b in the Euclidean division process, as long as this is possible. We want to consider here an algorithm similar to Lehmer's. This method appears to have no significant advantage or disadvantage relative to Lehmer's for rational integer greatest common divisor calculation. However, it generalizes in a more effective manner to Gaussian integers, and it also has the important advantage that one can better analyze the resulting algorithms. The idea was improved by Schönhage and carried over to polynomials by Moenck [31] and Strassen [38].

Let a and b be integers, $m = L_2(a)$, $n = L_2(b)$ ($m \geqslant n$) and $h < n$. Let $a' = \lfloor a/2^h \rfloor$, $b' = \lfloor b/2^h \rfloor$. The quotient of a' divided by b' will be roughly the same as that of a divided by b. Let $a'_1, a'_2, \ldots, a'_{t+1}$ and $q'_1, \ldots, q'_{t-1}$ be the Euclidean remainder and quotient sequences of a' and b'; let $u'_1, \ldots, u'_{t+1}$ and $v'_1, \ldots, v'_{t+1}$ be the associated cosequences. If $q'_i = q_i$ for $1 \leqslant i < j$, then $u'_i = u_i$ and $v'_i = v_i$ for $1 \leqslant i \leqslant j + 1$, so we can compute a_j and a_{j+1} using the formula $u_i \cdot a + v_i \cdot b = a_i$. Since a' and b' are smaller than a and b, a considerable amount of computation will be saved if h and j are sufficienty large. The problem is to determine the largest possible value of j for which $q'_j = q_j$. It turns out that

$$\min(a'_{i+2}, a'_{i+1} - a'_{i+2}) > 4|v'_{i+2}|$$

is a sufficient condition for $q'_i = q_i$, provided that $q'_1 = q_1, \ldots, q'_{i-1} = q_{i-1}$. The correctness proof — which can be found in [11] — is too lengthy to be presented here. The following algorithm uses the above ideas:

$c \leftarrow \text{IGCDLKS}(a, b)$

[integer greatest common divisor algorithm, Lehmer-Knuth-Schönhage. a and b are integers, $a > b > 0$, $c = \gcd(a, b)$]

(1)   Set $a_1 \leftarrow \text{IABS}(a)$, $a_2 \leftarrow \text{IABS}(b)$.

(2)   Compute the remainder sequence:
      as long as $a_2 \neq 0$ do steps (3)–(9).

(3)   If a_1 is a β-integer, return $c \leftarrow \text{IGCDE}(a_1, a_2)$.

(4)   Set $m \leftarrow L_2(a_1)$, $n \leftarrow L_2(a_2)$.

(5)   If $m - n$ is "too big" compute $a_4 = \lfloor a_1/a_2 \rfloor$ by IQR and set $a_3 \leftarrow a_2$.
      Go to step (9).

(6)   Choose h as large as possible so that $n > h$ and the resulting
      $a' = \lfloor a_1/2^h \rfloor$, $b' = \lfloor a_2/2^h \rfloor$ are β-digits.

(7)   Set $(u_1, u_2, v_1, v_2) \leftarrow \text{DPCC}(a', b')$.

(8)   If $v_1 = 0$ then compute $a_4 = \lfloor a_1/a_2 \rfloor$ by IQR and set $a_3 \leftarrow a_2$.
      Otherwise set $a_3 \leftarrow a_1 \cdot u_1 + a_2 \cdot v_1$, $a_4 \leftarrow a_1 \cdot u_2 + a_2 \cdot v_2$ using ISUM
      and IPROD.

(9) $a_1 \leftarrow a_3$, $a_2 \leftarrow a_4$.

(10) Return $c \leftarrow a_1$.   ∎

In IGCDLKS a subalgorithm DPCC is used, which we describe in the sequel:

$(u, u', v, v') \leftarrow \text{DPCC}(a_1, a_2)$

[digit partial cosequence calculation. a_1 and a_2 are β-integers, $a_1 \geqslant a_2 > 0$. u, u', v, v' are the last cosequence elements of a_1 and a_2 which can be guaranteed to correspond to correct quotient digits]

(1)   Set $a \leftarrow a_1$, $a' \leftarrow a_2$, $u \leftarrow 1$, $u' \leftarrow 0$, $v \leftarrow 0$, $v' \leftarrow 1$.

(2)   Repeat steps (3)–(5) until the termination criterion (4) is met.

(3) $q \leftarrow a/a'$, $a'' \leftarrow a - q \cdot a'$,
      $u'' \leftarrow u - q \cdot u'$, $v'' \leftarrow v - q \cdot v'$,
      $v^* \leftarrow 4 \cdot |v''|$.

(4)   If $a'' \leqslant v^*$ or $(a' - a'') \leqslant v^*$ then return the current values of u, u', v, v'.

(5) $a \leftarrow a'$, $a' \leftarrow a''$, $u \leftarrow u'$, $u' \leftarrow u''$, $v \leftarrow v'$, $v' \leftarrow v''$.   ∎

The computing time of IGCDLKS is again proportional to $n \cdot (m - k + 1)$. IGCDLKS, however, will be about j^* times as fast as the ordinary Euclidean algorithm, where j^* is the average value of j, the number of Euclidean division steps which we can carry out only on the leading digits of a and b.

Algorithm IGCDLKS can also be extended to an algorithm that computes the cosequence elements in addition to the greatest common divisor ([11]).

Let us just note that there is a gcd-algorithm with computing time dominated by $m \cdot M(\min(n, m - k + 1))$, where $m = L(a)$, $n = L(b)$, $k = L(\gcd(a, b))$. This algorithm makes appropriate use of the fast multiplication and division algorithms. The method is derived from the work of R. Moenck [30], which in turn is derived from the earlier work of D. E. Knuth [24] and A. Schönhage [35].

The Rational Numbers

Now that we have algorithms for performing arithmetic operations (including gcd calculation) in $\mathbb{Z}$ it is relatively simple to develop algorithms for arithmetic operations on arbitrary rational numbers $\mathbb{Q}$, the fraction field or field of quotients of the integral domain $\mathbb{Z}$. Van der Waerden, in [39], §13 describes how to embed an integral domain I into its field of quotients $Q(I)$.

The problem of computing canonical forms for the elements of $Q(I)$ arises at this point. An investigation and solution of this problem may be found in the chapter on algebraic simplification in this volume.

Let $>$ be a linear order on I such that $a + b > 0$ and $a \cdot b > 0$ for $a, b \in I$, $a > 0$ and $b > 0$. I is then said to be an ordered domain (relative to $>$). A linear order $>'$ on $Q(I)$ can be defined by the condition $(a, b) >' (c, d)$ in case $(bd > 0$ and $ad > bc)$ or $(bd < 0$ and $ad < bc)$. $Q(I)$ is an ordered field relative to $>'$.

For describing algorithms for addition and multiplication in the field of the rational numbers $\mathbb{Q} = Q(\mathbb{Z})$ one need only to apply the definitions

$$(*) \qquad \frac{a}{b} + \frac{c}{d} = \frac{ad + bc}{bd} \qquad \text{and} \qquad \frac{a}{b} \cdot \frac{c}{d} = \frac{ac}{bd}$$

and reduce the result to lowest terms by a canonical simplifier. P. Henrici, in [17], however, introduced less obvious algorithms which are in general more efficient if the inputs are reduced. His method can be stated in the general context of a unique factorization domain I with effective gcd. Let $r_1, r_2, s_1, s_2 \in I$, and $\gcd(r_1, r_2) = \gcd(s_1, s_2) = 1$. If we let $d = \gcd(r_2, s_2)$, $r_2' = r_2/d$, $s_2' = s_2/d$ then

$$\gcd(r_1 s_2' + s_1 r_2', r_2 s_2') = \gcd(r_1 s_2' + s_1 r_2', d).$$

This suggests an algorithm for adding two fractions:

$t \leftarrow \text{RNSUM}(r, s)$
[Sum of rational numbers. $r = (r_1, r_2)$ and $s = (s_1, s_2)$ are rational numbers in canonical form. $t = r + s$, $t = (t_1, t_2)$ is in canonical form.]

(1) $[r = 0$ or $s = 0]$ If $r = 0$ then set $t \leftarrow s$ and return.
         If $s = 0$ then set $t \leftarrow r$ and return.

(2) $d \leftarrow \text{IGCD}(r_2, s_2)$.

(3) $[d = 1]$ If $d = 1$ then set $t_1 \leftarrow r_1 \cdot s_2 + r_2 \cdot s_1$,
$$t_2 \leftarrow r_2 \cdot s_2 \text{ and return.}$$

(4) $[d \neq 1]$ $r'_2 \leftarrow r_2/d$, $s'_2 \leftarrow s_2/d$.
$t'_1 \leftarrow r_1 \cdot s'_2 + s_1 \cdot r'_2$.
$t'_2 \leftarrow r_2 \cdot s'_2$.
If $t'_1 = 0$ then set $t_1 \leftarrow 0$, $t_2 \leftarrow 1$ and return.
$e \leftarrow \text{IGCD}(t'_1, d)$.
If $e = 1$ then set $t_1 \leftarrow t'_1$, $t_2 \leftarrow t'_2$ and return.
$t_1 \leftarrow t'_1/e$, $t_2 \leftarrow t'_2/e$.
Return.   ■

The computing time for RNSUM is dominated by $m \cdot n^{\alpha-1} \cdot L(m)$ where $m = \max(L(r), L(s))$, $n = \min(L(r), L(s))$, and $\alpha = \log_2 3$ for Karatsuba multiplication and $\alpha = 2$ in the classical case. Here $L(r)$ is defined as $\max(L(r_1), L(r_2))$. The improvement in computing time by the use of RNSUM instead of an algorithm based on $(*)$ is a factor of about 3 for $d = 1$ and 1.52 for $d \neq 1$ (see [11]). But for other choices of the domain I (compare the section on rational functions) the improvement may be much more significant.

One could write a subtraction algorithm analogous to RNSUM, or simply compute $r - s$ as $r + (-s)$ where it is assumed that an algorithm RNNEG for negation is given (using INEG).

For multiplication of fractions we use the following fact: if $r_1, r_2, s_1, s_2 \in I$,

$$\gcd(r_1, r_2) = \gcd(s_1, s_2) = 1, \qquad d_1 = \gcd(r_1, s_2), \qquad d_2 = \gcd(s_1, r_2),$$

$$r'_1 = r_1/d_1, \qquad r'_2 = r_2/d_2, \qquad s'_1 = s_1/d_2, \qquad s'_2 = s_2/d_1$$

then $\gcd(r'_1 s'_1, r'_2 s'_2) = 1$. From this fact we have the following multiplication algorithm:

$t \leftarrow \text{RNPROD}(r, s)$

[Product of rational numbers. $r = (r_1, r_2)$ and $s = (s_1, s_2)$ are rational numbers in canonical form. $t = r \cdot s$, $t = (t_1, t_2)$ is in canonical form.]

(1) $[r = 0 \text{ or } s = 0]$ If $r = 0$ or $s = 0$ then set $t_1 \leftarrow 0$, $t_2 \leftarrow 1$ and return.

(2) $d_1 \leftarrow \text{IGCD}(r_1, s_2)$, $d_2 \leftarrow \text{IGCD}(s_1, r_2)$.

(3)   [compute quotients]
If $d_1 = 1$ then $r'_1 \leftarrow r_1$, $s'_2 \leftarrow s_2$
                else $r'_1 \leftarrow r_1/d_1$, $s'_2 \leftarrow s_2/d_1$.
If $d_2 = 1$ then $s'_1 \leftarrow s_1$, $r'_2 \leftarrow r_2$
                else $s'_1 \leftarrow s_1/d_2$, $r'_2 \leftarrow r_2/d_2$.

(4)   [compute products] $t_1 \leftarrow r'_1 \cdot s'_1$, $t_2 \leftarrow r'_2 \cdot s'_2$.
Return.   ■

The computing time t_{RNPROD} is dominated by $m \cdot n^{\alpha-1} \cdot L(m)$.

The inverse r^{-1} of the non-zero fraction $r = (r_1, r_2)$ is (r_2, r_1) if r_1 is in the ample set A and (ur_2, ur_1) otherwise, where $ur_1 = \gcd(r_1, 0)$. The quotient of two rational numbers r, s can readily be computed as $r \cdot s^{-1}$.

It is clear how to design an algorithm for comparing two rational numbers r and s: if the signs of r_1 and s_1 are different, then we can immediately decide which one is the greater. Otherwise we have to compute and compare the integers $r_1 s_2$ and $s_1 r_2$.

Certain best approximative representations of rational numbers, the so-called "fixed slash" and "floating slash" representations, have been suggested by D. W. Matula [27] and have been further developed in [28, 29]. These papers also address the issue of reconstructing fractions from given approximations.

Modular Arithmetic

When we are doing arithmetic (especially a lot of multiplications and only a few divisions and comparisons, see [23], p. 269, 275) on large integers, it is often preferable to work instead modulo several moduli $m_1, \ldots, m_k$. So let us just review the basic facts about modular arithmetic (see also the chapter on computing by homomorphic images).

Arithmetic modulo a positive integer m is best understood as working in the residue class ring of $\mathbb{Z}$ modulo the ideal generated by m, i.e. in $\mathbb{Z}/(m)$ (such ideals are the only ones in $\mathbb{Z}$, since it is well known that $\mathbb{Z}$ is a principal ideal domain. See, for example, [39], §17). The only residue classes in this case are $\bar{0}, \bar{1}, \ldots, \overline{m-1}$ ($\bar{a}$ denotes the residue class of a, i.e. $\bar{a} = \{a + bm \mid b \in \mathbb{Z}\}$). Addition and multiplication are defined by

$$\bar{a} + \bar{b} = \overline{a + b}, \qquad \bar{a} \cdot \bar{b} = \overline{a \cdot b}.$$

The additive identity is $\bar{0}$, the multiplicative identity is $\bar{1}$.

There are two natural choices of a set of representatives for $\mathbb{Z}/(m)$, namely $\{0, 1, \ldots, m-1\}$ and $\{a \mid -m/2 < a \leqslant m/2\}$. If m is odd, the second set is symmetric with respect to zero, while if m is even it contains $m/2$ but not $-m/2$. Each of these choices has certain advantages and disadvantages; we choose the first and denote it by $\mathbb{Z}_m$. By H_m we shall denote the canonical simplifier associated with $\mathbb{Z}_m$.

$\mathbb{Z}_m$ is itself a ring isomorphic with $\mathbb{Z}/(m)$ in which addition and multiplication are defined by

$$(*) \qquad\qquad a +_m b = H_m(a + b), \qquad a \cdot_m b = H_m(a \cdot b)$$

(of course we also have $a -_m b = H_m(a - b)$).

An algorithm MIHOM that realizes the modular integer homomorphism H_m takes m and a as input and essentially applies one of the division algorithms discussed in the section on integer arithmetic for computing $b = \text{rem}(a, m)$.

MISUM, MIDIF, and MIPROD for computing $a +_m b$, $a -_m b$, and $a \cdot_m b$ can be based directly on $(*)$, using MIHOM as a subalgorithm. This approach, however, is rather inefficient for MISUM and MIDIF, since the quotient computed by the application of MIHOM can only be zero or one. So it is preferable to subtract the modulus m from the result until the desired representative is reached.

For the computing times we have the relations $t^+_{\text{MISUM}} \sim t^*_{\text{MISUM}} \sim t^+_{\text{MIDIF}} \sim t^*_{\text{MIDIF}} \sim L(m)$. t_{MIPROD} is dominated by $L(m)^\alpha$.

Modular exponentiation has many applications in number theory. For integer exponentiation all three methods (repeated multiplication, left-to-right and right-to-left binary exponentiation) have computing times proportional to each other. Not so for modular exponentiation, because the modulus m is a bound for all integers which are multiplied. The time to compute $a^n \pmod{m}$ by repeated multiplication will be proportional to $n \cdot L(m)^\alpha$ as opposed to $L(n) \cdot L(m)^\alpha$ for any of the binary methods.

Division of a by b in a commutative ring R is defined just in case there is a unique c such that $a = b \cdot c$, and then c is the quotient a/b. In the case of $R = \mathbb{Z}_m$, b is a zero divisor (and therefore c is not uniquely determined) if $\gcd(m, b) \neq 1$. On the other hand if $\gcd(m, b) = 1$, then b is a unit and we can compute b^{-1} (thus the quotient $c = a/b$ is $a \cdot b^{-1}$). If $\gcd(m, b) = 1$ then there exist u, v such that $m \cdot u + b \cdot v = 1$, thus $b^{-1} = H_m(v)$. So an algorithm MIDIV for dividing a by b modulo m will essentially compute $d = \gcd(m, b)$ together with the cofactors u and v (e.g. by algorithm IGCDEXE). If $d \neq 1$ then a is not divisible by b, otherwise $a/b = a \cdot H_m(v)$. The computing time of MIDIV is dominated by $L(m)^\alpha \cdot L(L(m))$.

For the case in which m is a prime p an alternative for computing the inverse b^{-1} would be the use of Fermat's theorem (see [39], §43), according to which $b^{-1} = b^{p-1}$ in $\mathbb{Z}_p$. It follows from the time bound for modular exponentiation that the computing time of an algorithm using Fermat's theorem would be dominated by $L(m)^\alpha \cdot L(m)$ as opposed to the much better bound $L(m)^\alpha \cdot L(L(m))$ for an algorithm computing $\gcd(p, b)$. The following table compares both methods on a DEC10. The length of the prime p is close to the word size. The times are given in milliseconds.

Fermat	Euclid
3.795	0.042
3.787	0.053
3.797	0.050
3.800	0.058
3.773	0.034
3.786	0.052
3.782	0.043
3.783	0.029
3.789	0.024
3.768	0.042

The usual situation for the application of modular arithmetic is that one wants to replace a computation in $\mathbb{Z}$ (with possibly huge intermediate results) by a number of computations with respect to relatively prime moduli. The important question now is how we can convert fast into and out of the modular representation. Conversion from integer to modular representation is handled by the homomorphism H_m (computed by algorithm MIHOM). Next we shall consider algorithm MICRA1 for converting back from modular representation to the rational integers $\mathbb{Z}$. The mathematical background for the solution of this problem is provided by the Chinese remainder theorem which we need here for the Euclidean domain $\mathbb{Z}$. See the chapter on computing by homomorphic images in this volume, where in Section 2.2 the algorithm is presented.

The computing time t_{MICRA1} of an algorithm MICRA1 (modular integer Chinese remainder algorithm 1) based on these ideas is dominated by $L(m'_1) \cdot L(m'_2)^{\alpha-1} \cdot L(L(m'_2))$ where $m'_1 = \max(m_1, m_2)$ and $m'_2 = \min(m_1, m_2)$.

In order to obtain a "fast" Chinese remainder algorithm MICRA2, one must use a recursive algorithm, which divides its problem into two roughly equal subproblems. This is done by dividing the list of moduli $m_1, \ldots, m_k$ into two sublists $m_1, \ldots, m_{[k/2]}$ and $m_{[k/2]+1}, \ldots, m_k$. MICRA2 is applied to both sublists (if their lengths are greater than 2, otherwise MICRA1 is applied) and the solution of the given problem is computed from the solutions of the two subproblems by algorithm MICRA1. (Note that if $a \equiv b_1 \pmod{m_1 \cdot \cdots \cdot m_{[k/2]}}$ and $a \equiv b_2 \pmod{m_{[k/2]+1} \cdot \cdots \cdot m_k}$ and $b_1 \equiv a_i \pmod{m_i}$ for $1 \leqslant i \leqslant [k/2]$ and $b_2 \equiv a_i \pmod{m_i}$ for $[k/2] + 1 \leqslant i \leqslant k$ then $a \equiv a_i \pmod{m_i}$ for $1 \leqslant i \leqslant k$.)

The computing time of MICRA2 is dominated by $L(m)^{\alpha} \cdot L(L(m))^2$ where $m = m_1 \cdot \cdots \cdot m_k$.

When one is doing modular arithmetic, it is essential to be able to generate a large supply of prime numbers. Actually we only need relatively prime moduli. Knuth (in [23], p. 272) considers moduli of the form $2^e - 1$, for which relative primality can be checked by the simple rule $\gcd(2^e - 1, 2^f - 1) = 2^{\gcd(e,f)} - 1$. Whenever possible one is interested in single-precision moduli, because that makes the computation more efficient. If one used double-precision moduli, only half as many would be required for the solution of any given problem, but the calculations in $\mathbb{Z}_m$ for each modulus m would take more than twice as long.

An algorithm SPGEN (small prime generator) for generating prime numbers p in the range between m and $m + 4k$ for given m and k could be based on the following ideas: provide an array A of length k as workspace, $A[i]$ corresponding to the integer $n_i = m_1 + 4(i - 1)$ for $1 \leqslant i \leqslant k$ where $m_1 = m + 3 - \text{rem}(m, 4)$. Now successively consider all the possible divisors d of the integers n_i, $1 \leqslant i \leqslant k$, and mark each element of A which corresponds to a proper multiple of d. The integers whose corresponding fields in A remain unmarked by this process are the prime numbers in the given range. It suffices to consider as the possible divisors only the numbers $d = 3$ and $d = 6n \mp 1$, $n \geqslant 1$, since this set of numbers includes all odd primes. An upper bound for the divisors to be considered is $\sqrt{m_2}$, where $m_2 = m_1 + 4(k - 1)$. $d \leqslant \sqrt{m_2}$ can be decided by $d \leqslant [m_2/d]$.

In [11] $40\sqrt{n} + 250(\ln n)^2 + 1000 \ln n + 1500$ microseconds is given as an estimate of the execution time of SPGEN for generating 100 primes. The following table gives approximate values, in seconds, of this computing time function. In [11] it is also observed that for a Telefunken TR-440 computer the times of this table agree closely with observed times.

n	Time for computing 100 primes
2^{20}	0.1
2^{25}	0.3
2^{30}	1.4
2^{35}	7.6
2^{40}	42
2^{45}	238

Thus SPGEN is feasible for $n \leqslant 2^{48}$ but not for generating larger primes, where maybe probabilistic algorithms are acceptable ([23], Section 4.5.4).

Gaussian Integer Arithmetic

A Gaussian integer a is a number of the form $a_1 + a_2 i$ (a_1, a_2 rational integers, $i = \sqrt{-1}$). The components a_1 and a_2 of a are respectively called the real and imaginary part of a. The domain G of Gaussian integers constitutes a subring of the field of complex numbers. In fact, G is an integral domain. Addition and subtraction are defined component-wise. The ordinary algorithm for multiplication

$$(a_1 + a_2 i) \cdot (b_1 + b_2 i) = (a_1 b_1 - a_2 b_2) + (a_1 b_2 + a_2 b_1)i$$

can be speeded-up by Karatsuba's method (see the section on integer arithmetic), i.e. one multiplication can be traded for three additions/subtractions (see [7]).

It is clear that

$$t_{\text{GIADD}}(a, b) \leq L(a) + L(b),$$

$$t_{\text{GISUB}}(a, b) \leq L(a) + L(b),$$

$$t_{\text{GIMULT}}(a, b) \leq M'(L(a), L(b)).$$

(t_{GIADD}, t_{GISUB}, t_{GIMULT} times to add, subtract and multiply Gaussian integers; $L(a_1 + a_2 i) = L(a_1) + L(a_2)$ length of a Gaussian integer $a_1 + a_2 i$, M and M' are the dominating functions for multiplication of rational integers, see the section on integer arithmetic.)

For $a, b \in G$, "b divides a" may be decided by forming the quotient a/b according to the well-known formula

(DIV) $a/b = (a \cdot \bar{b}) \cdot (1/N(b))$ or, equivalently,

$$(a_1 + a_2 i)/(b_1 + b_2 i) = (a_1 b_1 + a_2 b_2)/(b_1^2 + b_2^2)$$
$$+ ((a_2 b_1 - a_1 b_2)/(b_1^2 + b_2^2))i$$

$(\bar{b} = \overline{b_1 + b_2 i} = b_1 - b_2 i$ the conjugate of b,

$N(b) = N(b_1 + b_2 i) := b \cdot \bar{b} = b_1^2 + b_2^2$ the norm of b),

and testing whether the real and imaginary part of a/b are both in G.

There is a method known to save 2 multiplications/divisions which is not known to be optimal (Alt van Leeuwen, private communication).

The units of G are $1 = i^0$, $i = i^1$, $-1 = i^2$ and $-i = i^3$. $S(a) := i^{5-j} \cdot a$ (j number of quadrant in which a lies) defines a canonical simplifier S for the equivalence relation "a and a' are associates" (compare the chapter on simplification in this volume). G is a Euclidean domain: the norm N is a function from G into $\mathbb{N}_0$ satisfying

(EUCL1) for all $a, b \in G$: $N(a \cdot b) = N(a) \cdot N(b)$ and
(EUCL2) for all $a, b \in G$, $b \neq 0$ there exist $q, r \in G$ such that $a = q \cdot b + r$,
$\qquad r = 0$ or $N(r) < N(b)$.

By (EUCL2), q and r are not uniquely determined. In fact, exactly all those $q \in G$ that satisfy $N(q - a/b) < 1$ together with $r := a - q \cdot b \in G$ are suitable in (EUCL2). In particular $q := \{a/b\}$ is such that $N(q - a/b) < 1$.

(For real numbers c_1, c_2, $\{c_1 + c_2 i\}$ is the nearest Gaussian integer to $c_1 + c_2 i$:

$$\{c_1 + c_2 i\} := \{c_1\} + \{c_2\}i,$$

where

$$\{c\} := \lceil a - 1/2 \rceil \text{ when } c \text{ is real.})$$

Hence, $q := \{a/b\}$ (the nearest quotient of a and b) and $r := a - q \cdot b$ (the least remainder of a and b) is a suitable definition of a Euclidean division procedure in G, on which a Euclidean algorithm for the computation of the greatest common divisor may be based in the usual way ([39]). If the computation of a/b in $q := \{a/b\}$ is based on formula (DIV), i.e. $q := \{c_1/d\} + \{c_2/d\}i$, where $c_1 + c_2 i = a \cdot \bar{b}$ and $d := N(b)$, and Newton's method is used for computing $\{c_1/d\}$ and $\{c_2/d\}$, then it can be shown, [7], that

$$t_{\text{GINQR}}(a, b) \leq M(L(b)) + M'(L(b), L(q))$$

(where t_{GINQR} is the time to compute q and r according to this method). The term $M(L(b))$ in this formula reflects the time to compute the norm of b. In case $L(q) < L(b)$, it cannot be absorbed by $M'(L(b), L(q))$. B. F. Caviness and G. E. Collins [7] have presented an alternative method for division in G that avoids computing $N(b)$ in a large percentage of cases when $N(q) < N(b)$. The time $t_{\text{GINQR}'}$ to compute q and r according to their method behaves as follows

$$t_{\text{GINQR}'}(a, b) \leq L(\max(L(b), L(q)) \cdot M(\max(L(b), L(q)))),$$

$$t^*_{\text{GINQR}'}(a, b) \leq M'(\max(L(b), L(q))), \min(L(b), L(q)),$$

i.e., roughly, in the average computing time the term $M(L(b))$ disappeared. The main idea of this method is trying to obtain $q := \{a/b\}$ from $q^* := \{a^*/b^*\}$ where a^*, b^* are approximations to a, b having only a few more bits than q. By a detailed case analysis it can be seen that the number $L^*(q)$ of bits in q can be estimated in advance:

$$q = 0, \quad \text{if} \quad L^*(a) - L^*(b) \leq -3,$$

and

$$L^*(a) - L^*(b) - 2 \leqslant L^*(q) \leqslant L^*(a) - L^*(b) + 3, \quad \text{otherwise.}$$

(Exact definition of L^*:

$$L^*(a_1 + a_2 i) := \max(L^*(a_1), L^*(a_2)), \quad \text{for } a_1 + a_2 i \in G,$$

$$L^*(a) := \begin{cases} [\log_2|a|] + 1, & \text{if } a \neq 0 \\ 0, & \text{otherwise.} \end{cases})$$

In fact, it turns out that in case $L^*(a) - L^*(b) > -3$ and $k \geqslant 6$

(APPDIV) $q = \{q^*/2^k\}$, with probability greater than $1 - 2^{-k+2}$
(where $q^* = \{a^*/b^*\}$,
$a^* = [a/2^{m-2h}]$, $b^* = [b/2^{n-h}]$,
$m = L^*(a)$, $n = L^*(b)$, $h = m - n + k$).

For computing $q^* = \{a^*/b^*\}$ the ordinary division method is applied. The following equivalence is an effective test for determining whether q is indeed equal to $\{q^*/2^k\}$:

$$q = \{q^*/2^k\} \Leftrightarrow |d_i| \neq 2^{k-1} \quad \text{for} \quad i = 1, 2$$

$$(\text{where } d_i := q_i^* - c_i \cdot 2^k, \ c_i := \{q_i^*/2^k\}).$$

Using this test (APPDIV) may be used iteratively for $k = k_0(\geqslant 6), 2k_0, 4k_0, 8k_0, \ldots$ until the test reveals that $q = \{q^*/2^k\}$ or $(m - n) + k \geqslant n$, in which case $q = \{a/b\}$ is completed by the ordinary method.

When Euclid's algorithm for computing the greatest common divisor of $a, b \in G$ is based on the above improved division algorithm a speed-up ratio of approximately two is reported in [7]. A Lehmer-type algorithm for Gaussian integer gcd [6] yields a further speed-up of two. The best Gaussian integer gcd algorithm known so far is in [7], p. 41. It again improves the computation time by a factor of two as has been shown by extensive test computations, [7], p. 42. It uses the idea of the original Lehmer algorithm (approximation of the quotient sequence by using the high-order, single-precision parts of the inputs), but avoids complex interval arithmetic. Very roughly for $a, b \in G$ this algorithm proceeds as follows:

$c \leftarrow \text{GIGCD} (a, b)$
[Gaussian integer greatest common divisor. $a, b \in G$, $c = \gcd(a, b)$].

(1)   If a and b are single-precision or one of them is zero then skip (2).

(2)   If $\{a/b\}$ may be multiple-precision
        [this can be effectively checked by considering the lengths of a and b]
        then
              $(a, b) \leftarrow (b, \text{rem}(a, b))$
        else
              $(a^*, b^*) \leftarrow$ high-order single-precision parts of a and b.
              Apply extended single-precision Euclidean algorithm to (a^*, b^*) until
              a certain termination criterion is satisfied. This yields u, u', v, v' as last
              elements of the first and second consequence.

$(a, b) \leftarrow (u \cdot a + v \cdot b, u' \cdot a + v' \cdot b)$
Go to step 1.

(3) $c \leftarrow GCD(a, b)$ using a single-precision gcd algorithm
(or $c \leftarrow S(a)$ if, for instance, $b = 0$). ∎

A suitable termination criterion in the loop in step 2 is

$$\frac{1}{6}\max(|a_1|, |a_2|) < |u_1| + |u_2| + |v_1| + |v_2|,$$

where $a_1 + a_2i, u_1 + u_2i, v_1 + v_2i$ are respectively the last element in the remainder sequence and first and second consequences starting from a^*, b^*. Normally, in step 2, the else-branch will be chosen. Essentially, this is the reason why the algorithm achieves a speed-up.

Polynomial Arithmetic

In this section we consider polynomials over a ring R in finitely many variables $x_1, \ldots, x_v$. We get univariate polynomials for $v = 1$, multivariate polynomials for $v > 1$, and elements of the ground ring R for $v = 0$. Since $R[x_1, \ldots, x_v]$ is isomorphic with $(\cdots(R[x_1])\cdots)[x_v]$ a polynomial $p \in R[x_1, \ldots, x_v]$ can be considered as a polynomial in x_v with coefficients in the ring $R[x_1, \ldots, x_{v-1}]$.

For

$$p(x_1, \ldots, x_v) = \sum_{i=1}^{m} p_i(x_1, \ldots, x_{v-1})x_v^{d_i},$$

where $d_1 > \cdots > d_m$ and $p_m \neq 0$, the degree of p (in the main variable x_v), written $\deg(p)$, is d_1. The degree of the zero polynomial is zero. For $1 \leq j \leq v$ the degree of p in x_j is defined as $\deg_j(p) = d_1$ for $j = v$ and $\deg_j(p) = \max_{1 \leq i \leq m} \deg_j(p_i)$ for $j < v$. $p_1(x_1, \ldots, x_{v-1})$ is the leading coefficient of p, written $\mathrm{ldcf}(p)$. By convention $\mathrm{ldcf}(0) = 0$. The trailing coefficient of p is $\mathrm{ldcf}(x_v^n \cdot p(1/x_v))$. The leading term of p is the polynomial

$$\mathrm{ldt}(p) = p_1(x_1, \ldots, x_{v-1})x_v^{d_1}$$

and the reductum of p is the polynomial

$$\mathrm{red}(p) = \sum_{i=2}^{m} p_i(x_1, \ldots, x_{v-1})x_v^{d_i}.$$

By convention, $\mathrm{ldt}(0) = \mathrm{red}(0) = 0$. The leading base coefficient of p belongs to the ring R and is defined as $\mathrm{lbcf}(p) = \mathrm{ldcf}(p)$ if $v = 1$ and $\mathrm{lbcf}(p) = \mathrm{lbcf}(\mathrm{ldcf}(p))$ if $v > 1$.

There are four major forms of representation for polynomials in $R[x_1, \ldots, x_v]$, depending on whether we represent the polynomials in a recursive or a distributive way and whether the polynomials are dense or sparse. In computer algebra multivariate polynomials are usually sparse.

Let $p \neq 0$ be a polynomial in $R[x_1, \ldots, x_v]$ in which the power products $x_1^{e_{1,1}} \cdots x_v^{e_{1,v}}, \ldots, x_1^{e_{m,1}} \cdots x_v^{e_{m,v}}$ have the respective coefficients $a_1, \ldots, a_m$ and all the other coefficients are zero then the sparse distributive representation of p is the list $(e_1, a_1, \ldots, e_m, a_m)$ where $e_i = (e_{i,1}, \ldots, e_{i,v})$ for $1 \leq i \leq m$. For most applications one wants the e_i to be ordered, e.g. according to the inverse lexicographical ordering. This representation makes nearly no distinction between the variables. We might, however, view the polynomial p as a polynomial in only one variable, say x_v, with coefficients in the ring $R[x_1, \ldots, x_{v-1}]$ (the same applies for the coefficients of p until the ground ring R is reached), since $R[x_1, \ldots, x_v]$ is isomorphic with $(\cdots(R[x_1])\cdots)[x_v]$. Thus if

$$p(x_1, \ldots, x_v) = \sum_{i=1}^{k} p_i(x_1, \ldots, x_{v-1}) \cdot x_v^{e_i},$$

where $e_1 > \cdots > e_k$ and $p_i \neq 0$ for $1 \leq i \leq k$, the sparse recursive representation of p is the list $(e_1, \pi_1, \ldots, e_k, \pi_k)$ where π_i is the sparse recursive representation of p_i for $1 \leq i \leq k$ (assuming that a representation for the coefficient ring R is available).

Let us just briefly mention possible corresponding dense representations. If a linear ordering is defined on the exponent vectors such that for each exponent vector e there are only finitely many other exponent vectors f with $f < e$ (e.g. the ordering which orders the exponent vectors according to their degree and inverse lexicographically within the same degree), $e_0 < e_1 < \cdots$, and

$$p(x_1, \ldots, x_v) = \sum_{i=0}^{h} a_i x^{e_i},$$

where $a_h \neq 0$ and x^{e_i} stands for $x_1^{e_{i,1}} \cdots x_v^{e_{i,v}}$, then the dense distributive representation of p may be the list $(e_h, a_h, \ldots, a_0)$.

If

$$p(x_1, \ldots, x_v) = \sum_{i=0}^{j} q_i(x_1, \ldots, x_{v-1}) \cdot x_v^i \qquad \text{and} \qquad q_j \neq 0$$

then the dense recursive representation of p may be the list $(j, q_j^*, \ldots, q_0^*)$, where q_i^* is the dense recursive representation of q_i for $1 \leq i \leq j$.

Addition and subtraction of polynomials is done by adding or subtracting the coefficients of like powers. One only has to be careful to delete "leading zeros" in the resulting polynomial. Polynomial arithmetic is similar to arithmetic on large integers, indeed simpler since there are no carries. The computing time for adding or subtracting two polynomials p, q by this method is dominated by

$$(L(p_0) + L(q_0)) \cdot \prod_{i=1}^{v} (\delta_i(p) + \delta_i(q) + 1),$$

where p_0 denotes the biggest coefficient in p and $\delta_i(p)$ is the degree of p in x_i.

The classical method for multiplying polynomials uses the formula

$$r(x) = p(x) \cdot q(x) = \sum_{l=0}^{m+n} \left(\sum_{i+j=l} p_i \cdot q_j \right) x^l,$$

where

$$p(x) = \sum_{i=0}^{m} p_i x^i, \qquad q(x) = \sum_{j=0}^{n} q_j x^j.$$

The computing time required by this classical algorithm applied to sparse polynomials is proportional to t^3, where t is the maximum number of terms in p and q. It can, however, be reduced to $t^2 \cdot \log_2 t$ (see [20, 21]) by the use of better sorting algorithms.

For the case of dense polynomials "fast" algorithms for multiplication have been developed. The computing time t_{MPC} of a recursive version of the classical method for multiplying two dense polynomials of degree n in each of v variables (a polynomial p in v variables is dense if every power product $x_1^{d_1} \cdots x_v^{d_v}$ has a non-zero coefficient in p for $d_1 + \cdots + d_v$ less or equal some degree d) is

$$t_{MPC}(n, v) = (n + 1)^2 \cdot t_{MPC}(n, v - 1) + n^2 \cdot (2n + 1)^{v-1},$$

where the first term comes from multiplying the coefficients and the second term from the required additions. Thus $t_{MPC}(n, v)$ is proportional to n^{2v} (more precisely, the time for multiplying two polynomials p, q is dominated by

$$L(p_0) \cdot L(q_0) \cdot \prod_{i=1}^{v} (\delta_i(p) + 1)(\delta_i(q) + 1)).$$

A first fast algorithm is based on the Karatsuba method: in order to multiply two polynomials p and q of degree n, p and q are represented as

$$p(x) = p_1(x) \cdot x^{\lceil n/2 \rceil} + p_0(x), \qquad q(x) = q_1(x) \cdot x^{\lceil n/2 \rceil} + q_0(x).$$

Then

$$r = p \cdot q = p_1 \cdot q_1 \cdot x^{2\lceil n/2 \rceil} + (p_1 \cdot q_1 + p_0 \cdot q_0 - (p_1 - p_0)(q_1 - q_0)) \cdot x^{\lceil n/2 \rceil}$$
$$+ p_0 \cdot q_0.$$

The computing time $t_{MPK}(n, v)$ for multiplying two dense polynomials of degree n in v variables by an algorithm based on the Karatsuba method is codominant with $n^{v \cdot \log_2 3}$.

A second fast algorithm is based on the fast Fourier transform. For this purpose we consider only polynomials with coefficients in a finite field $\mathbb{Z}_p$. This restriction can be removed by the use of the Chinese remainder theorem.

For a detailed description of the fast Fourier transform we refer to [33]. A precise description of an implementation is given in [3].

The computing time $t_{MPF}(n, v)$ for multiplying two polynomials of degree n in each of v variables by an algorithm using the FFT is proportional to $v \cdot n^v \cdot \log_2 n$.

For further details on fast multiplication algorithms for dense polynomials see [32]. The trade-off points for all these "fast" algorithms are rather high. In [32] R. T. Moenck presents a hybrid mixed basis FFT algorithm that achieves a trade-off point at degree 25 for the univariate case.

When we are given two polynomials $a(x), b(x)$ over a field, with $b(x) \neq 0$, division of a by b is possible, i.e. we can obtain a unique quotient q and remainder r such that

$$a(x) = q(x) \cdot b(x) + r(x) \qquad \text{and} \qquad r = 0 \qquad \text{or} \qquad \deg(r) < \deg(b).$$

The classical algorithm for this problem is given in [23], p. 402.

In many applications, however, (e.g. gcd computation of polynomials in $\mathbb{Z}[x]$) the underlying ring of coefficients is not a field and therefore the classical algorithm does not apply. So let I be a unique factorization domain, ufd, (e.g. $\mathbb{Z}$, $\mathbb{Z}[x]$ are ufds and $I[x]$ is a ufd if I is a ufd). For polynomials $a(x), b(x) \neq 0$ over I there exists a unique pseudoquotient $q(x) = \mathrm{pquo}(a(x), b(x))$ and pseudoremainder $r(x) = \mathrm{prem}(a(x), b(x))$ such that

$$\mathrm{ldcf}(b)^{\delta+1} \cdot a(x) = q(x) \cdot b(x) + r(x) \qquad \text{and} \qquad \deg(r) < \deg(b),$$

where $\delta = \deg(a) - \deg(b)$. An algorithm for pseudodivision is given in [23], p. 408.

Now assume that we are given polynomials a and $b \neq 0$ in $I[x]$ for some integral domain I. We want to compute the quotient $c = a/b$, if it exists. This is the problem of trial division. An algorithm for trial division is

$$c \leftarrow \mathrm{PTDIV}(a, b)$$
[polynomial trial division. a and b are polynomials over some integral domain. If $b \,|\, a$ then $c = a/b$, otherwise an error is reported]

(1)   Set $c \leftarrow 0$.

(2)   If $a = 0$ then return c.

(3)   If $m < n$, where $m = \deg(a)$ and $n = \deg(b)$, or $\mathrm{ldcf}(b) \nmid \mathrm{ldcf}(a)$ then b does not divide a and an error is reported.

(4)   Otherwise, $c_{m-n} \leftarrow \mathrm{ldcf}(a)/\mathrm{ldcf}(b)$ is the coefficient of x^{m-n} in the quotient c, if it exists.

(5) $a \leftarrow a - c_{m-n}x^{m-n} \cdot b$. Go to step (2). ■

In PTDIV we assume a trial division algorithm for I. Such an algorithm exists for $I = \mathbb{Z}$ and therefore we get trial division algorithms for $\mathbb{Z}[x]$, $\mathbb{Z}[x, y]$ etc.

G. E. Collins [10] observes that the algorithm PTDIV is efficient for cases in which the quotient exists. An analysis of the computing time, however, is very difficult.

Now let us consider the problem of computing the nth power of a polynomial p. The algorithms which can be efficiently used to solve this problem depend heavily on the "density" of the polynomial p. Investigations of algorithm complexity are carried out for the two extreme cases, where p is completely sparse or p is dense.

R. J. Fateman [13] gives comparative analyses of algorithms for the computation of integer powers of sparse polynomials. More precisely, all the following analyses are carried out for polynomials which are completely sparse to power N (for the exact definition we refer to [13], p. 145), where $N \geqslant n$, and n is the power to which p should be raised. All the polynomials considered have coefficients in a finite field.

Thereby we avoid the consideration of coefficient growth. By the use of the Chinese Remainder Algorithm this restriction can be removed (see [19]).

The most obvious algorithm, RMUL, successively computes $p^2 = p \cdot p$, $p^3 = p \cdot p^2, \ldots, p^n = p \cdot p^{n-1}$. The number of coefficient multiplications needed is

$$t \cdot \binom{t + n - 1}{t} - t,$$

where t is the number of non-zero terms in p.

A second algorithm, RSQ, for computing p^n uses the binary exponentiation method, based on the binary expansion of n (compare the corresponding algorithm for integer exponentiation). However, it turns out that RSQ is more expensive than RMUL. The reason for this complexity behaviour is, that it is less costly to multiply a large polynomial by a small one, than to multiply two "mediumsize" polynomials. For a more detailed comparison of RMUL and RSQ we refer to [15].

In [13] a number of further algorithms for computing the powers of a polynomial are considered, which rely on the multinomial or binomial expansion. The relation of the computing times for these algorithms to the computing time for RMUL is better than (or equal to) $t/(t + n - 1)$. The best of this class of algorithms seems to be BINB, an algorithm which can be described as follows:

$q \leftarrow \text{BINB}(p, n)$
[algorithm for raising the polynomial p to the nth power. Binomial expansion with half splitting. p is a polynomial with t non-zero terms, n is an integer, $q = p^n$]

(1)   Split p into $p_1 + p_2$, where the number of terms in p_1 is $\lfloor t/2 \rfloor$.

(2)   Compute $p_1^2, \ldots, p_1^n$ as well as $p_2^2, \ldots, p_2^n$.

(3)   Use the binomial theorem to compute

$$q = p^n = \sum_{i=0}^{n} \binom{n}{i} p_1^i \cdot p_2^{n-i}. \quad \blacksquare$$

The cost of BINB is

$$\binom{t + n - 1}{n} + t \cdot \binom{t/2 + n - 1}{n - 1} - 2 \cdot \binom{t/2 + n - 1}{n}$$
$$- (t/2) \cdot (t/4 - n - \log_2(t - 1) + 4).$$

For large n, and larger t BINB approaches $t^n/n! + O(t^{n-1}/2(n - 2)!)$.

The other extreme case is p being a dense polynomial, i.e. every power product $x_1^{d_1} \cdots x_v^{d_v}$ has a non-zero coefficient in p for $d_1 + \cdots + d_v$ less or equal to some degree d. In [14] R. F. Fateman gives an overview of exponentiation of dense polynomials. The number of coefficient multiplications for the method of repeated multiplication is bounded by $(d + 1)^{2v} \cdot n^{v+1}/(v + 1)$. Again there is the binary exponentiation method. If we use the Karatsuba algorithm for computing the intermediary results, the cost for the binary method is $O((nd)^{v \cdot \log_2 3})$.

Another algorithm, EVAL, computes the nth power of the polynomial p by computing $p(b_i)^n$ for $nd + 1$ integers $b_1 = 0, \ldots, b_{nd+1} = nd$, (where d is the degree of p) and then uses interpolation to compute $q = p^n$. For several variables $p(b_i)$ is a polynomial in one less variables and EVAL is applied recursively to compute $p(b_i)^n$. The number of multiplications required for computing p^n by the algorithm EVAL is $0((nd)^{(v+1)})$, which is, asymptotically, more efficient than the previous algorithms.

The binomial method can also be used in this case. It turns out that it is best to split p into p_1 and p_2 such that p_1 is a monomial. In this case the number of multiplications is $O(n^{2v}d^{2v-1} + n^{v+1}d^{2v})$. So the binomial method will be inferior to repeated multiplication and EVAL for large problems. Timings given in [14], however, show that the binomial method is quite good for many interesting problems.

Finally p^n may be calculated by the fast Fourier transform. The cost for exponentiation by the use of the FFT is $O(v \cdot (nd)^v \cdot \log_2(nd))$, which is the best known asymptotic bound for the dense multivariate case.

Now let us turn to the problem of evaluating a polynomial $p = \sum_{i=0}^{n} p_i x^i$ at some point x_0. An overview is given in [23], Section 4.6.4. The most obvious algorithm, which successively computes $p_0, p_1 x_0, \ldots, p_n x_0^n$ uses $2n - 1$ multiplications and n additions. An essential improvement in complexity is Horner's rule, by which $p(x_0)$ is computed as

$$((\cdots (p_n x_0 + p_{n-1})x_0 + \cdots)x_0 + p_0.$$

n multiplications and n additions are necessary for evaluating a polynomial p of degree n at some point x_0 by the Horner's rule. D. E. Knuth in [23] points out that the computation of $p(x_0)$ by Horner's rule is essentially the same as dividing $p(x)$ by $(x - x_0)$. Dividing $p(x)$ by another polynomial which has a root at x_0, namely $x^2 - x_0^2$, yields the second-order Horner's rule

$$p(x_0) = (\cdots (p_{2\lfloor n/2 \rfloor}x_0^2 + p_{2\lfloor n/2 \rfloor - 2})x_0^2 + \cdots)x_0^2 + p_0$$
$$+ ((\cdots (p_{2\lceil n/2 \rceil - 1}x_0^2 + p_{2\lceil n/2 \rceil - 3})x_0^2 + \cdots)x_0^2 + p_1)x_0.$$

$n + 1$ multiplications and n additions are necessary to evaluate p at x_0 by the second-order Horner's rule. So it is no improvement over the first-order Horner's rule, but if we have to evaluate $p(x_0)$ and $p(-x_0)$ at the same time, this can be achieved by just one more addition.

Polynomial evaluation is essential for performing the Taylor shift of a polynomial p, i.e. from given coefficients of p the coefficients of $p(x + x_0)$ should be computed:

$$p(x + x_0) = p(x_0) + p'(x_0)x + (p''(x_0)/2!)x^2 + \cdots + (p^{(n)}(x_0)/n!)x^n.$$

M. Shaw and J. F. Traub [37] investigate the complexity of algorithms for solving this problem. Successive use of Horner's rule for computing the coefficients of $p(x + x_0)$ requires $n(n + 1)/2$ multiplications and the same number of additions. In [37], p. 162 an algorithm for computing all coefficients of $p(x + x_0)$ by only $2n - 1$ multiplications and $n - 1$ divisions is presented:

$q \leftarrow \text{TST}(p, x_0)$

[Taylor shift. Algorithm of Shaw and Traub. p is a polynomial, x_0 a point in the range of p, $q = p(x + x_0)$]

(1)   Compute $t_i^{-1} \leftarrow p_{n-i-1} x_0^{n-i-1}$ for $i = n-1, \ldots, 1, 0$.

(2)   Compute $t_j^j \leftarrow p_n x_0^n$ for $j = 0, 1, \ldots, n$.

(3)   For $j = 0, 1, \ldots, n-1$, $i = j+1, \ldots, n$
       set $t_i^j \leftarrow t_{i-1}^{j-1} + t_{i-1}^j$.

(4) $q_j \leftarrow t_n^j / x_0^j$ for $j = 0, 1, \ldots, n$.   ■

TST is a special case of a family of splitting algorithms for computing the first m normalized derivatives $p^{(j)}/j!$ of a polynomial p, which is presented in [37].

A balance and conquer strategy due to Schönhage (private communication) splits $p(x) = p_1(x) \cdot x^r + p_0(x)$ similar to the Karatsuba split. Then the Taylor shift operator T is applied recursively over the degree

$$Tp = Tp_1 \cdot (x + x_0)^r + Tp_0,$$

where the basis is obvious and the powers $(x + x_0)^r$ are precomputed up to $r = \lfloor n/2 \rfloor$. For the multiplication a fast method can be useful.

The Rational Functions

In the section on rational numbers we have developed the basic concepts in a context general enough to contain rational functions as a subcase. Whenever I is an integral domain then $I[x_1, \ldots, x_v]$ is also an integral domain and therefore one can embed $I[x_1, \ldots, x_v]$ into its field of quotients $I(x_1, \ldots, x_v)$ (the elements of $I(x_1, \ldots, x_v)$ are called rational functions or quolynomials).

We are especially interested in the case where $I = \mathbb{Z}$. There exists a gcd-algorithm for $\mathbb{Z}[x_1, \ldots, x_v]$ (see the chapter on remainder sequences in this volume) and therefore unique representatives for the equivalence classes in $\mathbb{Z}(x_1, \ldots, x_v)$ can be computed.

Addition, subtraction, multiplication, inversion and division algorithms for $\mathbb{Z}(x_1, \ldots, x_v)$ are straightforward generalizations of the corresponding algorithms on rational numbers. The method of P. Henrici [17] and W. S. Brown [5] is used. For a detailed description of the algorithms see [9]. By carefully analyzing the points at which to compute gcd's it is also possible to design a Henrici-Brown algorithm for differentiating rational functions (see [9]).

A thorough computing time analysis of the Henrici-Brown algorithms for addition, multiplication and differentiation would be a considerable task. Let us just coarsly compare the Henrici-Brown multiplication algorithms to a classical algorithm, where the two rational functions to be multiplied are univariate. Most of the computing time for either algorithm will be devoted to gcd calculation, since the time to compute the gcd of two nth degree polynomials is approximately proportional to n^4 by [8]. Thus it follows easily that the Henrici-Brown algorithm will be faster than the classical algorithm by a factor of approximately 8. It is fairly clear that the advantage of the Henrici-Brown algorithms will increase as the number of variables increases.

Evaluation of rational functions reduces to evaluation of polynomials.

Power Series

A power series is a formal sum $U(x) = \sum_{n=0}^{\infty} u_n x^n$ whose coefficients u_n belong to a field. Of course, it is impossible to store all the infinitely many coefficients of a power series in a computer memory. As in the case of algebraic numbers, an initial segment $u_0, \ldots, u_N$ may be considered as an approximation of the power series, but conceptually an algorithm to compute for arbitrary N the next higher coefficients of the series is considered as the algebraic part of the power series representation.

Addition and subtraction are defined (and computed) component-wise. Multiplication of power series is done by the familiar Cauchy product rule. The quotient $Q(x) = U(x)/V(x)$ (for $v_0 \neq 0$) is the solution of $U = Q \cdot V$ for given U and V. The coefficients of Q are obtained by comparing like powers of x on both sides of this equation.

For computing the results of these operations it is necessary to have exact arithmetic in the underlying field, or, if the coefficients of the inputs are given only symbolically, to be able to simplify the occurring expressions. This establishes the connection between arithmetic on power series and other topics in computer algebra. Algebraic computations of power series coefficients are frequently used to increase the order of an approximation symbolically for subsequent numerical treatment (e.g. Runge-Kutta formulas).

D. E. Knuth, in [23], Section 4.7, presents algorithms for performing these operations. Furthermore he considers the problems of exponentiation (by a method suggested by J. C. P. Miller, see [18]) and reversion.

Finite Fields

Representations

If $q > 1$ is the power of some prime p there exists a finite field with q elements, this field is commutative and unique up to an isomorphism, it is denoted $\mathbb{F}_q$ or GF(q). Inversely if a finite field has q elements then q is a power of some prime number, say $q = p^k$. For our point of view the best source of information about finite fields is Berlekamp's book [2], Chap. 4. It is known that the nonzero elements of $\mathbb{F}_q$ are equal to the powers $\alpha^0, \alpha^1, \alpha^2, \ldots, \alpha^{q-2}$ of some element α called a primitive element ([2], Theorem 4.24).

The field $\mathbb{F}_q$ can be viewed in different ways:

(i) $\mathbb{F}_q = \{0\} \cup \{1, \alpha, \alpha^2, \ldots, \alpha^{q-2}\}$,

(ii) $\mathbb{F}_q$ is a vector space of dimension k over $\mathbb{F}_p$ with a given basis,

(iii) $\mathbb{F}_q = \mathbb{F}_p[X]/Q(X)$, where $Q(X)$ is a polynomial of $\mathbb{F}_p[X]$ of degree k and irreducible over $\mathbb{F}_p$.

See [2], Table 4.1, for an example of several representations of $\mathbb{F}_{16}$. Even if equivalent from the mathematical point of view, these representations are quite different in computer algebra because the algorithms of the arithmetical operations are very dependent on the representation, as we shall see below. Of course, representation (iii) is a special case of representation (ii).

Addition

In representation (ii) (and (iii)), if $(e_1, \ldots, e_k)$ is a basis of $\mathbb{F}_q$ as a vector field over $\mathbb{F}_p$ and if $u = u_1 e_1 + \cdots + u_k e_k, v = v_1 e_1 + \cdots + v_k e_k$ (where $u_1, \ldots, u_k, v_1, \ldots, v_k$ are integers modulo p) then

$$w = u + v = w_1 e_1 + \cdots + w_k e_k$$

where

$$w_j = (u_j + v_j) \bmod p, \qquad j = 1, \ldots, k.$$

In this case addition is very easy. But this is not at all the same in representation (i), if $u = \alpha^m$ and $v = \alpha^n$ then

$$w = u + v = \alpha^m + \alpha^n.$$

To obtain an integer l such that $w = \alpha^l$ there is no simple law and a table of the integers $a_0, a_1, \ldots$ such that

$$1 + \alpha^j = \alpha^{a_j}, \qquad j = 0, 1, \ldots, q - 2$$

is needed.

Multiplication

For this operation representation (i) is very practical:

$$\alpha^m \cdot \alpha^n = \alpha^l, \qquad \text{with} \qquad l = (m + n) \bmod (q - 1).$$

But now representation (ii) is not very good. To compute

$$(u_1 e_1 + \cdots + u_k e_k) \cdot (v_1 e_1 + \cdots + v_k e_k)$$

one needs a table of the products $e_i \cdot e_j$, $0 \leqslant i \leqslant j \leqslant k$. The special case of representation (iii) is better, then $e_i = X^{i-1}$, $i = 1, \ldots, k$ (more precisely, e_i is the image of X^{i-1} by the natural homomorphism $\mathbb{F}_p[X] \to \mathbb{F}_p[X]/Q(X)$!) and we need only to compute $X^i \bmod Q(X)$ for $i = k, \ldots, 2k - 2$.

Reciprocal

Again representation (i) is very practical

$$(\alpha^m)^{-1} = \alpha^{q-1-m} \qquad \text{for} \qquad m = 1, 2, \ldots, q - 2.$$

Representation (ii) is very bad. And if

$$u = u_1 + u_2 X + \cdots + u_k X^{k-1} = U(X)$$

the extended Euclidean algorithm in $\mathbb{F}_p[X]$ gives a relation

$$U(X)V(X) + Q(X)W(X) = 1$$

for some polynomials V and W over $\mathbb{F}_q$ and

$$u^{-1} = V(X) \bmod Q(X).$$

Some computational details of arithmetic in finite fields may be found in the chapter on computing in extensions in this volume.

p-adic Numbers

Let p be a prime number. On the set $\mathbb{Z}$ of rational integers we define the function

$$|a|_p = p^{-r} \qquad \text{if } p^r | a \text{ and } p^{r+1} \nmid a, \qquad a \neq 0,$$

and

$$|0|_p = 0.$$

This is a distance and $\mathbb{Z}_p$, the set of p-adic integers, is the completion of $\mathbb{Z}$ with respect to this distance. In $\mathbb{Z}_p$ a series

$$\sum_{i=0}^{\infty} c_i p^i, \qquad c_i \in \mathbb{Z}$$

is convergent. Moreover each element α of $\mathbb{Z}_p$ can be written uniquely in the canonical form

$$\alpha = \sum_{i=0}^{\infty} a_i p^i, \qquad a_i \in \{0, 1, \ldots, p-1\}, \qquad i \geqslant 0.$$

If

$$\beta = \sum_{i=0}^{\infty} b_i p^i$$

then

$$\alpha + \beta = \sum_{i=0}^{\infty} (a_i + b_i) p^i$$

(of course, in general this formula is not the canonical form but the canonical form can be obtained easily from it) and

$$\alpha\beta = \sum_{n=0}^{\infty} \left(\sum_{i+j=n} a_i b_j \right) p^n.$$

The formula to compute the canonical form of $-\beta$ is the following: suppose

$$\beta = b_k g^k + b_{k+1} g^{k+1} + \cdots, \qquad b_k \neq 0,$$

then

$$-\beta = 0 - \beta = (g \cdot g^k + (g-1)g^{k+1} + (g-1)g^{k+2} + \cdots) - \beta$$

so

$$-\beta = (g - b_k)g^k + (g - 1 - b_{k+1})g^{k+1} + (g - 1 - b_{k+2})g^{k+2} + \cdots$$

and this is a canonical expansion. Of course

$$\alpha - \beta = \alpha + (-\beta).$$

The ring $\mathbb{Z}_p$ is integral and its quotient field is called the field of p-adic numbers, $\mathbb{Q}_p$. Each non-zero element α of $\mathbb{Q}_p$ can be written uniquely in the form

$$\alpha = \sum_{n=r}^{\infty} a_n p^n, \qquad a_r \neq 0,$$

for some rational integer r. The formulas for addition, subtraction and multiplication in $\mathbb{Q}_p$ are similar to those for the same operations in $\mathbb{Z}_p$.

The formula for division in $\mathbb{Q}_p$ is the same as the formula for division of formal power series (this was also the case for addition and multiplication). Suppose we compute $\gamma = \alpha/\beta$, where

$$\alpha = a_0 + a_1 p + \cdots, \qquad \beta = b_0 + b_1 p + \cdots, \qquad b_0 \neq 0$$

(this normalization causes no loss of generality), then

$$\gamma = c_0 + c_1 p + c_2 p^2 + \cdots$$

with $c_0 = a_0/b_0$,

$$c_n = (a_n - c_0 b_n - c_1 b_{n-1} - \cdots - c_{n-1} b_1)/b_0, \qquad n \geq 1.$$

But in the previous expansion the c_i's are not rational integers in the general case, except if $b_0 = \pm 1$. To obtain the canonical expansion of α/β one can proceed as follows:

there is a digit $c_0' \neq 0$ such that

$$b_0 c_0' \equiv a_0 \pmod{p},$$

form

$$\alpha - c_0' \beta = a_1' p + a_2' p^2 + \cdots,$$

next there is a digit c_1' such that

$$c_1' \beta \equiv a_1' \pmod{p},$$

form

$$\alpha - c_0' \beta - c_1' p \beta = a_2'' p^2 + a_3'' p^3 + \cdots$$

etc....

A very interesting fact about $\mathbb{Q}_p$ is that some algebraic equations can be solved. Let us take an example. We try to solve the equation $x^2 + 1 = 0$ in $\mathbb{Q}_5$. Suppose $\alpha = a_0 + 5a_1 + 5^2 a_2 + \cdots$ is such a solution then

$$a_0^2 \equiv -1 \pmod{5}, \qquad \text{i.e.} \qquad a_0 = 2 \text{ or } 3.$$

Choose $a_0 = 2$. Then

$$0 = \alpha^2 + 1 = 5 + 4a_1 \cdot 5 + (4a_2 + a_1^2) \cdot 5^2 + \cdots.$$

These relations imply $1 + 4a_1 \equiv 0 \bmod 5$, hence $a_1 = 1$. Similarly we get $a_2 = 2$, $a_3 = 1, \ldots$ and

$$\alpha = 2 + 1 \cdot 5 + 2 \cdot 5^2 + 1 \cdot 5^3 + 2 \cdot 5^4 + \cdots.$$

For a detailed presentation of p-adic numbers we refer the reader to Mahler's book [26].

References

[1] Aho, A. V., Hopcroft, J. E., Ullman, J. D.: The Design and Analysis of Computer Algorithms. Reading, Mass.: Addison-Wesley 1974.

[2] Berlekamp, E. R.: Algebraic Coding Theory. New York: McGraw-Hill 1968.

[3] Bonneau, R. J.: Polynomial Operations Using the Fast Fourier Transform. Cambridge, Mass., MIT Dept. of Math., 1974.

[4] Brent, R. P.: Fast Multiple Precision Evaluation of Elementary Functions. J. ACM **23**, 242 – 251 (1976).

[5] Brown, W. S., Hyde, J. P., Tague, B. A.: The ALPAK System for Nonnumerical Algebra on a Digital Computer-II: Rational Functions of Several Variables and Truncated Power Series with Rational Function Coefficients. Bell Syst. Tech. J. **43**, No. 1, 785 – 804 (1964).

[6] Caviness, B. F.: A Lehmer-Type Greatest Common Divisor Algorithm for Gaussian Integers. SIAM Rev. **15**, No. 2, Part 1, 414 (April 1973).

[7] Caviness, B. F., Collins, G. E.: Algorithms for Gaussian Integer Arithmetic. SYMSAC **1976**, 36 – 45.

[8] Collins, G. E.: Computing Time Analyses of Some Arithmetic and Algebraic Algorithms. Univ. of Wisconsin, Madison, Comp. Sci. Techn. Rep. 36 (1968).

[9] Collins, G. E.: The SAC-1 Rational Function System. Univ. of Wisconsin, Madison, Comp. Sci. Techn. Rep. 135 (1971).

[10] Collins, G. E.: Computer Algebra of Polynomials and Rational Functions. Am. Math. Mon. **80**, No. 2, 725 – 755 (1973).

[11] Collins, G. E.: Lecture Notes in Computer Algebra. Univ. of Wisconsin, Madison.

[12] Collins, G. E., Musser, D. R.: Analysis of the Pope-Stein Division Algorithm. Inf. Process. Lett. **6**, 151 – 155 (1977).

[13] Fateman, R. F.: On the Computation of Powers of Sparse Polynomials. Stud. Appl. Math. **LIII**, No. 2, 145 – 155 (1974).

[14] Fateman, R. F.: Polynomial Multiplication, Powers and Asymptotic Analysis: Some Comments. SIAM J. Comput. **3**, No. 3, 196 – 213 (1974).

[15] Gentleman, W. M.: Optimal Multiplication Chains for Computing a Power of a Symbolic Polynomial. Math. Comput. **26**, No. 120 (1972).

[16] Hardy, G. H., Wright, E. M.: An Introduction to the Theory of Numbers, 5th ed. Oxford: Clarendon Press 1979.

[17] Henrici, P.: A Subroutine for Computations with Rational Numbers. J. ACM **3**, 6 – 9 (1956).

[18] Henrici, P.: Automatic Computations with Power Series. J. ACM **3**, 10 – 15 (1956).

[19] Horowitz, E.: The Efficient Calculation of Powers of Polynomials. 13th Annual Symp. on Switching and Automata Theory (IEEE Comput. Soc.) (1972).

[20] Horowitz, E.: A Sorting Algorithm for Polynomial Multiplication. J. ACM **22**, No. 4, 450 – 462 (1975).

[21] Johnson, S. C.: Sparse Polynomial Arithmetic. EUROSAM **1974**, 63 – 71.

[22] Karatsuba, A., Ofman, Yu.: Dokl. Akad. Nauk SSSR **145**, 293 – 294 (1962) [English translation: Multiplication of Multidigit Numbers on Automata. Sov. Phys., Dokl. **7**, 595 – 596 (1963)].

[23] Knuth, D. E.: The Art of Computer Programming, Vol. 2, 2nd ed. Reading, Mass.: Addison-Wesley 1981.

[24] Knuth, D. E.: The Analysis of Algorithms. Proc. Internat. Congress Math. (Nice, 1970), Vol. 3, pp. 269 – 274. Paris: Gauthier-Villars 1971.

[25] Lehmer, D. H.: Euclid's Algorithm for Large Numbers. Am. Math. Mon. **45**, 227 – 233 (1938).

[26] Mahler, K.: Introduction to p-adic Numbers and Their Functions. Cambridge Univ. Press 1973.

[27] Matula, D. W.: Fixed-Slash and Floating-Slash Rational Arithmetic. Proc. 3rd Symp. on Comput. Arith. (IEEE), 90–91 (1975).
[28] Matula, D. W., Kornerup, P.: A Feasibility Analysis of Binary Fixed-Slash and Floating-Slash Number Systems. Proc. 4th Symp. on Comput. Arith. (IEEE), 29–38 (1978).
[29] Matula, D. W., Kornerup, P.: Approximate Rational Arithmetic Systems: Analysis of Recovery of Simple Fractions During Expression Evaluation. EUROSAM 1979, 383–397.
[30] Moenck, R. T.: Studies in Fast Algebraic Algorithms. Ph.D. Thesis, Univ. of Toronto, 1973.
[31] Moenck, R. T.: Fast Computations of GCD's. Proc. 5th Symp. on Theory of Comput. (ACM), 142–151 (1973).
[32] Moenck, R. T.: Practical Fast Polynomial Multiplication. SYMSAC 1976, 136–148.
[33] Pollard, J. M.: The Fast Fourier Transform in a Finite Field. Math. Comput. 25, 365–374 (1971).
[34] Pope, D. A., Stein, M. L.: Multiple Precision Arithmetic. Commun. ACM 3, 652–654 (1960).
[35] Schönhage, A.: Schnelle Berechnung von Kettenbruchentwicklungen. Acta Inf. 1, 139–144 (1971).
[36] Schönhage, A., Strassen, V.: Schnelle Multiplikation großer Zahlen. Computing 7, 281–292 (1971).
[37] Shaw, M., Traub, J. F.: On the Number of Multiplications for the Evaluation of a Polynomial and Some of its Derivatives. J. ACM 21, No. 1, 161–167 (1974).
[38] Strassen, V.: The Computational Complexity of Continued Fractions. SYMSAC 1981, 51–67.
[39] van der Waerden, B. L.: Modern Algebra, Vol. 1. New York: Frederick Ungar 1948.

Prof. Dr. G. E. Collins
Computer Science Department
University of Wisconsin – Madison
1210 West Dayton Street
Madison, WI 53706, USA

Prof. Dr. M. Mignotte
Centre de Calcul de l'Esplanade
Université Louis Pasteur
5, rue René Descartes
F-67084 Strasbourg Cédex
France

Dipl.-Ing. F. Winkler
Institut für Mathematik
Johannes-Kepler-Universität Linz
Altenbergerstrasse 69
A-4040 Linz
Austria

Computer Algebra Systems

J. A. van Hulzen, Enschede, and **J. Calmet**, Grenoble

Abstract

A survey is given of computer algebra systems, with emphasis on design and implementation aspects, by presenting a review of the development of ideas and methods in a historical perspective, by us considered as instrumental for a better understanding of the rich diversity of now available facilities. We first indicate which classes of mathematical expressions can be stated and manipulated in different systems before we touch on different general aspects of usage, design and implementation, such as language design, encoding, dynamic storage allocation and a symbolic-numeric interface. Then we discuss polynomial and rational function systems, by describing ALTRAN and SAC-2. This is followed by a comparison of some of the features of MATHLAB-68, SYMBAL and FORMAC, which are pretended general purpose systems. Before considering giants (MACSYMA and SCRATCHPAD) and gnomes (muMATH-79), we give the main characteristics of TRIGMAN, CAMAL and REDUCE, systems we tend to consider as grown out special purpose facilities. Finally we mention some modern algebra systems (CAYLEY and CAMAC-79) in relation to recent proposals for a language for computational algebra. We conclude by stipulating the importance of documentation. Throughout this discussion related systems and facilities will be mentioned. Noticeable are ALKAHEST II, ALPAK, ANALITIK, ASHMEDAI, NETFORM, PM, SAC-1, SCHOONSCHIP, SHEEP, SMP, SYCOPHANTE and TAYLOR.

1. Mathematical Capabilities of Computer Algebra Systems

The pioneering work on differentiation by Kahrimanian [100] and Nolan [133] is often considered to be the first attempt to use a digital computer for formal algebraic manipulation. Now, almost 30 years later, it is estimated that about 60 systems exist for doing some form of computer algebra [136]. The mathematical capabilities of many of the well qualified systems of today are strongly related to the eminent, early successes of computer algebra, as achieved in integration [149, 126], Celestial Mechanics [91], General Relativity [10] and Quantum Electro Dynamics [80]. Some knowledge of the underlying goals and mathematical motives, leading to these results, can assist in understanding the present possibilities. They tended to model the classes of mathematical expressions of the different systems and the classes of operations on these expressions. Polynomial algebra, in some form, was considered as the basic requirement. All of the well-known elementary transcendental functions, naturally entering in the description of our (approximate) models of physical reality, were and are considered as intriguing objects (see Davenport [36] for a recent summary). However, algebraic systems of a less complicated type (Loos [114]) were often neglected, as reflected by the computer algebra courses, published in the past [25, 26, 52, 96]. A decade ago, Tobey [161], attempting to introduce and survey computer algebra (by him considered as an alchemistic discipline), employed the entities "user", "mathematician" and "designer/

implementer" to serve as a framework for his discussion. However, this distinction is often artificial. A good understanding of the mathematical capabilities of a system is essential for a "user" and a "designer/implementer" as well. For instance, a central issue like algebraic simplification is, citing Moses [127]: "The most pervasive process in algebraic manipulation". Then he continues: "It is also the most controversial. Much of the controversy is due to the difference between the desires of a user and those of a system designer". He suggested a classification of computer algebra systems, based on a single criterium: the degree to which a system insists on making a change in the representation of an expression as provided by a user.

Radical systems can handle a single well defined class of expressions (e.g. polynomials, rational functions, truncated power series, truncated Poisson series). Such systems radically alter the form of an expression in order to get it into its internal (canonical) form, implying that the task of the manipulating algorithms is well defined and lends itself to efficient manipulation.

New left systems arose in response to some of the difficulties with radical systems, such as caused by the automatic expansion of expressions (think of $(x + y)^{1000}$). Expansion, for instance, is brought under user control. Such systems can usually handle a wide variety of expressions with greater ease, though with less power than a radical system, by using labels for non rational (sub)expressions.

Liberal systems rely on a very general representation of expressions and use simplification transformations which are close in spirit to the ones used in paper and pencil calculations. Therefore a major disadvantage a liberal system has relative to a radical or new left system is its inefficiency, both in space and time requirements. An advantage might be that one can express problems more naturally.

Conservative systems are so unwilling to make inappropriate transformations that it essentially forces a user to write his own simplification rules. Their designers state that all simplification is determined by context and thus can better be based on the theoretical concept of Markov-algorithms.

Catholic systems use more than one representation for expressions and have more than one approach to simplification. The designers of such systems want to give the user the ease of working with a liberal system, the efficiency and power of a radical system and the attention to context of a conservative system. An implied disadvantage is therefore its size and organization.

2. Some General Aspects of Design and Implementation

These brief remarks, indicating the classes of mathematical expressions, to be stated and manipulated in various systems, suggest that a persistent user requires adequate documentation. His mathematical language, when using paper and pencil, is typically two-dimensional, using different alphabets and lots of different symbols, not provided by input devices. Most systems provide two dimensional output. Noticeable is William A. Martin's contribution [119, 67] to this important aspect of computer algebra. Gawlik's MIRA can even provide output in two colours in a very natural fashion [62, 63]. Illustrative examples for 10 different systems (ALTRAN, CAMAL, FORMAC, MACSYMA, muMATH-79, REDUCE, SAC-1, SCRATCHPAD, SYMBAL and TRIGMAN) are given by Yun and Stoutemyer [174]. They also provide a "consumer's guide", by briefly listing capabilities and availability conditions of these systems. Different systems allow command oriented, interactive access. Others require batchprocessing. Some combine both facilities. The first category requires simple instructions for reordering and dismantling of expressions, as well as substitution and pattern defining commands to allow simulating blackboard calculations. The last can be attractive for batchprocessing interactively tested programs, which (might) demand for excessive time and space requirements.

When a user types the command $(A + B) * (B + 2)$ he may get the answer (the meaning of this expression) in expanded form $(A * B + B ** 2 + 2 * A + 2 * B$ or $A * B + 2 * A + B ** 2 + 2 * B)$ in factored form $((A + B) * (B + 2))$ or, for instance, in recursive form $(A * (B + 2) + B ** 2 + 2 * B$ or $B ** 2 + B * (A + 2) + 2 * A)$, or in an equivalent one- or two-dimensional lay-out, depending on the output facilities of the system at hand. This simple example serves to indicate that the meaning of an expression can surprise a user. It is influenced by encoding and automatic or usercontrolled simplification actions, like ordering of operands with respect to an operator order; not only in the way the meaning is presented, but also in time and space in obtaining it. To illustrate this, let us (less formally) apply the rules for differentiation, as proposed by Tobey [161], on xe^{ax}, using a unary operator D, to denote differentiation with respect to x:

$$D(x \cdot e^{ax}) \rightarrow D(x) \cdot e^{ax} + x \cdot D(e^{ax})$$

$$\rightarrow 1 \cdot e^{ax} + x \cdot e^{ax} \cdot D(ax)$$

$$\rightarrow 1 \cdot e^{ax} + x \cdot e^{ax} \cdot \{D(a) \cdot x + a \cdot D(x)\}$$

$$\rightarrow 1 \cdot e^{ax} + x \cdot e^{ax} \cdot \{0 \cdot x + a \cdot 1\}.$$

A user will prefer to see $e^{ax} + ax e^{ax}$ or $(1 + ax)e^{ax}$, implying that further simplification is required and, eventually, additional user actions to obtain the factored form. Time and space spent are influenced by so-called intermediate expression swell. So an important aspect of encoding, if not a key consideration, is minimization of space requirements during execution. How does one deal with intermediate expression swell or how to avoid redundancy when the same subexpression occurs several times? Often some userfacilities are offered to free allocated space in combination with automated garbage collection, certainly available in LISP-based systems, or with an underlying reference count mechanism, as provided in SAC-1. How influential garbage collection can be is illustrated by Foderaro and Fateman [59] in describing their experiences when installing MACSYMA on the medium sized VAX-11 computer. It often essentially slows down interactive access in an environment with limited user core facilities. It stipulates Loos' remark [115]: "In some sense the only way to structure the garbage collection process and to limit its impact to a well defined level is to design a language around it in which all its aspects are bound together". It is just one illustration of Tobey's "alchemistic" view [161]: "Symbolic mathematical computation will become a science when its scope includes ... adequate techniques for measurement and validation".

This view on computer algebra was realistic, certainly by that time and perhaps even now, since must algebra systems evolved in parallel with the development of disciplines for design, programming, program verification and portability, as reflected by Barton and Fitch [11] in 1972: "Perhaps the most obvious difficulty that arises in the construction of an algebra system is that the hardware of a computer is not designed for the manipulation of algebraic expressions, and consequently the simple facilities of addition and multiplication between expressions must be provided by program. No algebraic facilities can exist on the computer at all until the program has been written, and the writing of the program, while it is in principle straightforward, has in practice required the use of advanced and in many cases recently developed programming techniques".

The only exceptions of the implicit presumption that with a computer algebra system some software package is meant are ANALITIK, available on the Soviet MIR-2 computer, and designed for numerical and analytical mathematical computations [88, 104] and its successor ANALITIK-74 running on the MIR-32 computer [105, 64]. ANALITIK has been implemented in hardware; actually the MIR-2 (from which the MIR-32 evolved) has been coined for the language rather then the language for the machine. ANALITIK is a compatible extension of the programming language MIR, an ALGOL-like language, more specifically designed for the solution of numerical problems. A similar approach was taken for the design of Formula ALGOL [137], which was a careful attempt to extend ALGOL-60 to include string processing and symbolic mathematical capabilities. But, like FAMOUS [50], a conservative system, and only used at one place [146]. This has been the misfortune of many other systems and facilities, like ABC-ALGOL [164], ALADIN [148], MIRA, POLMAN [168]. Reasons are often importability, language design or lack of significance when finally known outside its residence. POLMAN, for instance, is a polynomial system written in PL/I, like AMP [38], which shows a striking similarity with REDUCE. However, AMP seems to run more efficiently on its target IBM-machine. Formula ALGOL is, like SYMBAL and MACSYMA, an exception of how languages for computer algebra have been designed. Often, host languages are used via a preprocessor or a translator implying that, at some level, access to mathematical capabilities is provided through subroutine calls. FORTRAN as a hostlanguage can be argued for portability reasons, assuming ANSI-FORTRAN is used (ALTRAN, SAC-1). An easier choice, however, would have been LISP or a similar equivalent like AMBIT [27, 28, 29]. Intermediate expression swell, just to name one reason, demands dynamic storage allocation. Recursivity and listprocessing are natural instruments when attempting to automatize mathematics. Although Standard LISP [117] has been proposed, and is effectively used in describing and implementing REDUCE-2, well known other systems, like MACSYMA and SCRATCHPAD, have their own LISP base. This lack of standardization might be considered as a drawback. In this context it is worthwhile mentioning promising recent work of Jenks, Davenport and Trager [98, 37, 99] on the development of a language for computational algebra. Their primary goal in presenting such a language dealing with algebraic objects, is to take advantage of as much of the structure implicit in the problem domain as possible. The natural notions of domains extending one another, and collecting domains with common properties into categories is certainly a useful computational device. Their domains are created by functions (called functors). This is achieved by starting from some built-in functors (List, Vector, Struct and Union). By a domain (of computation) they mean a set of generic operations, a representation, describing a datatype used to represent the objects of the domain, a set of functions, which implement the operations in terms of the representation and a set of attributes, designating useful facts, such as axioms, for the operations as implemented by the functions. A category finally designates a class of domains with common operations and attributes, but with different functions and representations, i.e. it consists of a set of generic operations and a set of attributes designating facts which must be true for any implementation of the operations. Therefore, assuming to have defined the category of Euclidean domains a complete definition of the gcd function could be written

$$\gcd(x: R, y: R): R \ where \ R: \text{Euclidean Domain} = =$$

$$if \ x = 0 \ then \ y \ else \ \gcd \ (y, \text{remainder} \ (x, y)).$$

Even more convincing examples and details are given in [99].

3. The Symbolic-Numeric Interface

As stated above, one of the implicit objectives of designers of newleft, liberal and catholic systems can be described as "offering facilities for automated engineering mathematics to unknown users". But these general purpose packages thus introduced a communication problem with programming languages tailored for solving numerical problems. Although these problems are defined and described in an equivalent mathematical language, the intention is, however, completely different: production of (inaccurate) numbers, often using subroutine libraries. The last sentences of Rall's recent book on automatic differentiation [140] clearly illustrate this discrepancy: "The discussion up to now has been limited to software which will analyze formulas and produce code for evaluation of derivatives and Taylor coefficients of the functions considered. Software (and software systems) of the symbolic manipulation category is ... far beyond the scope of this book". As discussed in some detail by Brown and Hearn [19] two categories of communication problems can be distinguished: numerical evaluation of symbolic results and hybrid problems. Hybrid problems require a combination of symbolic and numerical techniques for their solution. Even systems like ALTRAN, based on FORTRAN, cannot offer the possibility of writing an efficient program which includes a symbolic computation step in it. FORTRAN is normally a compile, load and go system. Hence generating algebraic expressions during the go step must result in inefficient code, since compilation is impossible. So an approach to solve this communication problem has been to include auxiliary processors which can compile accurate and efficient subroutines for the numerical evaluation of their symbolic output, most of which are listed by Ng [132]. A recent example is VAXTRAN [109], an algebraic-numeric package running on the MACSYMA system under the VAX/VM/UNIX operating system. It allows users to create numeric FORTRAN subroutines, combining equations which have been created and manipulated in the MACSYMA system, templates of FORTRAN code, and FORTRAN library routines. But, as stated by Tobey [161] and Hearn [79] the code produced can be incomprehensible, in view of its length, and also often far from efficient. Tobey [159] and recently van Hulzen [167] advocated facilities for skeletal structure extraction to allow to generate optimized code. Another aspect is the possibility for removal of (numerically) insignificant portions of an expression, as argued by Moler [123]. Recently Smit [152] described how to achieve this in NETFORM [153], a REDUCE-program, designed for the symbolic analysis of electrical networks. Another approach has been to permit real computations to yield results of guaranteed accuracy [19, 47, 138, 147, 101].

4. Polynomial and Rational Function Systems

Many of the early systems in this category are surveyed by Sammet [145]. Tobey comparatively discussed the performance characteristics of some of them [160], i.e. Rom [144], Williams [171], Hartt [76], Gustavson [72] and more specifically Brown et al. [20] and Collins [32]. Among the reasons Tobey mentioned for focussing his attention on a comparison of Brown's ALPAK and Collin's PM were the extremely different approaches to dynamic storage allocation. Many of the early systems were programmed in FORTRAN. The runtime efficiency of FORTRAN is related to its static allocation of storage, thus bounding arrays at compile time. An expanded multivariate polynomial is easily represented by a multi-dimensional array of coefficients. Since, in practice, such polynomials are often sparse, such a dense representation, using exponents implicitly, contradicts the wish to minimize storage. This has lead to dynamic storage allocation schemes, based on a subtle use of arrays and explicit storage of exponents.

A good balance between time and space requirements was obtained by Brown, who used an assembly language to obtain more flexibility in encoding. The ALPAK encoding of a polynomial consists of a datavector of numerical elements (coefficients and exponents), a formatblock (carrying essential information about variables and exponents) and a headerblock (pointing to datavector and for-matblock, and carrying a count on the number of terms). The name of the pointer to the header is the name by which the polynomial is accessed by the user. The datavector elements alternate between coefficients and exponent combinations, such that a canonical form is obtained by ordering exponents in ascending order, subject to a variable order. Although Rom's encoding was similar, he disregarded a canonical ordering. Several pointer-words may point to the same header and it is a system requirement that polynomials which are to be manipulated together share the same formatblock. So, in fact, ALPAK utilizes sequential array storage, assumes a completely expanded form and has finite precision arithmetic. ALPAK provided sequential blocks of storage as needed and data compactification facilities [16]. Accessing of coefficients and exponents is simple; the basic datascan is an indexed loop.

PM, coded in REFCO III [32], a reference count list processing system for the IBM 7094 computer, and also designed by Collins, utilizes a canonical recursive form, i.e. PM treats polynomials as elements of $\mathbb{Z}[X_1][X_2] \cdots [X_n]$ and ALPAK as elements of $\mathbb{Z}[X_1, X_2, \ldots, X_n]$. The PM encoding of polynomials is based on lists with fixed nodes of two 7094 words. The first word contains an across pointer, a reference count and a type field (indicating whether the node is a list element or an atom). The second word contains a down pointer (to lists) or data if it is an atom. These data can be either a variable name and its exponent (thus limiting size of name and exponent) or portions of numerical coefficients. In contrast to ALPAK integers of indefinite length are allowed, by chaining the consecutive portions of these coefficients together. PM has efficient facilities for arbitrary precision arithmetic, since the least significant digits occur as the first element of the list. In addition one must begin with the least significant digits. It is also the best choice for the other arithmetic operations, due to their dependency on addition. PM obtains nodes, if needed, from the available space list, provided by REFCO III. It must trace through (redundant) list structures to find coefficients and exponents, i.e. the basic datascan is a down and/or across scan. The PM-list structures contained extraneous nodes, as recognized by Collins.

The performance of both systems not only depends on their space utilization and the rapidity of extracting coefficients and exponents, but also on inclusion of dynamic storage allocation, algorithm design and arithmetic. Tobey compared their performance in attempting to understand which encoding might be preferable for rational function integration. Basically, algorithms are required for multiplication, division and gcd computation of multivariate polynomials. If algorithms need to be recursive (division) and if unlimited arithmetic is required (gcd computation) it is unlikely that it would be very practical without the use of listprocessing. Tobey thought it to be plausible that the overhead incurred through the use of listprocessing (PM) is more than compensated for by the frequent moving of large blocks of data, required in systems not using it (ALPAK). Although for ordinary processing sequential data packing can be more efficient than liststructures, dynamic storage allocation and a frequent allocation of new arrays can be perturbing factors. These considerations lead him to an analysis of the multiplication algorithms of both ALPAK and PM. His conclusions were that PM will outperform ALPAK in multivariate polynomial multiplication involving 4 or more variables, while the ALPAK-approach is preferable for uni- and bivariate polynomial manipulation without gcd-extraction. This last remark was based on the then implemented facilities. It explains the interest of both Brown and Collins in developing better gcd algorithms.

ALPAK provided polynomial and rational function facilities, together with matrix-algebra and a package for truncated power series. Since ALPAK's arithmetic was very limited Brown tried to conquer coefficient growth in dropping the completely radical approach. In ALTRAN [18, 74], the successor of ALPAK and nicely summarized by Hall [73] and Feldman [48], factored rational expressions were introduced [17, 75], i.e. an algebraic (a polynomial or rational function) got the form $\mu v^{-1} \Pi_j \phi_j^{\varepsilon_j}$, where μ, v are relatively prime monomials, the ϕ_j distinct multinomials and the ε_j nonzero short integers. So an algebraic represents a formal product, being a block containing pointers to each of the factor headers. By syntax, rules are defined to declare the description of the formatblock. Several options allow a user to elect, problem dependent, factored or expanded numerators and/or denominators. Efficiency is gained by a large scale sharing of data, which is notably interesting when translating ALTRAN's internal form into FORTRAN [49]. Although the size of integers is brought under user control it is still limited to atmost 10 computer words, thus sometimes restricting its use [166]. The syntax of ALTRAN is a mixture of PL/I, FORTRAN and unique concepts. Although a FORTRAN based system which can be implemented on any sufficiently large machine supporting ANSI-FORTRAN, it allows recursive functions and sub-routines. To gain efficiency some primitives have to be coded in assembler. Its documentation is excellent, its design careful. However, since about 1975 ALTRAN's status is fixed.

PM's successor is SAC-1, a program system for performing operations on multivariate polynomials and rational functions with infinite precision coefficients. It is completely programmed in ANSI-FORTRAN, with the exception of a few primitives to improve efficiency. Its status by 1971 is summarized by Collins [33], more recent information is given in [34]. SAC-1 is organized as a series of modules, each module being designed to provide a functional capability or a class of capabilities. The first module is a general list processing system, the next deals with infinite precision integer arithmetic, etc. Its facilities cover, besides the usual operations, gcd and resultant calculation, factorization, exact polynomial real zero calculation, partial fraction decomposition, rational function integration, linear algebra, Gaussian integer and Gaussian polynomial operations (including calcu-lation of complex zeros) and facilities for polynomials over a real algebraic number field. Although its datastructures are inherited from PM the encoding is more efficient and tailored to specific modules when ever efficient. It is fair to state that the SAC-1 documentation is the largest collection of carefully designed and described algebraic algorithms, written in a style which easily permits real programming. This style of documentation inspired Collins and Loos in designing ALDES [115]. Currently ALDES is translated by a preprocessor in ANSI-FORTRAN; only direct recursion is automatically removed. Indirect recursion can be eliminated by intrinsic declarations, a restriction which may be lifted as soon as compilers are available. SAC-2, written in ALDES, is being developed to replace SAC-1. The SAC-1 reference counts have been replaced in SAC-2 by a garbage collection list processing system. At present not all capabilities of SAC-1 are available in SAC-2 [35, 21].

5. General Purpose Systems

Although ALTRAN is used to solve a variety of problems [18], its expression class limits its possibilities. And has not been an impetus for the development of SAC-2, via PM and SAC-1, Collins' interest in a quantifier elimination procedure for the first order theory of real closed fields (as nicely summarized by Arnon [2])? Although many systems evolved from computer facilities, designed for specific problem areas, some were developed to serve unknown users, i.e. are intentionally general purpose systems. Remarkable specimen are MATHLAB-68, SYMBAL, FORMAC and SYCOPHANTE, both in facilities and design.

MATHLAB-68 is an online system. It shares only its name with its predecessor the MATHLAB-program [42]. The real goal of its designers was, citing Engelman [43]: "The construction of a prototype system which could function as a mathematical aid to a scientist in his daily work. We would like the program to be so accessible that the scientist would think of it as a tireless slave to perform his routine, mechanical computations... A program intended for such intimate and continual use must obviously be available on line". Requisites are, according to Engelman [42], capabilities for ordinary numerical computations and a wide spectrum of symbolic computations, simple user commands, expandability by the expert and extensibility by the user. MATHLAB-68 is capable of performing differentiation, polynomial factorization, indefinite integration, direct and inverse Laplace transforms and the solution of differential equations with constant (symbolic) coefficients [44, 116]. Although available through DECUS, its capabilities are dated; Engelman reported its "death" in 1971 [44]. However, its design and facilities have been influential for at least IAM [29], MACSYMA, REDUCE, SCRATCHPAD and SIN [125]. MATHLAB-68 was written in LISP for the PDP-6 ITS timesharing system, such that all user data storage was in core. The most recent computed expression is thought of as occupying a construct named "workspace". Everything computed is preserved in a stack called "history of workspace", allowing to retrieve intermediate computations later on. Essential is the "catholic" concept of mixing general purpose simplification facilities with those for special purpose seminumeric RATional SIMPlification in the same system, with both families of transformations having access to the same data. This modular approach lead to the decision to store the data in the history-file in the form demanded as input by the two-dimensional display program, Millen's CHARYBDIS [122]. The general purpose facilities are due for the most part to Korsvold [106] and operate on prefix-expressions, being LISP S-expressions. RATSIMP considers the expressions rational and rewrites them with a single numerator and denominator, expanding all products and performing all available gcd cancellations. The encoding is similar to the SAC-conventions. This recursive structure is efficient for many of the MATHLAB-68 subroutines, like factorization (based on Kronecker's algorithm). MATHLAB-68 has, like ALPAK, finite precision arithmetic. Its gcd approach is as follows: start trying Collins' reduced prs algorithm. If overflow occurs try ALPAK's primitive prs algorithm. If overflow occurs again the user can gamble on the gcd being one, since in many situations claiming this is a fail-safe procedure.

SYMBAL is, like FORMAC, a batch system, belonging to the liberal family. Engeli's design goals for an Algebraic Processor, now known as SYMBAL, were [39] an utmost generality in the source language, high concentration of useful information in memory, while preserving a form suitable for efficiency in computation and economy in the use of space and time. This source language, originally called SYMBAL and described in [40], is an extension of ALGOL-60 using additional features of EULER. The SYMBAL compiler, primarily written in SCALLOP, an ALGOL-60 subset with some extensions to facilitate programming of non-numeric problems, is a one pass compiler, generating assembly language instructions. The complete processor consists of a translator, generating a syllable string for the source language program, an interpreter, reading this string and producing symbolic expressions, in obeying the instructions contained in it, a simplifier, a set of print routines and a collection of built-in standard functions (such as factorial, etc.) An interesting aspect of the source language is the array syntax; arrays are implemented as vectors of vectors, thus allowing to create complicated structures. Both encoding and simplification are remarkable [39]. Engeli used a subexpression list. Its elements are only accessible via an inventory vector. Each entry of this vector contains an across pointer to the subexpression, a reference count, used for garbage collection, and some further information. So,

$$Z := ((a + x)/(a - x) - 1)/(((a + x)/(a - x)) \uparrow (1/2) + 1)$$

is decomposed into

$$e_1 := a + x, \quad e_2 := a - x, \quad e_3 := e_1/e_2 - 1, \quad e_4 := e_1/e_2,$$
$$e_5 := e_4 \uparrow (1/2) + 1, \quad Z := e_3/e_5.$$

This indicates that the subexpression list must contain pointers to relevant inventory vector elements. To make garbage collection efficient the subexpression lists are headed with an erase flag and a lengthfield. This erase-flag allows to reset all erase-flags of list elements, originating from an erasable subexpression. Hence, compactification of non-erasable elements is possible in one pass. The garbage collection is fast, since the length fields allow to skip from element to element directly. The simplifier establishes a canonical form for any subexpression encountered, before it is actually stored. This has disadvantages, as discussed by Neidleman [130], since complete expressions are not stored. Further simplification is under user control, via a quite extensive output repertoire. SYMBAL was originally implemented for the CDC 1604 and the CDC 6000/CYBER series. In 1975 SYMBAL was improved and became also available on the ICL IV and the IBM 360/370-series [41]. Its present capabilities include indefinite-precision rational arithmetic, built-in functions for differentiation and for integration of certain classes of integrands, facilities for Taylor series (truncation is under user control), various trigonometric transformations and ALGOL- and FORTRAN-compatible output.

The basic concepts of FORMAC were developed by Sammet and Tobey in 1962. They felt the need to add a formal algebraic capability to an already existing numerical mathematical language, i.e. FORTRAN. In 1964 the first complete version of this FORTRAN-FORMAC [146] was running on an IBM 7090/94. A later version, PL/I-FORMAC [163, 173] on System/360 can be considered as a slight improvement. IBM decided to stop the FORMAC experiments at January 1, 1971. Although it had only a type III status the system was and is widely used and a variety of interactive versions was developed [165]. The PL/I-FORMAC interpreter consists of two modules of assembled routines, added to a system library: the preprocessor and the object time library. The preprocessor scans the source-program for FORMAC-statements and replaces them by PL/I-subroutine calls. After compilation of the thus rewritten program object time routines from both the FORMAC- and the PL/I-library are added during the link-edit step. The FORMAC-system gets free lists of 4 k bytes when needed. Garbage collection is under user control via an ATOMIZE-command. A SAVE-facility permits to use secondary storage. Some compactification in encoding is possible via a CS (common subexpression) header mechanism. FORMAC-expressions are built from constants (floating point numbers, rationals and integers (upto 2295 decimal digits) and π, e, $\sqrt{-1}$), indeterminates and bounded variables and a variety of built-in and/or user defined functions. FORMAC-routines, accepting these expressions as arguments, can handle simplification, substitution, differentiation, comparison and analysis of their arguments. An interface with PL/I permits to pass arguments between FORMAC and PL/I.

The FORMAC-approach to encoding and AUTomatic SIMplification [162] differs completely from Engeli's SYMBAL-design. Expressions are considered either as sums or as products. Their term(s) and factor(s) can be composite too. Internally expressions have a prefix notation allowing unary, binary and variary operators using a pseudo-canonical form to order operands with respect to the leading operators ($+$, $*$). These chains of operands, once in simplified form, have lexicographically ordered indeterminates, followed by functions (in a predefined order) and constants. A rich variety of AUTSIM rules is applied to achieve this, subject to some user options (EXPAND, for instance). A test problem, the Y_{2n}-functions [22], has been run by Sundblad for six different systems, including SYMBAL and FORMAC [158]. It showed SYMBAL's superiority over FORMAC (for this problem). An analysis of Bahr and Smit [6], using this problem, showed that AUTSIM's organization was poor, especially in lexicographical scans. Their improvements gave a 2.5 increase in execution speed. Their approach to measurement, an invocation count mechanism, proved to be a powerful tool for analyzing the performance of a system. Smit also used it for testing alternative algorithms for computing determinants of sparse matrices, as needed in NETFORM [153], a system born as an interactive FORMAC-facility [150].

Other improvements of Bahr [3, 4] also contributed to the new FORMAC-73 system [5]. However, advanced polynomial algebra facilities, such as factorization and integration, are not included in this latest version. It is doubtful if algorithms for factorization, requiring recursivity polynomial operations, can be implemented efficiently at all in systems like FORMAC and SYMBAL. But a renewed comparison of SYMBAL and FORMAC-73 might be interesting, albeit that such comparisons are rough instruments. For instance, the negative influence of the FORMAC integer representation, the reverse of the SAC-1 encoding, is strongly problem dependent.

The SYCOPHANTE-system, still in a more or less experimental stage, is also an intended general purpose system. As stated by Bergman [13] its design philosophy is based on unification and rewrite rules. The host language for this system, PROLOG, is a practical tool for "logical programming", based on predicate logic. The use of algebraic abstract data types for extensibility is currently being studied. For example, this interactive system provides sophisticated commands for generalized substitutions and pattern matching [102]. Its facilities include symbolic integration, similar to those provided by SIN [125], however, with some possibilities for limit calculations. Its modularity allows implementation on a mini-computer.

6. Special Purpose Systems

A number of systems originated from attempts to mechanize tedious, often straightforward, algebraic calculations, occurring in problems in CM, GR and QED. Some of these systems are grown out to general purpose facilities. Problems in CM belong to a restricted set of a rather well defined type, expressible in terms of Poisson series. As also argued by Jefferys [91] a canonical form can easily be provided. To obtain maximal cancellation it is best to store the polynomials P_i, occurring in such a (Fourier or) Poisson series

$$\Sigma_i P_i \, {\cos \atop \sin} \, (\Sigma_j \alpha_{ij} \phi_j),$$

where the α_{ij} are integers and the ϕ_j are angles, in expanded form, i.e. a radical ALPAK-like approach is profitable for a CM-system. But if the polynomial coefficients are floating point numbers or unlimited rationals some minor changes in encoding are profitable: lists of datablocks, each describing a term of a series and eventually pointing to lists representing these numbers. This indicates that systems like FORMAC and SYMBAL are wrong choices, in terms of time and space. Jefferys [91] surveys a number of CM-systems. Well known are his TRIGMAN [90, 92] and CAMAL [8, 9, 15].

TRIGMAN, written in FORTRAN for the CDC-6000 series, and originally only intended for Poisson-series calculations, can now be considered as a general purpose system. The major portion of the syntax is FORTRAN; a SNOBOL-preprocessor converts the source program in a sequence of subroutine calls. The TRIGMAN facilities include, besides the above indicated polynomial, Fourier and Poisson series capabilities, built-in functions for Fourier analysis. Truncation level and trigonometric expansion are under user control. FORTRAN- or ALGOL-compatible output is possible.

The CAMbridge ALgebra system has grown from a number of needs: CM [7, 14], and GR [51]. CAMAL is small, fast and powerful for solving the kind of problems it originated from. But one may consider it as a general purpose system, since it knows many applications from other disciplines [56]. This approach explains its structure and language design. Citing Fitch [54]: "We have hoped that getting answers overcame the contortions of programming, so we chose a simple autocode language which was easy to learn and use, with no complications like scope rules. The language was also designed to be easily changed, to allow new features to be added whenever necessary". In 1975, for instance Hiemstra [86] made a pre-editor, permitting recursive CAMAL-function procedures with arguments and local variables. The present version [53], written in BCPL [142], only allows the use of vectors, albeit of either type. A new version is in preparation and will include up to 4-dimensional arrays. CAMAL consists of a number of self-contained modules. Polynomials are transformed by a polynomial module. If one needs sines or other functions of polynomials a function module can be applied, which uses the polynomial module via an interface, protecting it from knowing about the polynomial structures. At present these function modules are three packaged systems, called F (Fourier series), E (complex exponential series) and H (a historic relic, standing for "Hump", a general purpose package). The polynomial structures are similar to those suggested above for TRIGMAN.

The purpose of using computer algebra in GR is to obtain a classification of geometries, described by field equations, in general a set of ten second order partial differential equations in ten unknown functions. The algebra involved is tedious [10]. CAMAL's underlying polynomial structures make this system efficient in this area. Different systems have been used for research in GR (see Fitch [56] for references to comparative studies). Important to mention is the LAM-family of systems, all of which are mainly intended for calculations in GR. The oldest member is ALAM, a LISP based system for an ATLAS-machine, written by d'Inverno, who also made a simplified version, entirely in LISP, called LAM. Russel-Clark implemented a much improved version, CLAM, for the CDC-6600-series; it was written in machine code. SHEEP, running on a PDP-10 and the latest member of the family, consists of three machine coded packages, mainly made by Frick [60, 61]. It is based on the basic idea's of LAM and borrowed some concepts from CLAM and REDUCE. The basic package of SHEEP consists of routines for I/O, simplification, substitution and differentiation. On top of this is a package of more user oriented commands, including routines to manipulate tensor components. Then on top of these facilities two application packages are made to compute the Einstein-tensor in different frames: CORD (for a coordinate frame) and FRAME (for a moving frame). Since no real need exists for polynomial division it is not implemented in SHEEP; but facilities are available to do such operations in REDUCE. Reasons for excluding it are implicitly given, when discussing ALPAK and PM. Hörnfeldt [87] made a package on top of SHEEP which can treat the data type "indicial formula", representing expressions containing tensors or spinors with symbolic indices. Recently Åman [1] made another extension, written in Standard LISP [117], to allow classification of many of the known vacuum metrics.

Another aspect of GR is the solution of the field equations. The only generally useful alternative for a numerical approach is likely to be some form of approximation technique or perturbation theory. This brings us to another aspect of CAMAL, since for approximation the simplification problem ceases to be dominant and characteristics are similar to those encountered in CM. CAMAL provides a flexible automatic series truncation capability, attractive for repeated approximation techniques [55]. This proved to be a useful instrument for the automatic derivation of periodic solutions to a class of weakly nonlinear differential equations, such as provided by ALKAHEST II [58], a program written in LISP, based on an earlier CAMAL-version, which generates programs, i.e. algorithms for solving such problems in a variety of languages (at present including CAMAL and REDUCE) by mimicking the actions of a human mathematician in solving equations. In relation to this we ought to mention a package called TAYLOR, made by Barton, Willers and Zahar [12] using a special purpose compiler based on CAMAL. It generates FORTRAN programs to solve, numerically, one-dimensional ordinary differential equations via the classical Taylor-series method. Norman and Willers give later extensions [134, 170].

The most widely used systems for doing calculations in QED are REDUCE and Veltman's SCHOONSCHIP [156, 157] originally written in COMPASS for the CDC 6000 and 7000 series. A nice specimen of a SCHOONSCHIP application is [143]. An IBM-version, based on PL/I, is made by Strubbe. A third system of some interest is ASHMEDAI, surveyed in [111, 112]. REDUCE, written in LISP, was originally designed as a batch system for solving QED problems [77]. It used a distributive polynomial encoding, i.e. expressions were considered as sums of products of so-called kernels and constants. The 4-types of kernels, generalized variables, which were introduced smoothed algorithm description in QED. In later, interactive, versions [78, 81] this ALPAK-like encoding was replaced by the recursive PM-representation. All expressions were internally represented as so-called standard quotients [135]. The kernel concept was generalized, thus allowing arbitrary, eventually not yet specified, user defined functions. Expansion came under user control. Hence REDUCE entered the new left family. But these general purpose facilities had their price, as remarked by Åman and Karlhede [1]: REDUCE is 10 to 20 times slower than SHEEP for doing the same calculation (in GR). In 1976 ideas were published for a new MODE-REDUCE [83, 71]. A remarkable gcd algorithm [85] and factored rational expression facilities, following Brown's ALTRAN approach, [84] were published for this experimental version. REDUCE has a varied repertoire of substitution and pattern defining capabilities, close in spirit to those discussed by Yun and Stoutemyer [174], facilities for matrix algebra and high energy physics and a number of flags for controlling output style including one for FORTRAN-compatible output. The system consists of a hierarchy of modules: on top is the so-called algebraic mode with a simple ALGOL-like user language with some LISP reminiscences. It is implemented in Standard LISP [117], a pleasantly structured subset of LISP. This Standard LISP proved to be a powerful tool to provide portability. At present over 500 copies of REDUCE are distributed. However, the documentation is not very satisfactory [82]. At present a new manual is in preparation and a new updated version will become available for the IBM 360/370 series, the PDP 10/20 series and the CDC 7600. It will include a number of new packages for integration [57], for polynomial factorization and gcd calculation [124], for solving ordinary differential equations [169] and integral equations [154], for arbitrary precision floating point arithmetic [147] and for network analysis problems [153], including determinant calculations especially for sparse matrices.

Recently SMP [30, 31] was presented, developed by Cole, Wolfram et al. Their primary motivation for the construction of the program was the necessity of performing very complicated algebraic manipulations in certain areas of theoretical physics. They considered SCHOONSCHIP as the only system designed to handle the very large expressions encountered and MACSYMA as the only one of any generality. Their system, thus motivated, although still more or less experimental, is remarkable.

7. Giants and Gnomes

Both MACSYMA and SCRATCHPAD are catholic systems. However, there are many noticeable differences in their design and development as well as on their impact and influence on computer algebra. Implementation of Project MAC's SYmbolic MAnipulator, written in MACLISP, began in July 1969. It has been multipled in size since the first paper describing it [120] appeared. The original design decisions, made by Engelman, Martin and Moses, were based on their earlier experiences [44, 118, 125]. "The system was", citing Moses [128], "intended to be useful to a wide variety of users without losing much efficiency in running time and working storage space. The emphasis of the design decisions for MACSYMA were on ease of time shared interaction with batch operation available, and assuming that users with relatively small problems wanted a great deal of built-in machinery, while users with large problems were presumed willing to spend more time in optimizing their programs in order to achieve space and time efficiencies". MACSYMA therefore utilizes different internal representations. The default general representation offers great flexibility and is quite useful in interactive situations since the internal form of the expressions is quite close to the displayed form and the user's input. Other representations are, for instance, rational, power series and Poisson series. By 1969 many of the basic algorithms used in computer algebra had still a lack of generality or basic inefficiencies (MATHLAB-68's factorization for instance). The MACSYMA group has successfully concentrated on improving these algorithms and on a continued implementation of a wide variety of new features. In May 1972 the MACSYMA system was made available over ARPA network. This certainly influenced and stimulated this trend. The proceedings of both MACSYMA user's conferences clearly demonstrate how influential this system is on research and engineering in many areas. A drawback of this rapid growth might be that the published algorithm descriptions are less balanced and detailed as those provided by the SAC-group. As admitted by Moses [128] an effect of the large size and generality of the MACSYMA system has been the sizable number of bugs due to the interactions between modules. The present capabilities of MACSYMA are shortly summarized in [174]. The latest version of its manual [121] is from 1977 and will soon be updated again. MACSYMA is implemented for the PDP-10 using the ITS operating system and for the Honeywell 60 series 68/80 using the MULTICS operating system. Fateman made it available on DEC's VAX-11 [59].

In contrast to MACSYMA the SCRATCHPAD-system has never known a wide user's community; in fact, this interactive system, is not used outside IBM. It was originally implemented in LISP, using an experimental System/360 LISP system [68]. One of the major features of this system was its provision for accessing a sizable number of LISP compiled and assembled programs. This enabled to incorporate significant portions of CHARIBDIS, MATHLAB, REDUCE, SIN, Korsvold's simplification system and Martin's Symbolic Mathematical Laboratory [118]. At present it is running on an IBM 370/168. Its capabilities are described in [174, 69, 89]. An extensible language approach was adopted in the design of SCRATCHPAD's user language. The initial user language, the base language, contains a set of basic syntactic constructs, described by notations resembling those in conventional mathematics. These basic constructs may be extended by an individual user. The objectives in the design and implementation of the evaluator (of the user commands) were to achieve efficiency (like REDUCE 2) and yet provide generality and understandability (like FAMOUS). The design of this evaluator is built upon the design of "replacement rules" [95] and provides a basis for the manipulation of not only algebraic expressions, but also inequalities and logical expressions. In fact SCRATCHPAD is a very-high-level non procedural language. This in contrast with MACSYMA, which has the more traditional approach of recursive functions or subroutines with arguments and local variables. To accomplish the objectives of syntax extensions, i.e. user chosen specialized notations on line, a translator writing system, META/LISP [93], has been developed, together with an extendable language, META/PLUS [94]. Through META/LISP, input translators for several languages, resident in the system, have been produced. The META/LISP facility enables sophisticated users to interactively modify existing input translators. From this long experience in language design evolved MODLISP [37], currently being implemented above LISP/370 [89, 172] to facilitate development and experiments with the earlier mentioned language for computational algebra [99].

Turning to gnomes we start mentioning the first published microcomputer algebra system muMATH-79, which was developed by Rich and Stoutemyer [141, 155]. Versions are available for INTEL 80, INTEL 85 and ZILOG-80 based micro computers. The system is written in a high-level Structured IMPlementation language muSIMP-79, which evolved from muLISP-77, developed by Rich for his IMSAI 8080 computer. The muSIMP syntax is partly ALGOL, partly unique, with LISP-influence in the semantics. The capabilities of muMATH-79 are listed in [174], and resemble those of REDUCE and MACSYMA in a more limited setting. There are facilities for explicit saving and restoring expressions to and from secondary storage. Facilities are available for redefining and extending the syntax by introducing new prefix, infix and postfix operators together with their precedence properties. At present MAPLE, a future competitor for muMATH-79 is in development at the University of Waterloo by W. M. Gentleman's group. Without doubt, this is only the second member of a growing group of micro computer algebra systems.

8. Some Conclusions

It is interesting to read a proposed "taxonomy for algebraic computation" [70] and J. Neubüser's comment on it [131] after studying the efforts of Jenks et al. in developing a language for computational algebra. Cannon [24] calls the systems reviewed in the previous sections Classical Algebra systems to contrast them with Modern Algebra systems. These MA-systems ought to enable a user to compute in a wide range of different structures. Are Jenks et al. creating a home for CAYLEY [23, 24], being an MA-system primarily designed for computing with discrete groups? And what should one think of CAMAC [139, 110], of which the original goal was to be able to handle groups and combinatorial structures in a unified way? The wishful thinking of Estabrook and Wahlquist [46] is, in this context, also of interest. "The arsenal we use consists of ideals of forms, vectors, the operations of exterior product, of exterior differentiation, of Lie-derivation and the various identities relating these. It is a fine set of weapons... These operations are so systematic and algorithmic that we even have hopes that it will be useful to analysts to have them implemented as part of the new symbol manipulation schemes for high speed computers". Recently Gragert [65, 66] used REDUCE to implement abstract Lie-algebra and the algebra of differential forms together with exterior differentiation for research in prolongation theory. These facilities were used by Kersten [103] to undertake the kind of work advocated by Estabrook [45], i.e. computation of infinitesimal symmetries for (extended) vacuum Maxwell equations. These are facilities, citing Barton and Fitch [10], which "force a user to adopt ad hoc techniques", i.e. the use of a "polynomial-algebra interface" to accomplish computations in a user domain of a less complicated algebraic type. These indications might underline the importance of the work of Jenks et al. in assisting to develop a unified view of computer algebra.

If, according to Jenks et al. the category "Field" designates a class of domains ("fields") with common operations and attributes but with different functions and representations, one can wonder if R have to be excluded. Assuming the representation of elements of R is a pair ([multiple precision] floating point number, [relative or absolute] error) how to implement functions to control the accuracy of the operations, subject to which attributes? How instrumental is the recently published work of Kulisch and Miranker on computer arithmetic [107] in this context? This might be an important research area as to improve the usability of computer algebra for numerical mathematicians.

The history of MACSYMA and REDUCE shows that a large users community is fertile but burdening to those who have to maintain and update, if not upgrade, a system. The publication tradition for numerical algorithms and the standardization of ALGOL-60 and FORTRAN made subroutine libraries possible. The variety of user- and implementation languages in computer algebra is not necessarily a handicap in exchanging new algorithms and packages, as indicated by Loos' experiments with Jenks' META/LISP [113]. More important, if not crucial, is standardization of terminology and algorithm description. The dedication of Collins and his group has shown that a careful description helps to debug and almost guarantees validation too. Such "management of computer algebra" is of tremendous importance in view of the expected impact of micro computer algebra systems. J. Moses' banquet address during the SYMSAC'81 conference was even entitled: "Algebraic Computation for the Masses" [129]. In this respect Kung's suggestions about the possible use of VLSI in computer algebra [108] are also confirmative. He stressed the importance of design, i.e. the definition and understanding of the different tasks, in view of concurrency and communication.

References

[1] Åman, J., Karlhede, H.: An Algorithmic Classification of Geometries in General Relativity. SYMSAC **1981**, 79 – 84.

[2] Arnon, D. S.: Algorithms for the Geometry of Semi-Algebraic Sets. Ph.D. Thesis, Univ. of Wisconsin, Madison, 1981.

[3] Bahr, K. A.: Utilizing the FORMAC Novelties. SIGSAM **9**/1, 21 – 24 (1975).

[4] Bahr, K. A.: Basic Algorithms in FORMAC: Design and Verification. SYMSAC **1976**, 189 – 197.

[5] Bahr, K. A.: FORMAC 73 User's Manual. Darmstadt: GMD/IFV.

[6] Bahr, K. A., Smit, J.: Tuning an Algebraic Manipulation System through Measurements. EUROSAM **1974**, 17−23.

[7] Barton, D.: A New Approach to the Lunar Theory. Ph.D. Thesis, University of Cambridge, 1966.

[8] Barton, D., Bourne, S. R., Fitch, J. P.: An Algebra System. Comput. J. **13**, 32−39 (1970).

[9] Barton, D., Bourne, S. R., Horton, J. R.: The Structure of the Cambridge Algebra System. Comput. J. **13**, 243−247 (1970).

[10] Barton, D., Fitch, J. P.: General Relativity and the Application of Algebraic Manipulation Systems. SYMSAM **1971**, 542−547.

[11] Barton, D., Fitch, J. P.: A Review of Algebraic Manipulation Programs and Their Application. Comput. J. **15**, 362−381 (1972).

[12] Barton, D., Willers, I. M., Zahar, R. V. M.: The Automatic Solution of Systems of Ordinary Differential Equations by the Method of Taylor Series. In: Proceedings Mathematical Software (Rice, J., ed.), pp. 369−390. New York: Academic Press 1971.

[13] Bergman, M.: The SYCOPHANTE System. CALSYF **1**, 1981 (unpublished Bull.). (Bergman, M., Calmet, J., eds.)

[14] Bourne, S. R.: Automatic Algebraic Manipulation and its Application to the Lunar Theory. Ph.D. Thesis, Univ. of Cambridge, 1970.

[15] Bourne, S. R., Horton, J. R.: The Design of the Cambridge Algebra System. SYMSAM **1971**, 134−143.

[16] Brown, W. S.: An Operating Environment for Dynamic-Recursive Computer Programming Systems. Commun. ACM **8**, 371−377 (1965).

[17] Brown, W. S.: On Computing with Factored Rational Expressions. EUROSAM **1974**, 26−34.

[18] Brown, W. S.: ALTRAN User's Manual, 4th ed. Murray Hill, N. J.: Bell Laboratories 1977.

[19] Brown, W. S., Hearn, A. C.: Application of Symbolic Mathematical Computations. Comp. Phys. Comm. **17**, 207−215 (1979).

[20] Brown, W. S., Tague, B. A., Hyde, J. P.: The ALPAK System for Numerical Algebra on a Digital Computer. Bell Syst. Tech. J. **42**, 2081−2119 (1963); **43**, 785−804 (1964); **43**, 1547−1562 (1964).

[21] Calmet, J.: A User's Presentation of SAC2/ALDES. CALSYF **1**, 1981 (unpublished Bull.). (Bergman, M., Calmet, J., eds.)

[22] Campbell, J.: Problem #2 − The Y_{2n} Functions. SIGSAM Bull. **22**, 8−9 (1972).

[23] Cannon, J. J.: A Draft Description of the Group Theory Language CAYLEY. SYMSAC **1976**, 66−84.

[24] Cannon, J. J.: The Basis of a Computer System for Modern Algebra. SYMSAC **1981**, 1−5.

[25] Caviness, B. F.: SAM Course Outlines. SIGSAM Bull. **8**/4, 15−25 (1974).

[26] Caviness, B. F., Collins, G. E.: Symbolic Mathematical Computation in a Ph.D. Computer Science Program. SIGSAM Bull. **23**, 25−28 (1972).

[27] Christensen, C.: On the Implementation of AMBIT, a Language for Symbol Manipulation. Commun. ACM **9**, 570−573 (1966).

[28] Christensen, C.: An Introduction to AMBIT/L, a Diagrammatic Language for Listprocessing. SYMSAM **1971**, 248−260.

[29] Christensen, C., Karr, M.: IAM, a System for Interactive Algebraic Manipulation. SYMSAM **1971**, 115−127.

[30] Cole, C. A., Wolfram, S.: SMP−A Symbolic Manipulation Program. SYMSAC **1981**, 20−22.

[31] Cole, C. A., Wolfram, S., et al.: SMP-Handbook, Version 1. Cal. Inst. of Techn. 1981.

[32] Collins, G. E.: PM, a System for Polynomial Manipulation. Commun. ACM **9**, 578−589 (1966).

[33] Collins, G. E.: The SAC-1 System: An Introduction and Survey. SYMSAM **1971**, 144−152.

[34] Collins, G. E.: SAC-1 Availability Notice. SIGSAM Bull. **10**/2, 14−15 (1976).

[35] Collins, G. E.: ALDES and SAC-2 Now Available. SIGSAM Bull. **14**/2, 19 (1980).

[36] Davenport, J.: Effective Mathematics: The Computer Algebra View Point. In: Constructive Mathematics (Richman, F., ed.). Lecture Notes in Mathematics, Vol. 873, pp. 31−43. Berlin-Heidelberg-New York: Springer 1981.

[37] Davenport, J., Jenks, R. D.: MODLISP. SIGSAM Bull. **15**/1, 11−20 (1981).

[38] Drouffe, J. M.: AMP User's Manual, Version 6. Gif-sur-Yvette, CEN Saclay 1981.

[39] Engeli, M. E.: A Language and Liststructure for an Algebraic Manipulation System. IFIP **1966**, 103–115.
[40] Engeli, M. E.: Formula Manipulation – the User's Point of View. In: Advances in Information Systems Science (Tou, J. T., ed.), Vol. 1, pp. 117–171. New York: Plenum Press 1969.
[41] Engeli, M. E.: An Enhanced SYMBAL System. SIGSAM Bull. **9**/4, 21–29 (1975).
[42] Engelman, C.: MATHLAB: A Program for On-Line Machine Assistance in Symbolic Computations. Spartan Books **1**, 413–421. Washington, D. C.: 1965.
[43] Engelman, C.: MATHLAB-68. Proc. IFIP 68, pp. 462–467. Amsterdam: North-Holland 1969.
[44] Engelman, C.: The Legacy of MATHLAB-68. SYMSAM **1971**, 29–41.
[45] Estabrook, F. B.: Differential Geometry as a Tool for Applied Mathematics. Proc. Scheveningen Conf. (Martini, R., ed.). Lecture Notes in Mathematics, Vol. 810, pp. 1–22. Berlin-Heidelberg-New York: Springer 1980.
[46] Estabrook, F. B., Wahlquist, H. D.: Prolongation Structures, Connection Theory and Bäcklund Transformation. In: Non Linear Evolution Equations Solvable by the Spectral Transform (Calogero, F., ed.). Research Notes Mathematics, Vol. 26, pp. 64–83. San Francisco-London-Melbourne: Pitman 1978.
[47] Fateman, R. J.: The MACSYMA "Big Floating-Point" Arithmetic System. SYMSAC **1976**, 209–213.
[48] Feldman, S. I.: A Brief Description of ALTRAN. SIGSAM Bull. **9**/4, 12–20 (1975).
[49] Feldman, S. I., Ho, J.: A Rational Evaluation Package. Murray Hill, N. J.: Bell Lab.'s C. S. Tech. Report **34** (1975).
[50] Fenichel, J.: An On-Line System for Algebraic Manipulation, Ph.D. Thesis, Harvard Univ. Cambridge, Mass., 1966.
[51] Fitch, J. P.: An Algebraic Manipulator. Ph.D. Thesis, Univ. of Cambridge, 1971.
[52] Fitch, J. P.: Course Notes, SIGSAM **9**/3, 4–8 (1975).
[53] Fitch, J. P.: CAMAL User's Manual. Univ. of Cambridge, Comput. Lab., 1975.
[54] Fitch, J. P.: The Cambridge Algebra System – An Overview. Proc. SEAS Anniversary Meeting, Dublin 1975.
[55] Fitch, J. P.: Mechanizing the Solution of Perturbation Problems. MAXIMIN **1977**, 93–98.
[56] Fitch, J. P.: The Application of Symbolic Algebra in Physics – A Case of Creeping Flow. EUROSAM **1979**, 30–41.
[57] Fitch, J. P.: User Based Integration Software. SYMSAC **1981**, 245–248.
[58] Fitch, J. P., Norman, A. C., Moore, M. A.: The Automatic Derivation of Periodic Solutions to a Class of Weakly Nonlinear Differential Equations. SYMSAC **1981**, 239–244.
[59] Foderaro, J. K., Fateman, R. J.: Characterization of VAX MACSYMA. SYMSAC **1981**, 14–19.
[60] Frick, I.: The Computer Algebra System SHEEP, What It Can and Cannot Do in General Relativity. Univ. of Stockholm: USIP Report **77**-14, 1977.
[61] Frick, I.: SHEEP User's Manual. Univ. of Stockholm 1977.
[62] Gawlik, H. J.: MIRA. SIGSAM Bull. **26**, 28–32 (1973).
[63] Gawlik, H. J.: The Further Development of MIRA. SIGSAM Bull. **11**/2, 22–28 (1977).
[64] Glushkov, V. M., et al.: ANALITIK-74. Kibernetika **5**, 114–147 (1978) (in Russian).
[65] Gragert, P.: Algebraic Operator, a Powerful Feature of REDUCE and Its Application in Non Commutative Algebras (Abstract). SIGSAM Bull. **14**/4, 18 (1980).
[66] Gragert, P.: Symbolic Computations in Prolongation Theory. Ph.D. Thesis, Twente Univ. of Technology, 1981.
[67] Griesmer, J. H.: William A. Martin, Some Personal Reflections. SIGSAM Bull. **15**/1, 1 (1981).
[68] Griesmer, J. H., Jenks, R. D.: SCRATCHPAD/1, an Interactive Facility for Symbolic Mathematics. SYMSAM **1971**, 42–58.
[69] Griesmer, J. H., Jenks, R. D., Yun, D. Y. Y.: SCRATCHPAD User's Manual. Yorktown Heights: IBM Research, Report RA **70**, 1975.
[70] Griesmer, J. H., Jenks, R. D., Yun, D. Y. Y.: A Taxonomy for Algebraic Computation. SIGSAM Bull. **12**/1, 25–28 (1978).
[71] Griss, M. L.: The Definition and Use of Datastructures in REDUCE. SYMSAC **1976**, 53–59.
[72] Gustavson, F. G.: On Constructing Formal Integrals of a Hamiltonian System Near an Equilibrium Point. Am. Astronomial Society Space Flight Mechanics Specialist Conf. Univ. of Denver: 1966.

[73] Hall, A. D.: The ALTRAN System for Rational Function Manipulation – A Survey. SYMSAM **1971**, 153 – 157, and Commun. ACM **14**, 517 – 521 (1971).

[74] Hall, A. D.: ALTRAN Installation and Maintenance. Murray Hill, N. J.: Bell Lab.'s 1972.

[75] Hall, A. D.: Factored Rational Expressions in ALTRAN. EUROSAM **1974**, 35 – 45.

[76] Hartt, K.: Some Analytic Procedures for Computers and their Applications to a Class of Multidimensional Integrals. J. ACM **4**, 416 – 421 (1964).

[77] Hearn, A. C.: Computation of Algebraic Properties of Elementary Particle Reactions Using a Digital Computer. Commun. ACM **9**, 573 – 577 (1966).

[78] Hearn, A. C.: REDUCE: A User Oriented Interactive System for Algebraic Simplification. In: Interactive Systems for Experimental Applied Mathematics (Klerer, M., Reinfeldt, J., eds.), pp. 79 – 90. New York-London: Academic Press 1968.

[79] Hearn, A. C.: The Problem of Substitution. Proc. of the 1968 Summer Inst. on Symb. Math. Comp. (Tobey, R. G., ed.), pp. 3 – 20. Cambridge, Mass.: IBM 1969.

[80] Hearn, A. C.: Applications of Symbol Manipulation in Theoretical Physics. SYMSAM **1971**, 17 – 21 and Common. ACM **14**, 511 – 516 (1971).

[81] Hearn, A. C.: REDUCE 2: A System and Language for Algebraic Manipulation. SYMSAM **1971**, 128 – 133.

[82] Hearn, A. C.: REDUCE User's Manual. Univ. of Utah: Report UCP-**19**, 1973.

[83] Hearn, A. C.: A New REDUCE Model for Algebraic Simplification. SYMSAC **1976**, 46 – 52.

[84] Hearn, A. C.: An Improved Factored Polynomial Representation, Proc. Hawaii Int. Conf. System Sci. **1977**, 155.

[85] Hearn, A. C.: Non Modular Computation of Polynomial GCDs Using Trial Division. EUROSAM **1979**, 227 – 239.

[86] Hiemstra, B.: A Pre-Editor for CAMAL. SIGSAM Bull. **9**/2, 30 – 34 (1975).

[87] Hörnfeldt, L.: A System for Automatic Generation of Tensor Algorithms and Indicial Tensor Calculus, Including Substitution of Sums. EUROSAM **1979**, 279 – 290.

[88] Husberg, N., Seppänen, J.: ANALITIK: Principle Features of the Language and Its Implementation. EUROSAM **1974**, 24 – 25.

[89] IBM: LISP/370, Program Description/Operations Manual. Doc. SH 20-2076-0. White Plains, N.Y.: IBM 1978.

[90] Jefferys, W. H.: FORTRAN-Based Listprocessor for Poisson-Series. Celest. Mech. **2**, 474 – 480 (1970).

[91] Jefferys, W. H.: Automated Algebraic Manipulation in Celestial Mechanics. SYMSAM **1971**, 328 – 331, and Commun. ACM **14**, 538 – 541 (1971).

[92] Jefferys, W. H.: A Precompiler for the Formula Manipulation System TRIGMAN. Celest. Mech. **6**, 117 – 124 (1972).

[93] Jenks, R. D.: META/LISP: An Interactive Translator Writing System. IBM Research Rep. RC 2968. Yorktown Heigths, N.Y.: 1970.

[94] Jenks, R. D.: META/PLUS: The Syntax Extension Facility for SCRATCHPAD. IFIP **1971**, Booklet TA-3, 61 – 63. Amsterdam: North-Holland 1971.

[95] Jenks, R. D.: The SCRATCHPAD-Language. SIGSAM Bull. **8**/2, 16 – 26 (1974).

[96] Jenks, R. D.: Course Outline, Yale University, New Haven. SIGSAM Bull. **9**/3, 9 – 10 (1975).

[97] Jenks, R. D.: Reflections of a Language Design. SIGSAM Bull. **13**/1, 16 – 26 (1979).

[98] Jenks, R. D.: MODLISP – An Introduction. EUROSAM **1979**, 466 – 480.

[99] Jenks, R. D., Trager, B. M.: A Language for Computational Algebra. SYMSAC **1981**, 6 – 13.

[100] Kahrimanian, H. G.: Analytic Differentiation by a Digital Computer. MA Thesis, Temple Univ. Phil., PA., 1953.

[101] Kanada, Y., Sasaki, T.: LISP Based "Big-Float" System is Not Slow. SIGSAM Bull. **15**/2, 13 – 19 (1981).

[102] Kanoui, H., Bergman, M.: Generalized Substitutions. MAXIMIN **1977**, 44 – 55.

[103] Kersten, P. H. M.: The Computation of the Infinitesimal Symmetries for Vacuum Maxwell Equations and Extended Vacuum Maxwell Equations. Twente University of Technology: TW memorandum **365** (1981).

[104] Korpela, J.: General Characteristics of the ANALITIK Language. SIGSAM Bull. **10**/3, 30 – 48 (1976).

[105] Korpela, J.: On the MIR-Series of Computers and Their Utilization for Analytic Calculations. MAXIMIN **1977**, 80 – 91.

[106] Korsvold, K.: An On-Line Algebraic Simplify Program. Stanford Univ.: Art. Int. Project Memorandum **37** (1965) and Commun. ACM **9**, 553 (1966).

[107] Kulisch, U. W., Miranker, W. L.: Computer Arithmetic in Theory and Practice. New York-London: Academic Press 1981.

[108] Kung, H. T.: Use of VLSI in Algebraic Computation. SYMSAC **1981**, 218 – 222.

[109] Lanam, D. H.: An Algebraic Front-End for the Production and Use of Numerical Programs. SYMSAC **1981**, 223 – 227.

[110] Leon, J. S., Pless, V.: CAMAC 79. EUROSAM **1979**, 249 – 257.

[111] Levine, M. J., Roskies, R.: ASHMEDAI and a Large Algebraic Problem. SYMSAC **1976**, 359 – 364.

[112] Levine, M. J., Roskies, R.: ASHMEDAI. MAXIMIN **1977**, 70 – 92.

[113] Loos, R.: Algebraic Algorithm Descriptions as Programs. SIGSAM Bull. **23**, 16 – 24 (1972).

[114] Loos, R.: Toward a Formal Implementation of Computer Algebra. EUROSAM **1974**, 1 – 8.

[115] Loos, R.: The Algorithmic Description Language ALDES (Report). SIGSAM Bull. **10**/1, 15 – 39 (1976).

[116] Manove, M., Bloom, S., Engelman, C.: Rational Functions in MATHLAB. IFIP **1966**, 86 – 102.

[117] Marti, J., et al.: Standard LISP Report. SIGSAM Bull. **14**/1, 23 – 43 (1980).

[118] Martin, W. A.: Symbolic Mathematical Laboratory. Ph.D. Thesis, M.I.T., Cambridge, Mass., Report MAC-TR-36, 1967.

[119] Martin, W. A.: Computer Input/Output of Mathematical Expressions. SYMSAM **1971**, 78 – 89.

[120] Martin, W. A., Fateman, R. J.: The MACSYMA-System. SYMSAM **1971**, 59 – 75.

[121] MATHLAB Group: MACSYMA Reference Manual. M.I.T., Cambridge, Mass.: Lab. of Comput. Sci. 1977.

[122] Millen, J. K.: CHARIBDIS: A LISP Program to Display Mathematical Expressions on Type Writer Like Devices. In: Interactive Systems for Experimental Applied Mathematics (Klerer, M., Reinfelds, J., eds.), pp. 155 – 163. New York-London: Academic Press 1968.

[123] Moler, C. B.: Semi-Symbolic Methods in Partial Differential Equations. SYMSAM **1971**, 349 – 351.

[124] Moore, P. M. A., Norman, A. C.: Implementing a Polynomial Factorization and GCD Package. SYMSAC **1981**, 109 – 116.

[125] Moses, J.: Symbolic Integration. Ph.D. Thesis, Math. Dept., M.I.T., Cambridge, Mass., 1967.

[126] Moses, J.: Symbolic Integration: The Stormy Decade. SYMSAM **1971**, 427 – 440, and Commun. ACM **14**, 548 – 560 (1971).

[127] Moses, J.: Algebraic Simplification: A Guide for the Perplexed. SYMSAM **1971**, 282 – 304, and Commun. ACM **14**, 572 – 537 (1971).

[128] Moses, J.: MACSYMA – the Fifth Year. EUROSAM **1974**, 105 – 110.

[129] Moses, J.: Algebraic Computation for the Masses. SIGSAM Bull. **15**/3, 4 – 8 (1981).

[130] Neidleman, L. D.: A User Examination of the Formula Manipulation Language SYMBAL. SIGSAM Bull. **20**, 8 – 24 (1971).

[131] Neubüser, J.: Some Remarks on a Proposed Taxonomy for Algebraic Computation. SIGSAM Bull. **14**/1, 19 – 20 (1980).

[132] Ng, E. W.: Symbolic-Numeric Interface: A Review. EUROSAM **1979**, 330 – 345.

[133] Nolan, J.: Analytic Differentiation on a Digital Computer. MA. Thesis, Math. Dept., M.I.T., Cambridge, Mass., 1953.

[134] Norman, A. C.: TAYLOR's User's Manual. Univ. of Cambridge: Computing Service 1973.

[135] Norman, A. C.: Symbolic and Algebraic Modes in REDUCE. REDUCE-Newsletter **3**, 5 – 9. Univ. of Utah: Symb. Comput. Group. 1978.

[136] Pavelle, R., Rothstein, M., Fitch, J.: Computer Algebra (Preprint). To appear in Scientific American.

[137] Perlis, A. J., Iturrigia, R., Standish, T. A.: A Definition of Formula ALGOL. Carnegie-Mellon Univ., Pittsburg: Computer Center 1966.

[138] Pinkert, R. J.: SAC-1 Variable Floating Point Arithmetic. Proc. ACM **1975**, 274 – 276.

[139] Pless, V.: CAMAC. SYMSAC **1976**, 171 – 176.

[140] Rall, L. B.: Automatic Differentiation: Techniques and Applications. Lecture Notes in Computer Science, Vol. 120. Berlin-Heidelberg-New York: Springer 1981.

[141] Rich, A., Stoutemyer, D. R.: Capabilities of the muMATH-79 Computer Algebra System for the INTEL-8080 Microprocessor. EUROSAM **1979**, 241 – 248.

[142] Richards, M.: The BCPL Programming Manual. Univ. of Cambridge: Comput. Lab. 1973.

[143] Rochon, A. R., Strubbe, H.: Solution of SIGSAM Problem No. 5 Using SCHOONSCHIP. SIGSAM Bull. **9**/4, 30 – 38 (1975).

[144] Rom, A. R.: Manipulation of Algebraic Expressions. Commun. ACM **4**, 396 – 398 (1961).

[145] Sammet, J. E.: Survey of Formula Manipulation. Commun. ACM **9**, 555 – 569 (1966).

[146] Sammet, J. E.: Programming Languages: History and Fundamentals. New York: Prentice-Hall 1969.

[147] Sasaki, T.: An Arbitrary Precision Real Arithmetic Package in REDUCE. EUROSAM **1979**, 358 – 368.

[148] Siret, Y.: A Conversational System for Engineering Assistance: ALADIN. SYMSAM **1971**, 90 – 99.

[149] Slagle, J. R.: A Heuristic Program that Solves Symbolic Integration Problems in Freshman Calculus. J. ACM **10**, 507 – 520 (1963).

[150] Smit, J.: Introduction to NETFORM. SIGSAM Bull. **8**/2, 31 – 36 (1974).

[151] Smit, J.: New Recursive Minor Expansion Algorithms, a Presentation in a Comparative Context. EUROSAM **1979**, 74 – 87.

[152] Smit, J.: A Cancellation Free Algorithm, with Factoring Capabilities, for the Efficient Solution of Large, Sparse Sets of Equations. SYMSAC **1981**, 146 – 154.

[153] Smit, J., Hulshof, B. J. A., Van Hulzen, J. A.: Netform and Code Optimizer Manual. Twente Univ. of Tech.: TW Memorandum **373** (1981).

[154] Stoutemyer, D. R.: Analytic Solution of Integral Equations, Using Computer Algebra. Univ. of Utah: Report UCP-34 (1975).

[155] Stoutemyer, D. R.: PICOMATH-80, an Even Smaller Computer Algebra Package. SIGSAM Bull. **14**/3, 5 – 7 (1980).

[156] Strubbe, H.: Presentation of the SCHOONSCHIP System. EUROSAM **1974**, 55 – 60.

[157] Strubbe, H.: Manual for SCHOONSCHIP. Comput. Phys. Commun. **8**, 1 – 30 (1974).

[158] Sundblad, Y.: One User's One-Algorithm Comparison of Six Algebraic Systems on the Y_{2n}-Problem. SIGSAM Bull. **28**, 14 – 20 (1973).

[159] Tobey, R. G.: Experience with FORMAC Algorithm Design. Commun. ACM **9**, 589 – 597 (1966).

[160] Tobey, R. G.: Algorithms for Anti-Differentiation of Rational Functions. Ph.D. Thesis, Harvard Univ., Cambridge, Mass., 1967.

[161] Tobey, R. G.: Symbolic Mathematical Computation – Introduction and Overview. SYMSAM **1971**, 1 – 16.

[162] Tobey, R. G., Bobrow, R. J., Zilles, S. N.: Automatic Simplification in FORMAC. Spartan Books **11**, 37 – 52. Wash., D.C.: 1965.

[163] Tobey, R. G., et al.: PL/I-FORMAC Symbolic Mathematics Interpreter. IBM, Proj. Number 360D-03.3.004., Contributed Program Library. IBM 1969.

[164] Van de Riet, R. P.: ABC ALGOL, a Portable Language for Formula Manipulation Systems. MC Trackts **46**/47. Amsterdam: Math. Centre 1973.

[165] Van Hulzen, J. A.: FORMAC Today, or What Can Happen to an Orphan. SIGSAM Bull. **8**/1, 5 – 7 (1974).

[166] Van Hulzen, J. A.: Computational Problems in Producing Taylor Coefficients for the Rotating Disk Problem. SIGSAM Bull. **14**/2, 36 – 49 (1980).

[167] Van Hulzen, J. A.: Breuer's Grow Factor Algorithm in Computer Algebra. SYMSAC **1981**, 100 – 104.

[168] Verbaeten, P., Tuttens, W.: A Subroutine Package for Polynomial Manipulation. Kath. Univ. Leuven. Applied Math. and Prog. div. Report TW12 (1973).

[169] Watanabe, S.: A Technique for Solving Ordinary Differential Equations Using Riemann's P-Functions. SYMSAC **1981**, 36 – 43.

[170] Willers, I. M.: A New Integration Algorithm for Ordinary Differential Equations Based on Continued Fraction Approximation. Commun. ACM **17**, 508 – 509 (1974).

[171] Williams, L. H.: Algebra of Polynomials in Several Variables for a Digital Computer. J. ACM **9**, 29 – 40 (1962).

[172] Wite, J. L.: LISP/370: A Short Technical Description of the Implementation. SIGSAM Bull. **12**/4, 23–27 (1978).

[173] Xenakis, J.: The PL/I-FORMAC Interpreter. SYMSAM **1971**, 105–114.

[174] Yun, D. Y. Y., Stoutemyer, R. D.: Symbolic Mathematical Computation. In: Encyclopedia of Computer Science and Technology (Belzer, J., Holzman, A. G., Kent, A., eds.), Vol. 15, pp. 235–310. New York-Basel: Marcel Dekker 1980.

J. A. van Hulzen
Department of Applied Mathematics
Twente University of Technology
P.O. Box 217
NL-7500 AE Enschede
The Netherlands

Dr. J. Calmet
IMAG, Laboratoire d'Informatique et de
Mathématiques Appliquées de Grenoble
BP 53
F-38041 Grenoble Cédex
France

Computer Algebra Applications

J. Calmet, Grenoble, and **J. A. van Hulzen,** Enschede

Abstract

A survey of applications based either on fundamental algorithms in computer algebra or on the use of a computer algebra system is presented. Since many survey articles are previously published, we did not attempt to be exhaustive. We discuss mainly recent work in biology, chemistry, physics, mathematics and computer science, thus again confirming that applications have both engineering and scientific aspects, i.e. apart from delivering results they assist in gaining insight as well.

1. Introduction

Applications have always played an important role in computer algebra (CA). Many systems evolved directly from applications, certainly in some areas of physics and in celestial mechanics and to some extent in (applied) mathematics. In fact, applications have often been presented as an existence theorem for the field itself.

Although this situation is different today, the programs of all major conferences on algebraic and symbolic manipulation traditionally include sessions on applications. Hence a rather precise and complete description of many of the more important applications can be found in the proceedings of these conferences as well as in those of the MACSYMA user's conferences. In addition it ought to be mentioned that designers of some of the well known systems, REDUCE and ALTRAN for instance, maintain bibliographies on applications of their products [13, 80]. The first bibliography on computer algebra, including applications, was composed by Jean E. Sammet [59]; some later extensions were published by Wyman [76]. Later survey articles, also concentrating on applications, are, for example, written by Barton and Fitch [5, 6], Brown and Hearn [14], Ng [55], Fitch [24], Yun and Stoutemyer [77], Pavelle [56], d'Inverno [39] and Calmet and Visconti [16].

The application areas are quite diverse and numerous, ranging from pure mathematics (projective geometry, for instance) to technology (arm motion of a robot or design of a plant for uranium mining). Our intention is not only to give an illustration of how CA-algorithms are actually used, but also to survey the most important contributions of CA in different fields of science and engineering. We start by considering those domains which only have loose links with CA, such as biology and chemistry. Then the impact on physics (including often reviewed areas like quantum physics, general relativity and celestial mechanics) is sketched. Finally the influence on computer science is indicated. We realize that this approach is inherently subjective. We do not pretend to be exhaustive, since hundreds of published papers report the use of one of the many, now existing CAS(ystems).

A striking feature of many of the most important applications is the simple and straightforward character of the underlying computational methods, such as polynomial arithmetic. These calculations are often tedious and lengthy. This inherent character of mechanized mathematics is one of the reasons we have chosen for a field-wise classification of applications rather than for an algorithm related characterization. In relation to this it ought to be mentioned that most systems are well enough debugged to be considered as errorfree, i.e. this mechanization leads to reliable results. In fact CAS have been used to eventually improve handmade calculations. With ANALITIK, running on the Soviet MIR-2 computer, for instance, several "misprints" in the more recent Russian editions of Gradshteyn and Ryzhik's Collection of integrals were found [43].

Another feature of applications is the way they rely on computer resources. In physics, for instance, systems are widely used, including some specifically written for this field. In many areas however, methods rather than systems have been adopted. It may be of interest to notice that LISP has often been used in the past [5]; a trend which has been vanished today.

Any description of the concept "Application" is controversial. Nevertheless, we ought to clarify our "intuitive" approach. We intend to distinguish between *system independent* and *system driven applications*. For the former category the underlying computational concepts, used to obtain a meaningful result, the application, are basically algorithms which (might) play an essential role in CA, even though CAS are not necessarily required or used. The computation of determinants of matrices with symbolic entries might serve as an illustrative example. Bossi et al. [10] discuss a PASCAL-program for this purpose. It is used to obtain structural identifiability tests for compartmental models, which play an important role in modelling different physico-chemical processes, as occurring in biology, medicine and ecology. Smit [63, 64] implemented such facilities in NETFORM [65], a REDUCE-program, designed for the symbolic analysis of electrical networks. The latter deals with the set of meaningful results which are essentially achieved by using the capabilities of existing CAS. However, this category needs a further distinction between results not leading to the necessity of implementing new algorithms in an existing CAS and those which are conceptually too heuristic. Hence Gragert's approach for doing symbolic computations in prolongation theory [29] belongs to the first sub-category as long as the elements of the Lie-algebra are rational functions; it might switch to the first category when the element definition is changed. An attempt, for example, to calculate elliptic integrals as an extension of the Rish algorithm to transcendental functions is not considered as an application. This in contrast to Ng's implementation of Carlson's non-algorithmic methods to achieve this [54], which belongs to the second sub-category.

This choice implies the presumption that the reader is more or less familiar with the revival of interest for mathematical problems, studied quite long ago, mainly in classical algebra, caused by the very existence of CA and its quests for efficient algebraic algorithms. In addition, we state that we do not underestimate the predictable impact of portable mini-CAS for every day use at highschool and undergraduate level, as already illustrated by public discussions [40, 69] and marketing papers [70, 71]. However, inclusion of such aspects of CA would open a limitless field of applications.

2. Biology and Chemistry

Here CA has almost no impact at all. Although very respectable papers were published in this area, which can be considered as CA applications, it can hardly be stated that CA caused any breakthrough. From the rare examples which we found in literature we discuss those we consider as representative for this area.

An existing system, MACSYMA, is used in biology to compute generating functions to study the secondary structure of RNA when specific pairing rules are imposed [37]. Essentially, a standard calculation of recursion functions is required. The same system was used for a study of urine concentration of some mammals [50]. A naive reader gets the impression that the technical description of the problem of how to obtain the solution for a multinephron kidney model is similar to those found in hydrodynamics. A statistical study of the relationships between amino acids in protein sequences [25], which demands a symbolic approach to the computation of trees, belongs to this area also, as well as the earlier indicated work of Bossi et al. [10].

PL/I-FORMAC was used to automatically generate programs, required for certain numerical calculations in chemical engineering [48]. Starting from a mathematical description of the problem some model dependent FORTRAN-routines are generated, mainly using PL/I to facilitate structuring these routines. FORMAC-facilities were only used for equation processing. An application in organic chemistry which can be qualified as "book-keeping" consists in fact of a deciphering technique for linear notations for the WLN (Wiswerser linear notation) code [58]. To achieve this, a database is created, giving information about some products (like existence, properties and bibliography), as to assist in synthesizing new products from those already known. It is worth mentioning that the construction of similar databases in organic chemistry is an important field of computer applications. From work in progress a study in chemical equilibria ought to be mentioned [68]. Here the problem-description demands the solution of a system of non-linear equations. This, however, can be reduced to the integration of a system of differential equations of the form $J(d\mathbf{x}/dt) = -f(\mathbf{x}_0)$ with the initial condition $\mathbf{x} = \mathbf{x}_0$. J and f depend on the investigated chemical process. Simple cases can be solved with a mini-CAS, such as muMATH-79. An application, finally, which belongs both to chemistry and biology is the calculation of association constants from binding data, leading to the need of solving ill-conditioned systems of linear equations [68].

3. Physics

This is a domain where CAS are extensively and routinely used. Many review papers have been devoted either to the entire field or to specific subdomains [5, 14, 16, 24, 56, for instance]. They provide numerous references to applications in high energy physics, celestial mechanics and general relativity in particular. Therefore, we only try to demonstrate why the impact of CA in these areas has been so drastic rather than discussing specific examples. Finally we mention in a separate subsection a number of other subjects.

3.1 High Energy Physics

For a long time already scientists in this field are well aware of the existence of CAS, such as REDUCE, SCHOONSCHIP and ASHMEDAI. They have learned that a specific package has been added to MACSYMA [11], may have heard of the availability of a new CAS, called SMP, and learned of the work in progress in ALDES/SAC 2 [75]. To be of any use in quantum physics, a CAS must offer the capability of performing the so-called Dirac algebra [8]. In addition, it must be suitable for pattern matching and polynomial manipulation. Depending on the strength of the interaction among the particles active in a given physical process, a distinction is made between theories of strong interactions, weak interactions and quantum electrodynamics (QED). QED is dealing with the interaction of leptons (electron, muon, ...) and photons. A recent unification scheme gives rise to quantum chromodynamics (QCD), which will eventually unify the three interactions. The name QCD originates from the fact that the quarks required by this theory are coming with at least three different "colours" (where the colour represents a quantum number). Originally CAS have only been used in QED. The advent of gauge theories made computerization of weak interaction problems possible. At present QCD calculations can be handled. It is fair to state that only the evolution of the physical theory made this trend possible.

Let us now describe a typical, oversimplified but still realistic, computation in QED. A more extensive discussion and references are given in [16]. Details pertaining to the physical theory involved can be found in any textbook, such as [8]. This description is in principle valid for QCD as well. In [26] for example, it is shown how REDUCE can be made instrumental for gauge theories, via changing a few instructions. Such a QED process is only computable through a perturbative series expansion of the form

$$P = A_1 \left(\frac{\alpha}{\pi} \right) + A_2 \left(\frac{\alpha}{\pi} \right)^2 + A_3 \left(\frac{\alpha}{\pi} \right)^3 + \cdots,$$

where α is the so-called fine structure constant. The calculation of each A_i requires the evaluation of several Feynman diagrams, which graphically represent the physical process. Many programs to generate the contributing diagrams up to a given order have been written. But this computational step is in fact only necessary if the calculation is supposed to be completely automized. This holds for the next step too, being the recognition of divergences in diagrams. The divergences are either of an infra-red type (when due to the null mass of the photon) or of an ultra-violet type (when resulting from the integration over the internal momenta). The cancellation of these divergences is a well defined step of the renormalization procedure. Each diagram is transformed into an algebraic expression, by applying the so-called Feynman rules, which have the general form

$$\int \frac{d^4 k_1 \cdots d^4 k_n}{a_1 a_2 \cdots a_n} * \mathrm{NUM}$$

if the a_i are defined by

$$a_i = \left\{ \sum_r (\zeta_r p_r + \eta_r k_r) \right\}^2 + m^2,$$

where $\zeta_r, \eta_r \in \{0, +1, -1\}$ and if p_r and k_r are the momenta of the external and internal particles involved in the process, respectively, and if m is a mass. This integral is parametrized by introducing the Feynman parameters via the identity

$$\frac{1}{a_1 a_2 \cdots a_n} = (n-1)! \int_0^1 d\alpha_1 \cdots \int_0^1 d\alpha_n \frac{\delta(\sum_{i=1}^n \alpha_i - 1)}{(\sum_{i=1}^n \alpha_i a_i)^n}.$$

The numerator NUM is typically a product of Dirac matrices γ_μ and of monomials of the form $(\sum_1^4 q^\nu \gamma_\nu - m)$, where the q^ν are the components of a momentum, i.e. a linear combination of p's and k's. The Dirac matrices are defined by the relationship

$$\gamma_\mu \gamma_\nu + \gamma_\nu \gamma_\mu = 2g_{\mu\nu},$$

where $g_{\mu\nu}$ is the metric tensor of the space considered. Usually this metric is $(+, +, +, -)$. For some well defined diagrams (closed loops) the trace of the numerator must be taken. For this purpose a set of formulas is available [8]. However, considering the complexity of the algorithm, derived from these formulas, it is shown that improvements are possible [75].

The main motivation for using a CAS in this field can be found in the structure of the defining formula of the γ-matrices: the numerator blows up very quickly, thus easily generating a few thousand terms during the calculation. To keep this growth under control the well-known Kahane-algorithm, for summation over the repeated indices, is available [41]. Once the algebra, required for the computation of the numerator, is performed only some pattern matching steps remain to be done. Essentially the renormalization procedure demands only the transformation (via an integral identity) of expressions of the form $(1/a^n - 1/b^n)$, where a and b are rather complicated polynomials. Then the integration over the internal momenta is also easily done by applying another identity. Finally, to specify the contribution of a particular diagram to a given physical process it is only necessary to either apply a projective operator or to square the expansion at hand. All these steps can be performed by a computer program. Even in the more complicated cases it requires only a few minutes. Hand calculations, except in trivial cases, are almost impossible. What finally remains to be done is the integration over the Feynman parameters, resulting in a number as answer to the problem. However, until now, nobody has been able to make use of the symbolic integrators, at present available in different CAS. The only symbolic approach, in computing significant contributions, is based on a table-look-up method [15]. The measure for computing times for this last phase of the computation is no longer the minute, but the hour, also when the integration is accomplished with numerical methods [16]. To conclude, it can be stated that a CAS, specialized in this field, will present the particular feature of handling Dirac matrices. It is, in fact, rather straightforward to implement such a package. Many impressive results, in particular the electron anomaly at high order, were only obtainable via CAS and are among the most important applications of CA.

3.2 Celestial Mechanics

This was also one of the first domains where CA has been used. The now classical example, mentioned in all review papers, is the recomputation by Deprit et al. [21] of Delaunay's work. The introduction of a new technique (the Lie transform) allows them to recompute the reduced Hamiltonian of the main problem in lunar theory. Delaunay spent 20 years to complete his calculation by hand. Deprit et al. used only 20 hours on a small computer, using Rom's CAS for the recomputation. It is a tribute to the accuracy of Delaunay that his in 1867 published results were correct up to order 9, with only one exception in an addition of order 7. This calculation of the orbit of the moon demands a careful consideration of the combination of many physical effects, such as the non-sphericity of the earth, the tilted ecliptic and the influence of the sun. The Hamiltonian describing the physical system requires several pages; up to 600 transformations are to be applied on each term. The reason that celestial mechanics is so well suited for mechanized computations is that the objects to be handled can be expressed in terms of Poisson-series. Such a series is in fact a Fourier series of the form

$$Q(x, y) = \sum_i \{P_i(x)\cos(i \cdot y) + R_i(x)\sin(i \cdot y)\},$$

where x, y and i are vectors of polynomial variables, trigoniometric variables and integers, respectively, and P_i and R_i polynomials. When no side relations occur these series can be written in a unique canonical form. This pleasant feature, a characteristic property of calculations in celestial mechanics, explains why specialized CAS have been designed. Apart from Rom's already mentioned system, well known and qualified systems are both Jeffreys' TRIGMAN and CAMAL. It is also the reason that one of the internal representations in MACSYMA is tailored for Poisson-series. Since this lunar orbit calculation, probably none has been of the same importance. But glancing over the journal "Celestial Mechanics" one almost immediately realizes that CA is now one of the basic tools in this field. This is in particular true for the estimation and prediction of artificial satellite orbits and for the construction of physical models for the gravitational field of the earth [78]. Essentially, an orbit calculation demands the integration of an equation of motion of the Newton-type in presence of side relations due to the presence of the solar and lunar masses and to the non-sphericity of the earth. In general, the partial differential equation, defining the total acceleration of an object, is a Keplerian ellipse with constant orbital elements. By introducing Gauss' perturbation equations it is, for instance, possible to derive the time variation of some Keplerian orbital elements [3].

3.3 General Relativity

This is the third domain where CAS are extensively used. As a consequence many reviews are already available [39, 56]. The fundamental object of general relativity is the metric tensor $g_{\mu\nu}$ which appears in the formula defining the length of a line element in this space:

$$ds^2 = g_{ij}(x) dx^i dx^j \qquad (i, j = 0, 1, 2, 3).$$

The metric tensor plays a central role because it enables to compute various objects such as the Riemann tensor, R_{ijkl}, through the Christoffel symbols, then the Ricci tensor, R_{ij}, and the Ricci scalar (R) by applying the equations:

$$R_{ij} = g^{\mu\nu} R_{\mu i\nu j},$$

$$R = g^{\mu\nu} R_{\mu\nu}.$$

The next step is the calculation of the Einstein tensor:

$$G_{\alpha\beta} = R_{\alpha\beta} - \tfrac{1}{2} R g_{\alpha\beta}.$$

The field equations are

$$G_{\mu\nu} = T_{\mu\nu},$$

where $T_{\mu\nu}$ is the energy momentum tensor. These simple equations are at the heart of any computation in general relativity. Among the questions which have been studied, one can mention the search for an alternative formulation of Einstein's vacuum equation [56]:

$$R_{ij} = 0.$$

A standard calculation consists of going through the above mentioned equations, while considering different metrics including some indeterminate variables. Until recently only diagonal metrics have been considered. Now, some work on non-diagonal metrics is in progress. To illustrate the need of CAS in this field it suffices to mention that a Ricci tensor in a 4-dimensional space has almost 10.000 diagonal terms and more than 13.000 of diagonal terms. The number of field equations is 10. One of the most difficult problems in general relativity is how to decide whether two different metrics describe the same gravitational field or not. An algorithmic approach to the classification of geometries in general relativity [2] is already producing results by classifying a number of Harrison metrics. If a complete classification of the metrics proves to be possible this certainly can be qualified as one of the major successes of the application of CAS in physics. In this area many systems were and are used: ALTRAN, FORMAC, SYMBAL, REDUCE, MACSYMA, CAMAL, ALAM, LAM, CLAM and SHEEP. This list, albeit probably not complete, proves that specialized CAS are not really necessary in this field if efficiency is not a priority. Indeed, the same calculation may require very different computing times depending on the availability of a specialized indicial tensor package [24].

3.4 Other Fields

Nowadays *electron optics* is essential and instrumental for, among others, particle accelerators, electron microscopy and the microfabrication technique for large scale integrated circuits. CA-tools for research in this area have been REDUCE [67] and CAMAL [32]. The basic physical problem encountered in the calculation of the general trajectory equation of an electron travelling in a cavity in presence of electric and magnetic fields. One wishes to use high frequency fields and to treat superimposed focusing and deflecting fields of both types. Mathematically the solution of this problem amounts to solve linear homogeneous second order differential equations. When aberrations are taken into account the algebra becomes cumbersome and a CAS must be used. Relativistic formulas have been obtained for chromatic and geometrical aberrations up to different orders. The study of microwave cavities has also been made possible. To illustrate the need of CA it suffices to mention that some of these formulas cover more than 50 pages.

Molecular physics gives a nice and illustrative example of what CA can perform. For a long time the calculation of the structure factor for ionic solutions in which one ionic specy is large enough to dominate many of the solution properties, was only possible with numerical techniques. It took in fact 15 years to obtain the first analytic solution [33]. The problem is to determine the quantity $c(x)$ in the equation

$$h(x) = c(x) + n\sigma^3 \int h(|\mathbf{x} - \mathbf{y}|)c(y) \, d\mathbf{y},$$

where $h(x)$ is the total correlation factor, n the particle number density and σ the diameter of the dominating ion. $c(x)$ is the product of the inverse temperature in energy units and the potential between two identical ions. The structure factor is obtained by a Fourier transformation of the above equation. Almost ten years ago the techniques to solve this equation were in principle already established. But finally REDUCE was used to achieve an analytic result.

An important contribution to *electrical engineering* has been made by Smit with his NETFORM system [65], which is a package, written in REDUCE, designed for the symbolic analysis of electrical networks. It operates on a simple graph description of a specific network as input. This input is used to construct the associated incidence matrix, describing the structure equations in accordance with Kirchhoff's laws. This structural information allows to obtain transfer functions, essentially via determinant calculations, i.e. NETFORM has also an impact on the study of the solution of large sparse sets of linear equations [63, 64]. NETFORM is able to perform several types of analysis, such as hybrid, state and mixed hybrid-state. Additional features for code optimization for a symbolic-numeric front-end exist [38] and are, for example, applied to analyze a transmission line network, as used in microwave technology [66].

In *plasma physics* CAS have assisted in studying the containment, the stability and the heating of plasmas. A standard calculation in this field starts from the equations of motion which either are or are transformed into large systems of partial or ordinary differential equations. Approximate solutions are obtained by both MACSYMA and REDUCE. Many references and examples are available in the literature [5, 24, MACSYMA 1977, 1979].

The mathematical problems found in *fluid mechanics* are similar. A recent application, for example, deals with the solution of the Navier-Stokes equations for a rotating fluid above an infinite disk [72]. It will be interesting to see whether the prediction of Fitch [24], stating that this field will see an extensive use of CAS materializes or not in the near future.

Many other fields of physics might be added. But to conclude with two areas are mentioned, where a practical motivation to apply CAS might be apparent:

industrial mechanics and *aeronautics*. An example of the former category is the use of REDUCE in studying a chain of rigid bodies, such that adjacent bodies have at least a common point [9]. A specimen of the latter is the application of CA in studying mathematical models of dynamical systems such as teaching a piloted simulator to respond to a command, requiring the determination of the location and the state of a vehicle [35]. Essentially, a set of at least 12 equations, expressing some features of the model (force, moment, Euler angles, ...), has to be solved. A detailed description of these problems can be found in [36].

4. Mathematics

In this domain the borderline between application and genuine work is hard to draw. This probably explains the title of one of the few review papers about the use of CA in this field [60]. According to our criteria we at least ought to mention the following subfields.

4.1 Number Theory

It is well known that computational problems in algebraic number theory require lengthy calculations, which can hardly be performed without the assistance of a computer. This explains why, already for a long time, CAS have been instrumental for research in number theory [79]. Many programs and algorithms are available, e.g. for the computation of the continued fraction expansion of a real algebraic number, for sign determination and rational operations in real algebraic fields, for the determination of the maximal order in real algebraic fields and for the construction of the Galois-group of the splitting field of an equation. In addition it ought to be mentioned that considerable efforts have been spent in developing a primality-test for very large integers. Although this is not a CA-application itself, the methods are very similar to those of CA. The existence of very efficient algorithms for polynomial multiplication are of key importance. A comprehensive exposure of the problem is presented in [45].

4.2 Projective Geometry

The rather ambitious project of investigating the feasibility of algebraic manipulation techniques in projective geometry, using MACSYMA, is being done at present at the University of Bergen, Norway. Holme [34] reported the implementation of algorithms for the computation of the Chern polynomial of a tensor product in terms of the Chern polynomials of the factors, the Chern and Segré classes of Grassmanians and embedding and duality properties of projective varieties. For details and definitions we refer to [34]. Future plans concentrate on problems related to generating varieties and classes of projective varieties.

4.3 Padé-Approximants

Applications of Padé-approximants can be found in a diversity of disciplines, ranging from physics to numerical analysis. An algorithm to compute these approximants for power series with polynomial coefficients over an arbitrary integral domain is due to Geddes [27]. It is easily implementable in any CAS.

Another interesting method is based on the possibility of using the extended Euclidean algorithm to compute entries in the Padé table. This approach to the calculation of Padé approximants is a consequence of the study of how to solve Toeplitz systems of linear equations [12].

4.4 Special Functions

This heading attempts to indicate research undertaken to learn how to handle recurrence relations, hypergeometric and transcendental functions. A remarkable specimen of these activities is the work, conducted by the MACSYMA group, on hypergeometric series, as discussed by Lafferty [44]. It illustrates the possibilities of the pattern matching capabilities of CAS. The aim of his research was to reduce hypergeometric functions to elementary or special functions by using different recurrence relations, of which over 200 are known. His work can be considered as part of the attempts to develop a general theory for the computerization of the treatment of the higher transcendental functions of mathematicians. In contrast to this approach, SAC-1 was once used by Verbaeten [73] for the automatic construction of pure recurrence relations for a special class of functions. His rather intuitive method naturally leads to the need of solving a system of linear equations.

4.5 Ordinary Differential Equations

Several groups are investigating the possibility of implementing an ODE-solver in an existing CAS. The ODE-solver of MACSYMA, for instance, relies on its integration package so as to produce closed form solutions. Such programs are often rather heuristic; a trend is to attempt to computerize solution strategies, which are known to be efficient [17]. Instead of discussing the many papers published in this field (references in all conference proceedings) we mention some recent contributions, thus attempting to indicate the present possibilities. Schmidt wrote a PL/I program for solving first order and first degree ODE's [61]. In his program, based on heuristics, the following stages can be distinguished: recognition of elementary types, calculation of integrating factors, particular solutions, substitutions for transforming a given ODE into a type, solvable by a previous step. Schmidt claims that his program is able to solve almost all known ODE's of the first order and first degree. This naive-user-approach does not, however, provide much insight for mathematicians. Ongoing research of Della Dora and Tournier to obtain formal solutions of high order DE's in the neighbourhood of singular points, using a Frobenius like method [20], is based on a completely opposite motivation. Similar arguments hold for the investigation of movable critical points of (second and) third order ODE's and of nth order systems of ODE's by Hajee [31]. His work might contribute to increasing insight in a possible connection between nonlinear evolution equations and ODE's of Painlevé-type [1]. In view of the many potential applications, the need for powerful ODE- and PDE-solvers is apparent, i.e. it is safe to qualify the (hoped for) future proof that the ODE problem is decidable as a real breakthrough. For additional remarks on ODE's see also the chapter on integration in finite terms in this volume.

4.6 Finite Algebras, Finite Fields, Rings and Groups

A survey on this area is given in the chapter entitled "Computing with Groups and Their Character Tables" in this volume. A few selected references are [19, 22, 23, 46, 49, 52, 53].

5. Computer Science

Here CA-methods and theorems rather than CAS have been applied. An illustrative example is Good's idea to use the Chinese remainder theorem for computing *discrete Fourier transforms* [28], as extended by Winograd [74] to the case of n points, when n is not a power of a prime. Winograd also emphasized in [74] the usability of the DFT-algorithm for the FFT-identity. Another very nice application of the Chinese remainder theorem can be noticed in *algebraic coding theory* when studying how to manage cryptographic keys [62]. Essentially the problem is that a secret, shared by n people, each of which has only partly and equally access (a key) to this information, can only be made public when at least $k(< n)$ keys are known. Hence, when only less than k keys can be collected it remains inaccessible. Shamir's solution of this problem [62], based on evaluation and interpolation of polynomials, allowed Mignotte [51] to demonstrate that using the Chinese remainder theorem is profitable. His proof is based on the property that n ought to satisfy at least k congruence relations (n and k are integers). Factorization of polynomials, defined over $GF_q[x]$ also plays an important role in algebraic code theory. During decoding a B-C-H code (see Berlekamp [7] for definitions and technical details) an important step consists of determining the roots of the error-locator polynomial. This problem is equivalent to finding the linear factors in the decomposition field of the polynomial when the Galois prime is 2 (i.e. when q is a power of 2). The performance and reliability of such algorithms is crucial, since this decoding might require real time processing. Group theory can be instrumental too in algebraic coding theory [46]. ALTRAN has been useful in this respect, for instance in studying the classification and enumeration of self-dual codes [57].

CAS can play a role in *software validation*, as illustrated by London and Musser [47], who used and extended REDUCE for implementing an interactive program verification system, based on inductive assertions, and thus requiring the implementation of transformations on relational and boolean expressions. The advent of abstract data type specification methods in program verification systems was related to the adaptation of algebraic algorithms for this purpose [30]. At present, these interactions are actively studied, possibly leading to the implementation of abstract data type analysis facilities in CAS. In that respect it can no longer be qualified as an application.

A rather curious but interesting application of MACSYMA is the analysis of *the reliability of on-line computer systems* [18]. The kernel of this problem is the computation of the core matrix of a semi-Markov process.

Some experiments, using MACSYMA, with Khachian's famous algorithm, used in *linear programming*, for deciding the consistency of a set of linear inequalities lead to an improved version of it [4].

Finally, the relationship between CA and *numerical analysis* ought to be mentioned. This is recently surveyed by Ng [55], who considers his article as supplementary to one written by Brown and Hearn [14]. This interaction is not only restricted to the generation of programs for numerical purposes (as shortly discussed in the chapter on CAS). That is nicely illustrated in a paper by Keller and Pereyra [42], dealing with the symbolic generation of finite difference formulas, often used as a tool to generate consistent approximations. Although only Taylor-series expansions are required to produce these formulas, this process can be tedious. The main motivation of these authors was, however, the study of corrections to ordinary boundary value problems.

References

[1] Ablowitz, M. J., Romani, A., Segur, H.: A Connection Between Nonlinear Evolution Equations and Ordinary Differential Equations of *P*-Type. I. J. Math. Phys. **21**/4, 715 – 721 (1980).

[2] Åman, J., Karlhede, H.: An Algorithmic Classification of Geometries in General Relativity. SYMSAC **1981**, 79 – 84.

[3] Anderson, J. D., Lau, E. L.: Application of MACSYMA to First Order Perturbation Theory in Celestial Mechanics. MACSYMA **77**, 395 – 404 (1977).

[4] Anderson, N., Wang, P. S.: MACSYMA Experiments with a Modified Khachian-Type Algorithm in Linear Programming. SIGSAM Bull. **14**/1, 8 – 14 (1980).

[5] Barton, D., Fitch, J. P.: A Review of Algebraic Manipulation Programs and Their Application. Comp. J. **15**, 362 – 381 (1972).

[6] Barton, D., Fitch, J. P.: Applications of Algebraic Manipulative Programs in Physics. Reports on Progress in Physics **35**, 235 – 314 (1972).

[7] Berlekamp, E. R.: Algebraic Coding Theory. New York: McGraw-Hill 1968.

[8] Bjorken, J. D., Drell, S. D.: Relativistic Quantum Fields. New York: McGraw-Hill 1964.

[9] Bordoni, L., Colagrossi, A.: An Application of REDUCE to Industrial Mechanics. SIGSAM Bull. **15**/2, 8 – 12 (1981).

[10] Bossi, A., Colussi, L., Cobelli, C., Romanin Jacur, G.: Identifiability of Compartmental Models: Algorithms to Solve an Actual Problem by Means of Symbolic Calculus. SIGSAM Bull. **14**/4, 35 – 40 (1980).

[11] Brenner, R. L.: The Evaluation of Feynman Diagrams in MACSYMA. MACSYMA **1979**, 1 – 24.

[12] Brent, R. P., Gustavson, F. G., Yun, D. Y. Y.: Fast Solution of Toeplitz Systems of Equations and Computation of Padé-Approximants. Yorktown Heights, N. Y.: IBM Research, Report RC 8173 1980.

[13] Brown, W. S.: ALTRAN User's Manual, 4th ed. Murray Hill, N. J.: Bell Laboratories 1977.

[14] Brown, W. S., Hearn, A. C.: Application of Symbolic Mathematical Computations. Comput. Phys. Commun. **17**, 207 – 215 (1979).

[15] Caffo, M., Remiddi, E., Turrini, S.: An Algorithm for the Analytic Evaluation of a Class of Integrals. EUROSAM **1979**, 52 – 57.

[16] Calmet, J., Visconti, A.: Computing Methods in Quantum Electrodynamics. In: Fields Theory, Quantization and Statistical Physics (Tirapegui, E., ed.), pp. 33 – 57. Dordrecht: Reidel 1981.

[17] Char, B.: Using Lie Transformation Groups to Find Closed Form Solutions to First Order Ordinary Differential Equations. SYMSAC **1981**, 44 – 50.

[18] Chattergy, R.: Reliability Analysis of On-Line Computer Systems Using Computer Algebraic Manipulations. University of Hawaii: Report A 76-1, 1976.

[19] Computers in Non-Associative Rings and Algebra (Beck, R. E., Kolman, B., eds.). New York-London: Academic Press 1977.

[20] Della Dora, J., Tournier, E.: Formal Solutions of Differential Equations in the Neighbourhood of Singular Points. SYMSAC **1981**, 25 – 29.

[21] Deprit, A., Henrard, J., Rom, A.: Lunar Ephemeris: Delaunay's Theory Revisited. Science **168**, 1569 – 1570 (1970).

[22] Feldman, H. A.: Some Symbolic Computations in Finite Fields. SYMSAM **1966**, 79 – 96.

[23] Felsch, V.: A KWIC Indexed Bibliography on the Use of Computers in Group Theory and Related Topics. SIGSAM Bull. **12**/1, 23 – 86 (1978).

[24] Fitch, J. P.: The Application of Symbolic Algebra in Physics – a Case of Creeping Flow. EUROSAM **1979**, 30 – 41.

[25] Franchi-Zannettacci, M. P.: Evaluation d'Orbre Pour un Calcul Formel. (Sixth CAAP' 81.) Lecture Notes in Computer Science, Vol. 111. Berlin-Heidelberg-New York: Springer 1981.

[26] Gastmans, R., Van Proeyen, A., Verbaeten, P.: Symbolic Evaluation of Dimensionally Regularized Feynman Diagrams. Comput. Phys. Commun. (to appear).

[27] Geddes, K. O.: Symbolic Computation of Padé Approximants. ACM Trans. Math. Software **5**, 218 – 233 (1979).

[28] Good, J. J.: The Interaction of Algorithms and Practical Fourier Series. J. Royal Statist. Soc. **B20**, 361 – 372 (1958).

[29] Gragert, P.: Symbolic Computations in Prolongation Theory. Ph.D. Thesis, Twente Univ. of Technology, 1981.

[30] Guttag, J. V., Horowitz, E., Musser, D. R.: Abstract Data Types and Software Validation. Commun. ACM **21**, 1048 – 1064 (1978).

[31] Hajee, G.: The Investigation of Movable Critical Points of ODE's and Nth Order Systems of ODE's Using Formula Manipulation. MS Thesis, Dept. of Appl. Math., Twente Univ. of Technology, 1982.

[32] Hawkes, P. W.: Computer Calculation of the Aberration Coefficients of High-Frequency Lenses. Proc. Sixth Int. Conf. on High Voltage Electron Microscopy, Antwerp 80. Electron Microscopy **4**, 46 – 49 (1980).

[33] Hayter, J. B., Penfold, J.: An Analytic Structure Factor for Macro Ion Solutions. Molecular Phys. **42**, 109 – 118 (1981).

[34] Holme, A.: Some Computing Aspects of Projective Geometry I. University of Bergen: Preprint 1980.

[35] Howard, J. C.: MACSYMA as an Aid to Simulation Modelling. MACSYMA **1979**, 483 – 521.

[36] Howard, J. C.: Practical Applications of Symbolic Computation. Guilford, Surrey, England: LPC Sci. and Techn. Press 1980.

[37] Howell, J. A., Smith, J. F., Waterman, M. S.: Using MACSYMA to Compute Generating Functions for RNA. MACSYMA **1979**, 126 – 139.

[38] Hulshof, B. J. A., Van Hulzen, J. A., Smit, J.: Code Optimization Facilities, Applied in the NETFORM Context. Twente Univ. of Technology: TW Memorandum **368** (1981).

[39] D'Inverno, R. A.: A Review of Algebraic Computing in General Relativity. Einstein Commemorative Volume (Held, A., et al., eds.). New York: Plenum Press 1980.

[40] Jenks, R. D.: Push 80. SIGSAM Bull. **13**/4, 3 – 6 (1979).

[41] Kahane, J.: An Algorithm for Reducing Contracted Products of γ-Matrices. J. Math. Phys. **9**, 1732 (1968).

[42] Keller, H. B., Pereyra, V.: Symbolic Generation of Finite Difference Formulas. Math. of Comput. **32**, 955 – 971 (1978).

[43] Krupnikov, E. D.: Private Communication.

[44] Lafferty, E. L.: Hypergeometric Function Reduction – an Adventure in Pattern Matching. MACSYMA **1979**, 465 – 481.

[45] Lenstra Jr., H. W.: Primality Testing Algorithms. Seminaire Bourbaki, 33-ième année 1980-81, exposé 576. Paris: Ecole Normale Sup. 1981.

[46] Leon, J. S., Pless, V.: CAMAC 79. EUROSAM **1979**, 249 – 257.

[47] London, R. L., Musser, D. R.: The Application of a Symbolic Mathematical System to Program Verification. Proc. ACM 1974, 265 – 273. New York: ACM 1974.

[48] Mah, R. S. H., Rafal, M.: Automatic Program Generation in Chemical Engineering Computation. Trans. Inst. Chem. Eng. **49**, 101 – 108 (1971).

[49] Maurer, W. D.: Computer Experiments in Finite Algebra. SYMSAM **1966**, 598 – 603.

[50] Mejia, R., Stephenson, J. L.: Symbolics and Numerics of a Multinephron Kidney Model. MACSYMA **1979**, 597 – 603.

[51] Mignotte, M.: Introduction à la cryptographie, Comptes-rendus des V-ièmes Journées Informatiques, University of Nice: IMAN 1980.

[52] Mora, F.: Computation of Finite Categories. University of Genova: Unpublished Note 1979.

[53] Neubüser, J., Bülow, R., Wondratschek, H.: On Crystallography in Higher Dimensions I, II, III. Acta Cryst. **A27**, 517−535 (1971).

[54] Ng, E. W.: A Study of Alternative Methods for the Symbolic Integration of Elliptic Integrals. SYMSAC **1976**, 44−50.

[55] Ng, E. W.: Symbolic-Numeric Interface: A Review. EUROSAM **1979**, 330−345.

[56] Pavelle, R.: Applications of MACSYMA to Problems in Gravitation and Differential Geometry. MACSYMA **1979**, 256−276.

[57] Pless, V., Sloane, N. J. A.: On the Classification and Enumeration of Self-Dual Codes. J. Comb. Theory **18**, 313−334 (1975).

[58] Ponsignon, J. P.: Decryptage des Notations Linéaires en Chimie Organique. Application à WLN. CALSYF **1**, 1981 (unpublished Bull.). (Bergman, M., Calmet, J., eds.)

[59] Sammet, J. E.: Revised Annotated Description Based Bibliography on the Use of Computers for Non-Numerical Mathematics. IFIP **1966**, 358−484.

[60] Sammet, J. E., Ivinsky, V.: Brief Annotated Bibliography/Reference List of Items Relevant to the Use of Computers in Formal Mathematics. SIGSAM Bull. **27**, 30 (1973).

[61] Schmidt, P.: Automatic Symbolic Solution of Differential Equations of First Order and First Degree. EUROSAM **1979**, 114−125.

[62] Shamir, A.: How to Share a Secret. Commun. ACM **22**, 612−613 (1979).

[63] Smit, J.: New Recursive Minor Expansion Algorithms, a Presentation in a Comparative Context. EUROSAM **1979**, 74−87.

[64] Smit, J.: A Cancellation Free Algorithm, with Factoring Capabilities, for the Efficient Solution of Large, Sparse Sets of Equations. SYMSAC **1981**, 146−154.

[65] Smit, J., Hulshof, B. J. A., Van Hulzen, J. A.: NETFORM and Code Optimizer Manual. Twente Univ. of Tech.: TW Memorandum **373**, 1981.

[66] Smit, J., Van Hulzen, J. A.: Symbolic-Numeric Methods in Microwave Technology. Twente Univ. of Technology: TW Memorandum **374**, 1981.

[67] Soma, T.: Relativistic Aberration Formulas for Combined Electromagnetic Focusing-Deflection System. Optik **49**, 255−262 (1977).

[68] Squire, W.: Private Communication.

[69] Stoutemyer, D. R.: Computer Symbolic Math. & Education: A Radical Proposal. MACSYMA **1979**, 142−158, and SIGSAM Bull. **13**/3, 8−24 (1979).

[70] Stoutemyer, D. R.: Computer Symbolic Math. in Physics Education. Am. J. Phys. **49**, 85−88 (1981).

[71] Stoutemyer, D. R.: The Coming Revolution in Scientific Computation. Comput. & Educ. **5**, 53−55 (1981).

[72] Van Hulzen, J. A.: Computational Problems in Producing Taylor Coefficients for the Rotating Disk Problem. SIGSAM **14**/2, 36−49 (1980).

[73] Verbaeten, P.: The Automatic Construction of Pure Recurrence Relations. EUROSAM **1974**, 96−98.

[74] Winograd, S.: On Computing the Discrete Fourier Transform. Math. Comp. **32**, 175−199 (1978).

[75] Woll, A.: γ-Algebra Algorithms for Canonical Hypercomplex Representations. EUROSAM **1979**, 546−557 (1979).

[76] Wyman, J.: Addition No. 3, 4 and 5 to J. E. Sammet [59]. SIGSAM Bull. **10** (1968), **12** (1969), **15** (1970).

[77] Yun, D. Y. Y., Stoutemyer, R. D.: Symbolic Mathematical Computation. In: Encyclopedia of Computer Science and Technology (Belzer, J., Holzman, A. G., Kent, A., eds.), Vol. 15, pp. 235−310. New York-Basel: Marcel Dekker 1980.

[78] Zeis, E., Cefola, P. J.: Computerized Algebraic Utilies for the Construction of Non-Singular Satellite Theories. MACSYMA **1979**, 277−294.

[79] Zimmer, H. G.: Computational Problems, Methods and Results in Algebraic Number Theory. Lecture Notes in Mathematics, Vol. 262. Berlin-Heidelberg-New York: Springer 1972.

[80] REDUCE-Newsletters. Univ. of Utah: Symb. Comp. Group (since 1978).

Dr. J. Calmet
IMAG, Laboratoire d'Informatique
et de Mathématiques Appliquées de Grenoble
BP 53 X
F-38041 Grenoble Cédex
France

J. A. van Hulzen
Department of Applied Mathematics
Twente University of Technology
P.O. Box 217
NL-7500 AE Enschede
The Netherlands

Some Useful Bounds

M. Mignotte, Strasbourg

Abstract

Some fundamental inequalities for the following values are listed: the determinant of a matrix, the absolute value of the roots of a polynomial, the coefficients of divisors of polynomials, and the minimal distance between the roots of a polynomial. These inequalities are useful for the analysis of algorithms in various areas of computer algebra.

I. Hadamard's Inequality

Hadamard's theorem on determinants can be stated as follows:

Theorem 1. *If the elements of the determinant*

$$D = \begin{vmatrix} a_{11} & \cdots & a_{1n} \\ \vdots & & \vdots \\ a_{n1} & \cdots & a_{nn} \end{vmatrix}$$

are arbitrary complex numbers, then

$$|D|^2 \leqslant \prod_{h=1}^{n} \left(\sum_{j=1}^{n} |a_{hj}|^2 \right)$$

and equality holds if and only if

$$\sum_{h=1}^{n} a_{hj} \bar{a}_{hk} = 0 \quad for \quad 1 \leqslant j < k \leqslant n,$$

where $\bar{a}_{hk}$ is the conjugate of a_{hk}.

We do not give a proof of this classical result, it can be found in many textbooks on linear algebra (for example: H. Minc and M. Marcus, Introduction to Linear Algebra, Macmillan, New York, 1965).

II. Cauchy's Inequality

The following result gives an upper bound for the modulus of the roots of a polynomial in terms of the coefficients of this polynomial.

Theorem 2. *Let*

$$P(X) = a_0 X^d + a_1 X^{d-1} + \cdots + a_d, \quad a_0 \neq 0, \quad d \geqslant 1, \tag{*}$$

be a polynomial with complex coefficients. Then any root z of P satisfies

$$|z| < 1 + \frac{\text{Max}\{|a_1|, \ldots, |a_d|\}}{|a_0|}.$$

Proof. Let z be a root of P. If $|z| \leqslant 1$ the theorem is trivially true so we suppose $|z| > 1$. Put

$$H = \max\{|a_1|, \ldots, |a_d|\}.$$

By hypothesis z satisfies

$$a_0 z^d = -a_1 z^{d-1} - \cdots - a_d,$$

so that

$$|a_0| |z|^d \leqslant H(|z|^{d-1} + \cdots + 1) < \frac{H|z|^d}{|z| - 1},$$

and

$$|a_0|(|z| - 1) < H.$$

This proves the result. ∎

Corollary. *Let P be given by (∗) and $a_d \neq 0$. Then any root z of P satisfies*

$$|z| > \frac{|a_d|}{|a_d| + \text{Max}\{|a_0|, |a_1|, \ldots, |a_{d-1}|\}}.$$

Proof. If z is a root of P then z^{-1} is a root of the polynomial

$$a_d X^d + a_{d-1} X^{d-1} + \cdots + a_0.$$

Applying the theorem to this polynomial gives the result. ∎

There are many other known bounds for the modulus of the roots of a polynomial, most of which can be found in the book of Marden [3].

III. Landau's Inequality

Cauchy's inequality gives an upper bound for the modulus of *each* root of a polynomial. Landau's inequality gives an upper bound for the product of the modulus of *all* the roots of this polynomial lying outside of the unit circle. Moreover this second bound is not much greater than Cauchy's.

Theorem 3. *Let P be given by (∗). Let $z_1, \ldots, z_d$ be the roots of P. Put*

$$M(P) = |a_0| \prod_{j=1}^{d} \text{Max}\{1, |z_j|\}.$$

Then

$$M(P) \leqslant (|a_0|^2 + |a_1|^2 + \cdots + |a_d|^2)^{1/2}.$$

To prove this theorem a lemma will be useful. If $R = \sum_{k=0}^{m} c_k X^k$ is a polynomial we put

$$\|R\| = \left(\sum_{k=0}^{m} |c_k|^2 \right)^{1/2}.$$

Lemma. *If Q is a polynomial and z is any complex number then*

$$\|(X + z)Q(X)\| = \|(\bar{z}X + 1)Q(X)\|.$$

Proof. Suppose

$$Q(X) = \sum_{k=0}^{m} c_k X^k.$$

The square of the left hand side member is equal to

$$\sum_{k=0}^{m} (c_{k-1} + z\bar{c}_k)(\bar{c}_{k-1} + \bar{z}c_k) = (1 + |z|^2)\|Q\|^2 + \sum_{k=0}^{m} (zc_k\bar{c}_{k-1} + \bar{z}\bar{c}_k c_{k-1})$$

where $c_{-1} = 0$.

It is easily verified that the square of the right hand side admits the same expansion. ∎

Proof of the Theorem. Let $z_1, \ldots, z_k$ be the roots of P lying outside of the unit circle. Then $M(P) = |a_0| |z_1 \cdots z_k|$. Put

$$R(X) = a_0 \prod_{j=1}^{k} (\bar{z}_j X - 1) \prod_{j=k+1}^{d} (X - z_j) = b_0 X^d + \cdots + b_d.$$

Applying k times the lemma shows that $\|P\| = \|R\|$. But

$$\|R\|^2 \geqslant |b_0|^2 = M(P)^2. \quad \blacksquare$$

IV. Bounds for the Coefficients of Divisors of Polynomials

1. An Inequality

Theorem 4. *Let*

$$Q = b_0 X^q + b_1 X^{q-1} + \cdots, \qquad b_0 \neq 0$$

be a divisor of the polynomial P given by (*). Then

$$|b_0| + |b_1| + \cdots + |b_q| \leqslant |b_0/a_0| 2^q \|P\|.$$

Proof. It is easily verified that

$$|b_0| + \cdots + |b_q| \leqslant 2^q M(Q).$$

But

$$M(Q) \leqslant |b_0/a_0| M(P)$$

and, by Landau's inequality,

$$M(P) \leqslant \|P\|. \quad \blacksquare$$

Another inequality is proved in [4], Theorem 2.

2. An Example

The following example shows that the inequality in Theorem 4 cannot be much improved.

Let q be any positive integer and

$$Q(X) = (X - 1)^q = b_0 X^q + b_1 X^{q-1} + \cdots + b_q;$$

then it is proved in [4] that there exists a polynomial P with integer coefficients which is a multiple of Q and satisfies

$$\|P\| \leqslant Cq(\text{Log } q)^{1/2},$$

where C is an absolute constant.

Notice that in this case

$$|b_0| + \cdots + |b_q| = 2^q.$$

This shows that the term 2^q in Theorem 3 cannot be replaced by $(2 - \varepsilon)^q$, where ε is a fixed positive number.

V. Isolating Roots of Polynomials

If $z_1, \ldots, z_d$ are the roots of a polynomial P we define

$$\text{sep}(P) = \min_{z_i \neq z_j} |z_i - z_j|.$$

For reasons of simplicity we consider only polynomials with simple zeros (i.e. square-free polynomials); for the general case see Güting's paper [1].

The best known lower bound for $\text{sep}(P)$ seems to be the following.

Theorem 5. *Let P be a square-free polynomial of degree d and discriminant D. Then*

$$\text{sep}(P) > \sqrt{3} \, d^{-(d+1)/2} |D|^{1/2} \|P\|^{1-d}.$$

Proof. Using essentially Hadamard's inequality, Mahler [2] proved the lower bound

$$\text{sep}(P) > \sqrt{3} \, d^{-(d+2)/2} |D|^{1/2} M(P)^{1-d}.$$

The conclusion follows from Theorem 3. ∎

Corollary. *When P is a square-free integral polynomial $\text{sep}(P)$ satisfies*

$$\text{sep}(P) > \sqrt{3} \, d^{-(d+2)/2} \|P\|^{1-d}.$$

Other results are contained in [4], Theorem 5. It is possible to construct monic irreducible polynomials with integer coefficients for which $\text{sep}(P)$ is "rather" small. Let $d \geqslant 3$ and $a \geqslant 3$ be integers. Consider the following polynomial

$$P(X) = X^d - 2(aX - 1)^2.$$

Eisenstein's criterion shows that P is irreducible over the integers (consider the prime number 2). The polynomial P has two real roots close to $1/a$: clearly

$$P(1/a) > 0$$

and if $h = a^{-(d+2)/2}$

$$P(1/a \pm h) < 2a^{-d} - 2a^2 a^{-d-2} = 0,$$

so that P has two real roots in the interval $(1/a - h, 1/a + h)$. Thus

$$\operatorname{sep}(P) < 2h = 2a^{-(d+2)/2}.$$

References

[1] Güting, R.: Polynomials with Multiple Zeros. Mathematika **14**, 181 – 196 (1967).
[2] Mahler, K.: An Inequality for the Discriminant of a Polynomial. Michigan Math. J. **11**, 257 – 262 (1964).
[3] Marden, M.: The Geometry of the Zeros of a Polynomial in a Complex Variable. Publ. AMS **1949**, Math. Surv. 3.
[4] Mignotte, M.: Some Inequalities about Univariate Polynomials. SYMSAC **1981**, 195 – 199.

Prof. Dr. M. Mignotte
Centre de Calcul
Université Louis Pasteur
7, rue René Descartes
F-67084 Strasbourg
France

Author and Subject Index

Computergesteuerter Fotosatz und Umbruch: Dipl.-Ing. Schwarz' Erben KG, A-3910 Zwettl NÖ. —
Reproduktion und Offsetdruck: Novographic, Ing. Wolfgang Schmid, A-1230 Wien.

DUE DATE